Birkhäuser

Martin Schechter

Critical Point Theory

Sandwich and Linking Systems

Birkhäuser

Martin Schechter
Brooklyn, NY, USA

ISBN 978-3-030-45605-4 ISBN 978-3-030-45603-0 (eBook)
https://doi.org/10.1007/978-3-030-45603-0

Mathematics Subject Classification: 35A15, 35B38, 49J40, 58E05, 70G75, 35-02, 47-02, 49-02, 58-02

This book is published under the imprint Birkhäuser, www.birkhauser-science.com by the registered company Springer Nature Switzerland AG.
The registered company address is: Gewerbestrasse 11, 6330 Cham, Switzerland

BS"D

To my wife, Deborah,
our children, our grandchildren,
our great grandchildren,
and our extended family.
May they all enjoy many happy years.

Preface

The Purpose of the Book

Many problems arising in science and engineering call for the solving of nonlinear partial differential equations or systems. These equations are difficult to solve, and there are very few general techniques that can be applied to solve them. Different equations seem to require different methods of solution. (Of course, some nonlinear equations have no solutions.) However, it was noticed that many of these equations are Euler equations of functionals (i.e., functions from some space to $\mathbb{R}$). This means that for each equation there is a C^1 functional $G(u)$ such that any solution u of the equation satisfies

$$G'(u) = 0. \tag{1}$$

(Here C^1 refers to continuously differentiable functions.) As an illustration, the equation

$$-\Delta u(x) = f(x, u(x))$$

is the Euler equation of the functional

$$G(u) = \frac{1}{2}\|\nabla u\|^2 - \int F(x, u(x))\, dx$$

on an appropriate space S, where

$$F(x, t) = \int_0^t f(x, s)\, ds, \tag{2}$$

and the norm is that of L^2. (Here

$$\Delta u(x) = \sum_{k=1}^{n} \frac{\partial^2 u(x_1, x_2, \cdots, x_n)}{\partial^2 x_k},$$

$$\nabla u = (\partial u(x)/\partial x_1, \partial u(x)/\partial x_2, \cdots, \partial u(x)/\partial x_n),$$

and

$$\|u\|^2 = \int_S |u(x)|^2.)$$

The solving of the Euler equation is tantamount to finding critical points of the corresponding functional. This has motivated researchers to study critical points of functionals in order to solve the corresponding Euler equations and systems. It has led to the development of several techniques to find critical points. In many cases, these techniques lead to results superior to those obtained by other methods of solving these equations. This volume is dedicated to the latest developments and applications of these techniques.

In order to explain the nature of the problem, we describe in detail the one-dimensional case.

A One-Dimensional Problem

Consider the problem of finding a solution of

$$-u''(x) + u(x) = f(x, u(x)), \quad x \in I = [0, 2\pi], \tag{3}$$

under the conditions

$$u(0) = u(2\pi), \; u'(0) = u'(2\pi). \tag{4}$$

We assume that the function $f(x,t)$ is continuous in $I \times \mathbb{R}$ and is periodic in x with period 2π. In order to solve it, you may first want to solve the linear problem corresponding to (3), (4), namely

$$-u''(x) + u(x) = f(x), \quad x \in I = [0, 2\pi], \tag{5}$$

under the conditions (4), where the function $f(x)$ is continuous in I and is periodic in x with period 2π. After a substantial calculation, one finds that there is a unique solution given by

$$u(x) = Ae^x + Be^{-x} + \int_0^x \sinh(t-x)\, f(t)\, dt, \tag{6}$$

where

$$2A = \frac{e^{2\pi}}{e^{2\pi}-1} \int_0^{2\pi} e^{-t} f(t)\, dt \tag{7}$$

and

$$2B = \frac{e^{-2\pi}}{1 - e^{-2\pi}} \int_0^{2\pi} e^t f(t)\, dt. \tag{8}$$

Can this solution be used to solve (3), (4)? It can if $f(x, t)$ is bounded for all x and t. For then, we can define

$$Tu(x) = A(u)e^x + B(u)e^{-x} + \int_0^x \sinh(t - x)\, f(t, u(t))\, dt, \tag{9}$$

where

$$2A(u) = \frac{e^{2\pi}}{e^{2\pi} - 1} \int_0^{2\pi} e^{-t} f(t, u(t))\, dt \tag{10}$$

and

$$2B(u) = \frac{e^{-2\pi}}{1 - e^{-2\pi}} \int_0^{2\pi} e^t f(t, u(t))\, dt. \tag{11}$$

Then a solution of (3), (4) will exist if we can find a function $u(x)$ such that

$$Tu(x) = u(x), \quad x \in I. \tag{12}$$

Such a function is called a **fixed point** of the operator T. All solutions of (3), (4) are fixed points of the operator T. In the present case, one can show that there is indeed a fixed point for the operator T when $f(x, t)$ is bounded. If more than one exists, there is no organized way of finding all of them.

However, there is another way of solving (3), (4) called the variational approach. If $u(x)$ is a solution, then we have

$$(u', v') + (u, v) = (-u'' + u, v) = (f(\cdot, u), v)$$

for all $v \in C^1(I)$ satisfying (4). Here,

$$(u, v) = \int_0^{2\pi} u(x)v(x)\, dx,$$

$C^1(I)$ is the set of continuously differentiable periodic functions on I, and we used the fact that no boundary terms arise in the integration by parts. The expression

$$(u, v)_H = (u', v') + (u, v) \tag{13}$$

is a scalar product corresponding to the norm

$$\|u\|_H = (\|u'\|^2 + \|u\|^2)^{1/2}. \tag{14}$$

Thus, a solution of (3), (4) satisfies

$$(u, v)_H = (f(\cdot, u), v) \tag{15}$$

for all $v \in C^1(I)$ satisfying (4). We let H be the completion of periodic functions in $C^1(I)$ with respect to the norm (14).

In searching for solutions of (3) and (4), we look for functions $u \in H$ which satisfy (15). If $\{u_k\}$ is a Cauchy sequence in H of functions in $C^1(I)$, then

$$\|u_j - u_k\| \to 0, \ \|u_j' - u_k'\| \to 0.$$

This means that there are functions $u, h \in L^2(I)$ such that

$$u_k \to u, \ u_k' \to h \quad \text{in} \ L^2(I). \tag{16}$$

Note that

$$(u_k, v') = -(u_k', v), \quad v \in C^1(I), \tag{17}$$

by integration by parts. Thus in the limit,

$$(u, v') = -(h, v), \quad v \in C^1(I). \tag{18}$$

If $u \in C^1(I), \ h \in L^2(I)$, and (18) holds, then $h = u'$ a.e.

Now, suppose that $u \in L^2(I)$ and there is an $h \in L^2(I)$ such that (18) holds. Note that h is unique. Even though u is not in $C^1(I)$ and we do not know whether or not it has a derivative at any point, we define the "weak" derivative of u to be h and denote it by u'. It behaves like a derivative with respect to integration by parts. Since H is the completion of $C^1(I)$ with respect to the norm given by (14), then every function in H has a weak derivative in $L^2(I)$. Conversely, every function in $L^2(I)$ that has a weak derivative in $L^2(I)$ is in H.

Although functions in H need not be in $C^1(I)$, they are in $C(I)$, the set of continuous functions on I. In fact, there is a constant K such that

$$|u(x)| \leq K\|u\|_H, \quad x \in I, \ u \in H. \tag{19}$$

Moreover,

$$u(0) = u(2\pi). \tag{20}$$

More precisely, every function $u \in H$ is almost everywhere equal to a function in $C(I)$. Inequality (19) holds for the continuous function equal to u a.e., i.e., it holds for u if we adjust it on a set of measure zero to make it continuous. The same is true of (20). Moreover, if the sequence $\{u_k\}$ converges in H, then it converges uniformly on I. We also have

Theorem 1 *If $f \in L^2(I)$, $u \in H$, and*

$$(u, v)_H = (f, v), \quad v \in C^1(I), \tag{21}$$

then $u' \in H$ and $u'' = (u')' = u - f$. In particular, u' is continuous in I and is the derivative of u in the usual sense. If, in addition, f is in $C(I)$, then u'' is continuous in I, and $u'' = u - f$ in the usual sense. In particular, it is a solution of (3), (4). Consequently, $u \in H$ is a solution of (3), (4) iff it satisfies (15) for all $v \in H$.

Assume that $f(x, t)$ is a continuous function on $I \times \mathbb{R}$ satisfying

$$|f(x, t)| \leq C(|t| + 1), \quad x \in I, \ t \in \mathbb{R}. \tag{22}$$

We let

$$G(u) = \frac{1}{2}\|u\|_H^2 - \int_0^{2\pi} F(x, u(x))\, dx, \tag{23}$$

where

$$F(x, t) = \int_0^t f(x, s)\, ds. \tag{24}$$

Note that $G(u)$ is defined on H.

Next, we calculate the derivative of G. We have

$$\begin{aligned} G(u + \varepsilon v) - G(u) &= \frac{1}{2}(\|u + \varepsilon v\|_H^2 - \|u\|_H^2) \\ &\quad - \int_0^{2\pi} [F(x, u + \varepsilon v) - F(x, u)]\, dx \\ &= \frac{1}{2}(\|u\|_H^2 + 2\varepsilon(u, v)_H + \varepsilon^2\|v\|_H^2 - \|u\|_H^2) \\ &\quad - \int_0^{2\pi}\int_0^1 \frac{d}{d\theta}F(x, u + \varepsilon\theta v)\, d\theta\, dx \\ &= \varepsilon(u, v)_H + \varepsilon^2\frac{1}{2}\|v\|_H^2 - \varepsilon\int_0^{2\pi}\int_0^1 f(x, u + \varepsilon\theta v)v\, d\theta\, dx. \end{aligned}$$

Hence,

$$[G(u + \varepsilon v) - G(u)]/\varepsilon - (u, v)_H + \int_0^{2\pi} f(x, u)v\, dx$$

$$= \frac{\varepsilon}{2}\|v\|_H^2 - \int_0^{2\pi}\int_0^1 [f(x, u + \varepsilon\theta v) - f(x, u)]v d\theta dx.$$

The derivative will exist and satisfy

$$(G'(u), v) = (u, v)_H - \int_0^{2\pi} f(x, u)v\,dx, \quad v \in H, \tag{25}$$

if we can show that the expression

$$\int_I \int_0^1 |f(x, u + \varepsilon\theta v) - f(x, u)|^2 d\theta\,dx$$

converges to 0 as $\varepsilon \to 0$. If this were not the case, there would be a sequence $\varepsilon_k \to 0$ such that

$$\int_I \int_0^1 |f(x, u + \varepsilon_k\theta v) - f(x, u)|^2 d\theta\,dx \geq \delta > 0. \tag{26}$$

But $u + \varepsilon_k\theta v \to u$ uniformly on I. Consequently, the integrand of (26) converges to 0 a.e. The integrand of (26) is majorized by constants depending on the H norms of $u + \varepsilon_k\theta v$ and u. Since these norms are bounded, the integrand is majorized by a constant. This implies that the integral converges to 0, contrary to assumption. Consequently, the derivative of G satisfies (25). As a result, $u \in H$ is a solution of (3), (4) iff $G'(u) = 0$. This reduces the search for solutions of (3) and (4) to the search for critical points of G. If we wish to compare the two methods, we must ask ourselves which is easier, finding fixed points of the operator T or critical points of the functional G.

Before we decide, let us turn to the n-dimensional problem

$$-\Delta u(x) + u(x) = f(x, u(x)),$$

where $x \in \Omega \subset \mathbb{R}^n$. It is extremely difficult to obtain a formula $Tu(x)$ for a solution of

$$-\Delta u(x) + u(x) = f(x)$$

satisfying given boundary conditions on the boundary $\partial\Omega$ of a region $\Omega \subset \mathbb{R}^n$. Finding fixed points is just as difficult if not more so. On the other hand, finding critical points of the corresponding functional

$$G(u) = \frac{1}{2}\|\nabla u\|^2 + \|u\|^2 - \int_\Omega F(x, u(x))\,dx$$

is doable. We explain this in the next section.

Palais–Smale and Cerami Sequences

Once it was known that solutions of Euler equations were critical points of the corresponding functionals, researchers began searching for critical points. The classical approach was to look for maxima or minima. Searching for extrema illustrates the difficulty of the problem. We are trying to find points where the gradient of the functional vanishes. But all we have to work with are inequalities involving the functional itself, and not its gradient. If the functional is bounded from below and one is looking for a minimum, one can obtain a minimizing sequence satisfying

$$G(u_k) \to a = \inf G. \tag{27}$$

If such a sequence converges or has a convergent subsequence, then we indeed obtain a minimum. However, in dealing with such sequences it is difficult, in general, to establish the convergence of a subsequence because there is very little with which to work.

Luckily, there is some help. In such a case, one can show that there is a sequence, called a Palais–Smale (PS) sequence, satisfying

$$G(u_k) \to a, \quad G'(u_k) \to 0, \tag{28}$$

where $a = \inf G$. It is much easier to establish the existence of a convergent subsequence of a PS sequence than of a minimizing sequence. In fact, a minimizing sequence may not have a convergent subsequence, while a PS sequence for the same functional does.

Actually, one can do better. If the functional $G(u)$ is bounded from below, then there exists a sequence (called a Cerami sequence) satisfying

$$G(u_k) \to a, \quad (1 + \|u_k\|)G'(u_k) \to 0 \tag{29}$$

for $a = \inf G$. As in the case of a PS sequence, if a Cerami sequence has a convergent subsequence, it will produce a minimum. The advantage of obtaining such a sequence is that the additional structure allows one to prove the convergence of a subsequence in cases where a corresponding PS sequence need not have a converging subsequence.

General Critical Points

When one wishes to solve Euler equations, one is not merely looking for extrema. One is interested in finding all critical points. The same is true even for semibounded functionals if one wishes to obtain critical points which are not extrema. However, when the functional is not semibounded, there is no organized procedure for producing critical points. In particular, how does one search for critical points when the functional is not semibounded? Is there anything that can be used to replace semiboundedness?

In this book, we present several approaches which produce sequences similar to (29) when one is searching for critical points whether or not they are extrema.

Outline of the Book

In the first six chapters of this book, we present different methods which can be used to find critical points. Each method works under certain circumstances, but not under all situations. They are based on a fundamental approach of obtaining two special sets A, B such that (29) results when the functional G has a certain relationship with the sets. We describe these methods below.

The remaining chapters of the book show how these methods can be used to solve various problems in mathematics and physics. In Chaps. 7–9, we discuss the nonlinear Schrödinger equation. The method applies to general semilinear partial differential equations.

In Chap. 10, we study second order Hamiltonian systems from mechanics.

In Chaps. 11 and 12, we apply the methods to semilinear partial differential equations in bounded domains. In Chap. 13, we consider elliptic systems of partial differential equations.

In Chap. 14, we consider the p-Laplacian problem in a bounded domain (cf. below).

We study the periodic semilinear wave equation in Chap. 15.

In Chap. 16, we consider light waves propagating in a photorefractive crystal.

We study periodic solutions of n-dimensional nonlinear wave equations in Chap. 17.

In our final chapter, Chap. 18, we show how our methods can be used to obtain multiple solutions.

One point we want to express is that these methods produce some of the strongest results for the given problems. We now give a detailed description of the chapters.

Linking

An important method of detecting critical points is called **linking.** It was discovered that there are pairs of sets A, B such that whenever they separate a functional G, i.e., satisfy

$$a_0 := \sup_A G \leq b_0 := \inf_B G,$$

one obtains a sequence of the form

$$G(u_k) \to c, \quad (1 + \|u_k\|)G'(u_k) \to 0, \tag{30}$$

where $b_0 \le c$. If A, B is such a pair, we say that A **links** B and call A, B a **linking set**.

The main question is to identify such pairs of sets. We show in Chap. 1 that there are quite a number of linking sets. We give a general sufficient condition as follows:

Let E be a Banach space. The set Φ of mappings $\Gamma(t) \in C(E \times [0, 1], E)$ is to have the following properties:

(a) for each $t \in [0, 1)$, $\Gamma(t)$ is a homeomorphism of E onto itself and $\Gamma(t)^{-1}$ is continuous on $E \times [0, 1)$,
(b) $\Gamma(0) = I$,
(c) for each $\Gamma(t) \in \Phi$, there is a $u_0 \in E$ such that $\Gamma(1)u = u_0$ for all $u \in E$ and $\Gamma(t)u \to u_0$ as $t \to 1$ uniformly on bounded subsets of E, and
(d) for each $t_0 \in [0, 1)$ and each bounded set $A \subset E$, we have

$$\sup_{\substack{0 \le t \le t_0 \\ u \in A}} \{\|\Gamma(t)u\| + \|\Gamma^{-1}(t)u\|\} < \infty.$$

We then prove

Theorem 2 *A subset A of E links a subset B of E if $A \cap B = \phi$ and, for each $\Gamma(t) \in \Phi$, there is a $t \in (0, 1]$ such that $\Gamma(t)A \cap B \neq \phi$.*

We show how to generate them, and we give a partial list.

Sandwich Pairs

What if we cannot find linking sets that separate the functional? Is there anything else that can be done?

There is another approach which works when one cannot obtain linking sets which separate the functional. This will be described in Chap. 2. In this approach, one looks for suitable sets A, B such that the functional is bounded from below on one set and bounded from above on the other with no requirement on the relationship between the bounds. In other words, one looks for suitable subsets A, B of a Banach space E, which are such that

$$a_0 := \sup_A G < \infty, \quad b_0 := \inf_B G > -\infty \tag{31}$$

holding for a given C^1 functional G on E implies the existence of a PS sequence

$$G(u_k) \to c, \quad G'(u_k) \to 0, \tag{32}$$

where $b_0 \leq c \leq a_0$. If A, B are such that (31) always implies (32), we say that A, B forms a **sandwich pair.** It will be shown in Chaps. 2 and 3 that sandwich pairs are not rare. They are not as plentiful as linking sets, but they exist and can be used when one cannot find linking sets which separate the functional. Just as in the case of linking sets, there is no organized way of obtaining sandwich pairs. In Chaps. 2 and 3, we present some methods of finding and using them. We give sufficient conditions. A typical theorem states the following:

Theorem 3 *Let N be a finite dimensional subspace of a Hilbert space E and let $M = N^\perp$. Then M,N form a sandwich pair.*

Note that (32) is not a Cerami sequence as was produced by a linking set that separates the functional. This raises the question as to whether or not there are sandwich pairs that produce Cerami sequences. Indeed there are, and we call them **strong sandwich pairs.** An example is given by the following:

Theorem 4 *Let N be a finite dimensional subspace of a Hilbert space E with complement $M \oplus \{v_0\}$, where v_0 is an element in $M^\perp$ having unit norm, and let δ be any positive number. Let $\varphi(t) \in C^1(\mathbb{R})$ be such that*

$$0 \leq \varphi(t) \leq 1, \ \ \varphi(0) = 1,$$

and

$$\varphi(t) = 0, \qquad |t| \geq 1.$$

Let

$$F(v+w+sv_0) = v+[s+\delta-\delta\varphi(\|w\|^2/\delta^2)]v_0, \quad w \in M, \ v \in N, \ s \in \mathbb{R}. \tag{33}$$

If $R_0 > \delta$, then $A = [N \oplus \{v_0\}]\backslash\mathcal{B}_{R_0}$ and $B = F^{-1}(\delta v_0)$ form a strong sandwich pair.

We exhibit and use strong sandwich pairs.

The Monotonicity Trick

Linking sets and strong sandwich pairs produce Cerami sequences. Many times this alone will imply that the sequence has a convergent subsequence. More often, one needs additional information. The crucial step is to show that the sequence is bounded or has a bounded subsequence. For if $\|u_k\| \to \infty$, there is no way we can get convergence of a subsequence. Moreover, in many cases, it is easy to prove the convergence of a subsequence once we know that the sequence is bounded. In Chap. 4, we discuss a technique for showing that some sequences obtained by linking and sandwich methods are bounded. Of course, there is a penalty to pay. The method is called the **monotonicity trick** and involves embedding the functional in a family of functionals depending on a parameter. A typical result is as follows:

Let E be a reflexive Banach space with norm $\|\cdot\|$. Suppose that $G \in C^1(E, \mathbb{R})$ is of the form: $G(u) := I(u) - J(u)$, $u \in E$, where $I, J \in C^1(E, \mathbb{R})$ map bounded sets to bounded sets. Define

$$G_\lambda(u) = \lambda I(u) - J(u), \quad \lambda \in \Lambda,$$

where Λ is an open interval contained in $(0, +\infty)$. Assume one of the following alternatives holds.

(H$_1$) $I(u) \geq 0$ for all $u \in E$ and $I(u) + |J(u)| \to \infty$ as $\|u\| \to \infty$.

(H$_2$) $I(u) \leq 0$ for all $u \in E$ and $|I(u)| + |J(u)| \to \infty$ as $\|u\| \to \infty$.

Furthermore, suppose that

(H$_3$) There is a sandwich pair A, B such that

$$a_0 := \sup_A G_\lambda < \infty, \qquad b_0 := \inf_B G_\lambda > -\infty \tag{34}$$

for each $\lambda \in \Lambda$.

Then we have

Theorem 5 *Assume that* **(H$_1$)** *(or* **(H$_2$)***) and* **(H$_3$)** *hold. Then for almost all $\lambda \in \Lambda$ there exists a bounded sequence $u_k(\lambda) \in E$ such that*

$$\|G'_\lambda(u_k)\| \to 0, \quad G_\lambda(u_k) \to a(\lambda).$$

Of course, you realize that we obtain a bounded PS (or Cerami) sequence only for almost every value of $\lambda \in \Lambda$. The boundedness usually helps one to prove the existence of a solution of $G'_\lambda(u) = 0$, but only for almost every value of $\lambda \in \Lambda$. However, we show that in many cases the addition of appropriate hypotheses implies that $G'_\lambda(u) = 0$ has a solution for **all** $\lambda \in \Lambda$.

Weak Linking

It should be noted that in all linking and sandwich examples one of the sets must be contained in a finite dimensional subspace. The reason for this is that verifying linking or sandwich properties requires the use of finite dimensional index theory. Yet, many of the applications cannot be subjected to such restrictions. There are several approaches to dealing with such situations. We present one of them in Chap. 5. We introduce the concept of **weak linking**. We make use of the following theorem:

Theorem 6 *Let N be a closed separable subspace of a Hilbert space E. Let G be a continuously differentiable functional on E such that*

$$v_n = Pu_n \to v \text{ weakly in } E, \quad w_n = (I - P)u_n \to w \text{ strongly in } E$$

implies

$$G'(v_n + w_n) \to G'(v + w) \text{ weakly in } E, \tag{35}$$

where P is the projection of E onto N. Assume

$$a_0 := \sup_N G < \infty, \quad b_0 := \inf_M G > -\infty. \tag{36}$$

Then there is a sequence $\{u_k\} \subset E$ such that

$$G(u_k) \to c, \ b_0 \le c \le a_0, \ G'(u_k) \to 0. \tag{37}$$

We introduce the notion of a **weak sandwich pair** and prove the following:

Theorem 7 *Let E be a reflexive Banach space with norm $\|\cdot\|$. Let G be a continuously differentiable functional on E such that*

$$v_n = Pu_n \to v \text{ weakly in } E, \quad w_n = (I - P)u_n \to w \text{ strongly in } E \tag{38}$$

implies

$$G'(v_n + w_n) \to G'(v + w) \text{ weakly in } E, \tag{39}$$

where P is the projection of E onto a closed subspace N. Suppose that G is of the form: $G(u) := I(u) - J(u), u \in E$, where $I, J \in C^1(E, \mathbb{R})$ map bounded sets to bounded sets. Define

$$G_\lambda(u) = \lambda I(u) - J(u), \quad \lambda \in \Lambda,$$

where Λ is an open interval contained in $(0, +\infty)$. Assume one of the following alternatives holds.

(H₁) *$I(u) \ge 0$ for all $u \in E$ and $I(u) + |J(u)| \to \infty$ as $\|u\| \to \infty$.*

(H₂) *$I(u) \le 0$ for all $u \in E$ and $|I(u)| + |J(u)| \to \infty$ as $\|u\| \to \infty$.*

Furthermore, suppose that

(H₃) *There is a weak sandwich pair A,B such that*

$$a_\lambda := \sup_A G_\lambda < \infty, \qquad b_\lambda := \inf_B G_\lambda > -\infty \tag{40}$$

for each $\lambda \in \Lambda$.

Then we have

(1) *For almost all $\lambda \in \Lambda$, there exists a constant $k_0(\lambda) := k_0$ (depending only on λ) such that for each $\delta > 0$ there exists a $K \in \mathcal{K}$ such that*

$$\sup_K G_\lambda \le a(\lambda) + \delta$$

and

$$\|u\| \le k_0 \quad \text{whenever} \quad u \in K \text{ and } G_\lambda(u) \ge a(\lambda) - \delta. \tag{41}$$

(2) *For almost all $\lambda \in \Lambda$, there exists a bounded sequence $u_k(\lambda) \in E$ such that*

$$\|G'_\lambda(u_k)\| \to 0, \quad G_\lambda(u_k) \to a(\lambda), \quad \text{as } k \to \infty.$$

Differential Equations

In proving the theorems, we are required to solve equations of the form

$$\sigma'(t) = g(\sigma(t)), \; t \in [t_0, T), \quad \sigma(t_0) = v \in E \tag{42}$$

and obtain estimates for the solution. In almost every case, we do not know the size of the interval in which the solution must exist. This requires us to insist that the function g be such that we get a unique solution in $[t_0, \infty)$. In Chap. 6, we prove the following:

Lemma 8 *Let $\rho(t) > 0$ be a locally Lipschitz continuous function on $[t_0, \infty)$ satisfying*

$$\int_{u_0}^{\infty} \frac{d\tau}{\rho(\tau)} = \infty, \tag{43}$$

where $u_0 > 0$ and $t_0 \in \mathbb{R}$. Then, there is a unique solution of

$$u'(t) = \rho(u(t)), \quad t \in [t_0, \infty), \quad u(t_0) = u_0, \tag{44}$$

which is positive in (t_0, ∞) and depends continuously on u_0. Moreover, if

$$\int_0^{u_0} \frac{d\tau}{\rho(\tau)} = \infty, \tag{45}$$

then there is a unique solution of

$$u'(t) = -\rho(u(t)), \quad t \in [t_0, \infty), \quad u(t_0) = u_0, \tag{46}$$

which depends continuously on u_0.

Theorem 9 *Let $g(t, x)$ be a continuous map from $\mathbb{R} \times E$ to E. Assume that for each point $(\hat{t}, \hat{x}) \in \mathbb{R} \times E$, there are constants $K, b > 0$ such that*

$$\|g(t,x)-g(t,y)\| \leq K\|x-y\|, \quad |t-\hat{t}| < b, \ \|x-\hat{x}\| < b, \ \|y-\hat{x}\| < b \tag{47}$$

and

$$\|g(t,x)\| \leq \rho(d(x,Q)), \quad x \in E, \ t \in [t_0, \infty), \tag{48}$$

where Q is a subset of E, with ρ nondecreasing or bounded. Assume the hypotheses of Lemma 8. Then for each $x_0 \in E$ and $t_0 \in \mathbb{R}$, there is a unique solution $x(t)$ of the equation

$$\frac{dx(t)}{dt} = g(t, x(t)), \quad t \in [t_0, \infty), \quad x(t_0) = x_0. \tag{49}$$

Moreover, $x(t)$ depends continuously on x_0 and satisfies

$$u_1(t) \leq d(x(t), Q) \leq u_2(t), \quad t \in [t_0, \infty), \tag{50}$$

where $u_1(t)$ is the solution of (46) in $[t_0, \infty)$ satisfying $u(t_0) = u_0 = d(x_0, Q)$, and $u_2(t)$ is the solution of (44) in $[t_0, \infty)$ satisfying $u(t_0) = u_0 = d(x_0, Q)$.

Applications

In Chaps. 7–9, we apply the methods developed to the study of the nonlinear Schrödinger equation. In Chap. 7, we consider the following theorem:

Theorem 10 *Let $\mathcal{A} = -\Delta + V(x)$ on $H^1(\mathbb{R}^n)$. Assume*

1. *V is continuous, 1-periodic in $x_1, \cdots, x_k$.*
2. *$(a, b) \subset \rho(\mathcal{A})$, with $a < 0 < b$.*
3. *$C_0^\infty(\mathbb{R}^n) \subset D := D(|\mathcal{A}|^{(1/2)}) \subset H^{2,2}(\mathbb{R}^n)$, where*

$$\|u\|_{2,2} \leq C\|u\|_D = \||\mathcal{A}|^{(1/2)}u\|, \quad u \in D. \tag{51}$$

4. *D is compact in $L^q_{loc}(\mathbb{R}^n)$ and*

$$\|u\|_q \leq C\|u\|_D, \quad u \in D, \tag{52}$$

where $q \in (2, 2^*)$, $2^* := 2n/(n-2)$, $n > 2$, $2^* := \infty$, $n \leq 2$, and $\|\cdot\|_q$ is the norm of $L^q(\mathbb{R}^n)$.

5. *$f(x,t)$ is continuous, 1-periodic in $x_1, \cdots, x_k$.*
6. *$|f(x,t)| \leq C(|t|^{q-1} + 1), \quad x \in \mathbb{R}^n, \ t \in \mathbb{R}$.*

7. $|f(x,t)| \le \sigma|t|, \ \ |t| < \delta, \ x \in \mathbb{R}^n, \ t \in \mathbb{R}$, *for some* $\sigma < min[-a, b], \ \delta > 0.$
8. $F(x,t) \ge 0, \quad x \in \mathbb{R}^n, \ t \in \mathbb{R}.$
9. $F(x,t)/t^2 \to \infty$ *as* $t^2 \to \infty$ *uniformly in* x.
10. $2F(x,t+s) - 2F(x,t) - (2rs - (r-1)^2 t)f(x,t) \ge -W(x), \quad x \in \mathbb{R}^n,$ $s, t \in \mathbb{R}, \ r \in [0,1]$, *where* $W(x) \in L^1(\mathbb{R}^n)$.

Then

$$\mathcal{A}u = f(x,u), \quad u \in D \tag{53}$$

has a ground state solution.

(The reason for the periodicity is to guarantee that the resolvent of $\mathcal{A}$ is not empty. In Theorem 10 above, it is assumed that $0 \in \rho(\mathcal{A})$. One wishes to find nontrivial solutions and the so-called "minimizing solutions." These are solutions that minimize the corresponding energy functional. If they are nontrivial, they are called "ground state solutions.") We actually consider equations which are more general than the Schrödinger equation, but the latter is the most popular. In Theorem 10, we assumed that $0 \in \rho(\mathcal{A})$. This means that 0 is an interior point of $\rho(\mathcal{A})$. This is of considerable help in solving the problem. In Chap. 8, we allow 0 to be a boundary point of $\rho(\mathcal{A})$. In this case, much of this help falls away, and the problem becomes more difficult. In particular, we are able to prove the following:

Theorem 11 *Let* $\mathcal{A} = -\Delta + V(x)$ *on* $H^1(\mathbb{R}^n)$. *Assume*

1. V *is continuous, 1-periodic in* $x_1, \cdots, x_k$.
2. $(0, b) \subset \rho(\mathcal{A})$.
3. $C_0^\infty(\mathbb{R}^n) \subset D := D(|\mathcal{A} - c|^{(1/2)}) \subset H^{2,2}(\mathbb{R}^n)$ *for some* $c \in (0, b)$.
4. D *is compact in* $L^q_{loc}(\mathbb{R}^n)$ *and*

$$\|u\|_q \le C\|u\|_D, \quad u \in D, \tag{54}$$

   where $q \in (2, 2^*)$.
5. $f(x,t)$ *is continuous, 1-periodic in* $x_1, \cdots, x_k$.
6. $|f(x,t)| \le C(|t|^{q-1} + 1), \quad x \in \mathbb{R}^n, \ t \in \mathbb{R}.$
7. $|f(x,t)| \le C|t|^{q-1}, \quad |t| < \delta, \ x \in \mathbb{R}^n, \ t \in \mathbb{R}$, *for some* $C > 0, \ \delta > 0$.
8. $2F(x,t) \ge 0, \quad x \in \mathbb{R}^n, \ t \in \mathbb{R}.$
9. $F(x,t)/t^2 \to \infty$ *as* $t^2 \to \infty$,
10. $2F(x,t+s) - 2F(x,t) - (2rs - (r-1)^2 t)f(x,t) \ge -W(x) \in L^1(\mathbb{R}^n),$ $x \in \mathbb{R}^n, \ s, t \in \mathbb{R}, \ r \in [0,1].$

Then

$$\mathcal{A}u = f(x,u), \quad u \in D \tag{55}$$

has a nontrivial solution and a minimum energy solution.

The periodicity assumptions made in Chaps. 7 and 8 guaranteed that the resolvent of $\mathcal{A}$ was not empty. They also guaranteed that there are no isolated eigenvalues. Therefore, hypotheses cannot involve corresponding subspaces. In Chap. 9, we study the equation

$$\mathcal{A}u = f(x, u), \quad u \in D \tag{56}$$

without making any periodicity assumptions on the potential or on the nonlinear term. But we must be assured that the linear operator has nonempty resolvent. To accomplish this, we make assumptions on $V(x)$ which guarantee that the essential spectrum of $-\Delta u + V(x)u$ is the same as that of $-\Delta u$. In other words, our assumptions are such that the potential does not change the essential spectrum of the linear operator. This results in $[0, \infty)$ being the absolutely continuous part of the spectrum of $\mathcal{A}$.

Concerning the function $V(x)$, we make the following assumptions:

(V_1)

$$\sup_y \int_{|x-y|<\delta} |V(x)|\, \omega_2(x-y)\, dx \to 0 \; as \; \delta \to 0,$$

and

(V_2)

$$\int_{|x-y|<1} |V(x)|\, \omega_2(x-y)\, dx \to 0 \; as \; |y| \to \infty,$$

where

$$\omega_s(x) = \begin{cases} |x|^{s-n}, & 0 < s < n \\ 1 - \ln |x|^2, & s = n \\ 1, & s > n. \end{cases}$$

These assumptions imply that there is a forms extension $\mathcal{A}$ of the operator

$$-\Delta u(x) + V(x)u(x)$$

on the space $H = H^{1,2}(\mathbb{R}^n)$ having essential spectrum equal to $[0, \infty)$ and a (possibly empty) discrete, countable negative spectrum consisting of isolated eigenvalues of finite multiplicity with a finite lower bound $-L$.

$$-\infty < -L \le \lambda_0 < \lambda_1 < \lambda_2 < \ldots < \lambda_l < \ldots < 0. \tag{57}$$

(cf. Sect. 9.8.) For the operator $\mathcal{A}$, there are three possibilities: (a) it has no negative eigenvalues, (b) it has only one negative eigenvalue, and (c) it has two or more negative eigenvalues. What is interesting is that each of these possibilities must be dealt with differently. We study all of them separately. The hypotheses on $f(x, u)$ depend on the situations (a), (b), or (c). We also obtain the least energy solutions.

In Chap. 10, we study second order Hamiltonian systems. We consider the system

$$-\ddot{x}(t) = B(t)x(t) + \nabla_x V(t, x(t)), \tag{58}$$

where

$$x(t) = (x_1(t), \cdots, x_n(t)) \tag{59}$$

is a map from $I = [0, T]$ to $\mathbb{R}^n$ such that each component $x_j(t)$ is a periodic function in H^1 with period T, and the function $V(t, x) = V(t, x_1, \cdots, x_n)$ is continuous from $\mathbb{R}^{n+1}$ to $\mathbb{R}$ with

$$\nabla_x V(t, x) = (\partial V/\partial x_1, \cdots, \partial V/\partial x_n) \in C(\mathbb{R}^{n+1}, \mathbb{R}^n). \tag{60}$$

For each $x \in \mathbb{R}^n$, the function $V(t, x)$ is periodic in t with period T.

We study this problem under several sets of assumptions. First, we make no assumption on $\nabla_x V(t, x)$ other than (60). This allows both sublinear and superlinear problems. The elements of the symmetric matrix $B(t)$ are to be real-valued functions $b_{jk}(t) = b_{kj}(t)$. Our assumption on $B(t)$ is as follows.

$(B1)$ Each component of $B(t)$ is an integrable function on I, i.e., for each j and k, $b_{jk}(t) \in L^1(I)$.

This assumption implies that there is an extension $\mathcal{D}$ of the operator

$$\mathcal{D}_0 x = -\ddot{x}(t) - B(t)x(t)$$

having a discrete, countable spectrum consisting of isolated eigenvalues of finite multiplicity with a finite lower bound $-L$

$$-\infty < -L \leq \lambda_0 < \lambda_1 < \lambda_2 < \ldots < \lambda_l < \ldots . \tag{61}$$

The energy functional corresponding to (58) is

$$G(x) = d(x) - 2\int_I V(t, x)\, dt, \tag{62}$$

where $d(x) = (\mathcal{D}x, x)$.

We introduce a parameter λ to make use of the monotonicity trick of Chap. 4. This allows us to obtain solutions under very weak hypotheses but only for almost every value of the parameter. We can then obtain solutions for all values of the parameter by introducing appropriate mild assumptions.

In Chap. 11, we study the equation

$$\mathcal{A}u = f(x, u), \quad u \in D(\mathcal{A}) \tag{63}$$

by analyzing the "core" function

$$H(x,t) := tf(x,t) - 2F(x,t),$$

where

$$F(x,t) := \int_0^t f(x,s)\,ds.$$

In particular, if there is a number τ satisfying $0 < \tau < 2$ such that $H(x,t)$ satisfies an inequality of the form

$$|H(x,t)| \le W(x)(|t|^\tau + 1), \quad x \in \Omega,\ t \in \mathbb{R}, \tag{64}$$

we have

Theorem 12 *If (64) holds, then $f(x,t)$ satisfies an inequality of the form*

$$|f(x,t)| \le V(x)(|t| + 1), \quad x \in \Omega,\ t \in \mathbb{R},$$

and there exist functions $\alpha_\pm(x)$ such that

$$f(x,t)/t \to \alpha_\pm(x) \text{ as } t \to \pm\infty.$$

Moreover, the function $f(x,t)$ is completely determined by the core function H(x,t) and the limits $\alpha_\pm(x)$.

We study this phenomenon in detail in the chapter.

In Chap. 12, we adjust our method of sandwich pairs to obtain results such as

Theorem 13 *Consider the problem*

$$-\Delta u = f(x,u),\ x \in \Omega; \quad u = 0 \text{ on } \partial\Omega, \tag{65}$$

where $\Omega \subset \mathbb{R}^n$ is a bounded domain whose boundary is a smooth manifold, and $f(x,t)$ is a continuous function on $\bar{\Omega} \times \mathbb{R}$. Assume the following.

(a_1) *There are constants $c_1,\ c_2 \ge 0$ such that*

$$|f(x,t)| \le c_1 + c_2|t|^s,$$

where $0 \le s < (n+2)/(n-2)$ if $n > 2$.

(a_2)

$$f(x,t) = o(t) \text{ as } t \to 0, \text{ uniformly in } x.$$

(a_3) *Either*

$$F(x,t)/t^2 \to \infty \text{ as } t \to \infty$$

or

$$F(x,t)/t^2 \to \infty \text{ as } t \to -\infty,$$

where

$$F(x,t) = \int_0^t f(x,s)\,ds.$$

Then the boundary value problem

$$-\Delta u = \beta f(x,u),\ x \in \Omega; \quad u = 0 \text{ on } \partial\Omega, \tag{66}$$

has a nontrivial solution for almost every positive β.

In proving such theorems, we introduce the concept of **custom sandwich pairs.** These are sandwich pairs such that

$$a_0 := \sup_A G < \infty, \qquad b_0 := \inf_B G > G(0) \tag{67}$$

implies the existence of a PS sequence. One example of such a pair is

Theorem 14 *Let A be a continuous curve in E connecting 0 and ∞, and let B the boundary of a bounded open set in E containing 0. Then A,B form a custom sandwich pair.*

In Chap. 13, we consider systems of the form

$$-\mathcal{A}v = f(x,v,w) \tag{68}$$

$$\lambda\mathcal{B}w = g(x,v,w), \tag{69}$$

where $\mathcal{A}$, $\mathcal{B}$ are positive, self-adjoint operators on $L^2(\Omega)$ with compact resolvents, $\Omega \subset \mathbb{R}^n$,

$$f(x,v,w) = \partial F/\partial v, \quad g(x,v,w) = \partial F/\partial w \tag{70}$$

are Carathéodory functions satisfying

$$|f(x,v,w)| + |g(x,v,w)| \le C_0(|v| + |w| + 1), \quad v,w \in \mathbb{R}, \tag{71}$$

and $F(x,v,w)$ is a Carathéodory function on $\Omega \times \mathbb{R}^2$. (The reason for the negative sign is that a positive sign leads to a simple minimization problem when $\lambda > 0$.) Let $\lambda_0(\mu_0)$ be the lowest eigenvalue of $\mathcal{A}(\mathcal{B})$. These are assumed positive. The theorems we obtain include the following.

Theorem 15 *Assume*

$$2F(x, s, 0) \geq -\lambda(x)s^2 - W_1(x), \quad x \in \Omega, s \in \mathbb{R} \tag{72}$$

and

$$2F(x, 0, t) \leq \mu(x)t^2 + W_2(x), \quad x \in \Omega, s, t \in \mathbb{R}, \tag{73}$$

where $W_i(x) \in L^1(\Omega)$ *and*

$$\lambda(x) \leq\not\equiv \lambda_0, \ \mu(x) \leq\not\equiv \mu_0, \quad x \in \Omega. \tag{74}$$

Then the system (68), (69) has a solution for a.e. $\lambda \geq \beta/\mu_0$, *where* $\beta = \sup \mu(x)$.

Theorem 16 *In addition, assume that the eigenfunctions of* λ_0 *and* μ_0 *are bounded and* $\neq 0$ *a.e. in* Ω, *and there is a* $q > 2$ *such that*

$$\|w\|_q^2 \leq Cb(w), \quad w \in M. \tag{75}$$

Assume that for some $\delta > 0$,

$$2F(x, s, t) \leq \beta t^2 - \lambda_0 s^2, \quad |t| + |s| \leq \delta, \tag{76}$$

where $\beta = \sup \mu(x)$. *Then the system (68), (69) has a nontrivial solution for a.e.* $\lambda \geq \beta/\mu_0$.

Of course, in proving these theorems we use the monotonicity trick.

In Chap. 14, we consider the p-Laplacian problem

$$\begin{cases} -\Delta_p u = f(x, u) & \text{in } \Omega \\ u = 0 & \text{on } \partial\Omega, \end{cases} \tag{77}$$

where Ω is a bounded domain in $\mathbb{R}^n$, $n \geq 1$, $\Delta_p u = \operatorname{div}\left(|\nabla u|^{p-2}\, \nabla u\right)$ is the p-Laplacian of u, $1 < p < \infty$, and f is a Carathéodory function on $\Omega \times \mathbb{R}$ with subcritical growth. Solutions of (77) coincide with the critical points of the C^1 functional

$$G(u) = \int_\Omega |\nabla u|^p - p\, F(x, u), \tag{78}$$

where $F(x, t) = \displaystyle\int_0^t f(x, s)\, ds$, defined on the Sobolev space $W_0^{1,\, p}(\Omega)$.

The nonlinear eigenvalue problem

$$\begin{cases} -\Delta_p u = \lambda\, |u|^{p-2}\, u & \text{in } \Omega \\ u = 0 & \text{on } \partial\Omega \end{cases} \tag{79}$$

has eigenvalues satisfying $0 < \lambda_1 < \lambda_2 \leq \cdots \to \infty$.

Setting $H(x,t) = p\,F(x,t) - tf(x,t)$, we prove existence theorems such as follows.

Theorem 17 *If* $\lambda_l < \lambda_{l+1}$ *and*

$$(\lambda_l + \varepsilon)\,|t|^p - W(x) \leq p\,F(x,t) \leq \lambda_{l+1}\,|t|^p + W(x), \quad t \in \mathbb{R} \tag{80}$$

for some $l, \varepsilon > 0$ *and* $W \in L^1(\Omega)$, *and*

$$H(x,t) \leq C\,(|t|^\alpha + 1), \qquad \overline{H}(x) := \limsup_{|t|\to\infty} \frac{H(x,t)}{|t|^\alpha} < 0 \ \text{a.e.}, \tag{81}$$

for some α *satisfying* $0 < \alpha \leq p$, *then* (77) *has a solution. If*

$$\lambda_l\,|t|^p - W(x) \leq p\,F(x,t) \leq (\lambda_{l+1} - \varepsilon)\,|t|^p + W(x), \quad t \in \mathbb{R} \tag{82}$$

for some $l, \varepsilon > 0$ *and* $W \in L^1(\Omega)$, *and*

$$H(x,t) \geq -C\,(|t|^\alpha + 1), \qquad \underline{H}(x) := \liminf_{|t|\to\infty} \frac{H(x,t)}{|t|^\alpha} > 0 \ \text{a.e.}, \tag{83}$$

then (77) *has a solution.*

In Chap. 15 we study periodic solutions of the Dirichlet problem for the semilinear wave equation:

$$\Box u - \mu u := u_{tt} - u_{rr} - \mu u = p(t,r,u), \quad t \in \mathbb{R}, \quad 0 < r < R, \tag{84}$$

$$u(t,R) = u(t,0) = 0, \quad t \in \mathbb{R}, \tag{85}$$

$$u(t+T,r) = u(t,r), \quad t \in \mathbb{R}, \quad 0 \leq r \leq R. \tag{86}$$

Our basic assumption is that the ratio R/T is rational. Thus, we can write

$$2R/T = a/b, \tag{87}$$

where a, b are relatively prime positive integers. We also assume

$$|p(t,r,s)| \leq C(|s| + 1), \quad s \in \mathbb{R}. \tag{88}$$

We have the following.

Theorem 18 *Under assumptions (85)–(87), the operator* $\Box$ *has a self-adjoint extension L having discrete spectrum except for the point* $\{0\}$. *Assume that*

$$-\alpha s^2 - W_1(t,r) \leq 2P(t,r,s) \leq \beta s^2 + W_2(t,r), \tag{89}$$

where

$$P(t,r,s) = \int_0^s p(t,r,\sigma)\,d\sigma, \tag{90}$$

$\mu \in (m_-, m_+)$ *is contained in the resolvent set of L,* $0 < \alpha < \min[m_+ - \mu, \mu - m_-]$, $\beta > 0$, *and the functions* W_1, W_2 *are in* $L^1(Q)$ *with* $Q = [0,T] \times [0,R]$. *We also assume that* $p(t,r,s)$ *is nondecreasing in s if* $m_- \geq 0$ *and nonincreasing in s if* $m_+ \leq 0$. *We let M be the subspace of* $E = D(|L|^{1/2})$ *on which* $L > \mu$, *and N the subspace on which* $L < \mu$. *For* $\lambda > 0$, *we define*

$$L_\lambda u = \lambda[L-\mu]w + [L-\mu]v, \quad u = v + w,\ w \in M,\ v \in N.$$

Then

$$L_1 = L - \mu.$$

We embed (84) in the problem

$$L_\lambda u = p(t,r,u), \quad u \in D(L),\ \lambda > 0. \tag{91}$$

Then (91) has at least one weak solution for a.e. $\lambda > \beta/(m_+ - \mu)$.

An important aspect of this theorem is that both subspaces M and N are permitted to be infinite dimensional. Moreover, all rational values of R/T are allowed and the hypotheses are given in terms of the primitive (90) of $p(t,r,s)$. In Chap. 16, we apply our methods to the study of light waves propagating in a photorefractive crystal. They are governed by a nonlinear Schrödinger equation. Steady-state solutions satisfy the following equation over a periodic domain $\Omega \subset \mathbb{R}^2$:

$$\Delta u = \frac{Pu}{1 + V(x) + |u|^2} + \lambda u, \tag{92}$$

where P, λ are parameters and $V(x)$ is a continuous, nonnegative function periodic in $\overline{\Omega}$. The solutions u are to be periodic in Ω with the same periods as those of Ω. This equation has the trivial solution $u = 0$. We show that there are many intervals of the parameters in which nontrivial solutions exist. The results are true in any dimension.

In Chap. 17, we study periodic solutions of the Dirichlet problem for the semilinear wave equation

$$\Box u := u_{tt} - \Delta u = f(t, x, u), \quad t \in \mathbb{R}, \quad x \in \mathcal{B}_R \tag{93}$$

$$u(t, x) = 0, \quad t \in \mathbb{R}, \quad x \in \partial\mathcal{B}_R \tag{94}$$

$$u(t + T, x) = u(t, x), \quad t \in \mathbb{R}, \quad x \in \mathcal{B}_R, \tag{95}$$

where

$$\mathcal{B}_R = \{x \in \mathbb{R}^n : |x| < R\}. \tag{96}$$

In this case we have

$$f(t, x, u) = f(t, |x|, u), \quad x \in \mathcal{B}_R.$$

The basic assumption is that the ratio R/T is rational. Thus, one can write

$$8R/T = a/b, \tag{97}$$

where a, b are relatively prime positive integers. We show that

$$n \not\equiv 3 \qquad (\text{mod } (4, a)) \tag{98}$$

implies that the linear problem corresponding to (93)–(95) has no essential spectrum. If

$$n \equiv 3 \qquad (\text{mod } (4, a)), \tag{99}$$

then the essential spectrum of the linear operator consists of precisely one point

$$\lambda_0 = -(n - 3)(n - 1)/4R^2. \tag{100}$$

Thus, the spectrum has at most one limit point. We consider the nonlinear case for $f(t, r, s)$ satisfying

$$|f(t, r, s)| \le C(|s| + 1), \quad s \in \mathbb{R},\ r = |x|. \tag{101}$$

Assume that

$$m_- s^2 - W_1(t, r) \le 2F(t, r, s) \le m_+ s^2 + W_2(t, r), \tag{102}$$

where

$$F(t, r, s) = \int_0^s f(t, r, \sigma)\, d\sigma, \tag{103}$$

(m_-, m_+) is contained in the resolvent set and the functions W_1, W_2 are in $L^1(Q, \rho)$ with $Q = [0, T] \times [0, R]$ and $\rho = r^{n-1}$. We also assume that

$$H(t, r, s) := sf(t, r, s) - 2F(t, r, s) \tag{104}$$

satisfies

$$\limsup_{|s|\to\infty} H(t, r, s)/|s| \leq h(t, r) < 0. \tag{105}$$

We have the following.

Theorem 19 *If (98) holds, then (93)–(95) has a weak rotationally invariant solution. If (99) holds and $m_- \geq \lambda_0$, assume in addition that there is a point $\mu \in (m_-, m_+)$ such that*

$$p(t, r, s) = f(t, r, s) - \mu s \tag{106}$$

is nondecreasing in s. If $m_+ \leq \lambda_0$, assume that there is such a point such that $p(t, r, s)$ is nonincreasing in s. Then (93)–(95) has a weak rotationally invariant solution.

This theorem should be compared to the results of Chap. 15. In Chap. 18, we consider the following problem:

$$\begin{cases} -\ddot{u}(t) + B(t)u = \mu \nabla_u V(t, u), \\ \\ u(T) - u(0) = \dot{u}(T) - \dot{u}(0) = 0, \end{cases} \tag{107}$$

where B is a nonnegative symmetric matrix-valued function and μ is a positive real parameter and $u : [0, T] \to \mathbb{R}^n$ is a vector-valued function. We will assume the following conditions on the matrix-valued function B in $I = [0, T]$:

(*B*1) *$B(t) = \big(b_{ij}(t)\big)$ is a symmetric matrix with each $b_{ij}(t) \in L^1(I)$.*
(*B*2) *There exists a function $\gamma \in L^1(I)$ such that $\gamma \geq 0$, $\gamma \not\equiv 0$, and*

$$B(t)x \cdot x \geq \gamma(t)|x|^2$$

for every $x \in \mathbb{R}^n$ and a.e. t in I.

The assumptions imply that there is an extension $\mathcal{D}$ of the operator

$$\mathcal{D}_0 x = -\ddot{x}(t) - B(t)x(t)$$

having a discrete, countable spectrum consisting of isolated eigenvalues of finite multiplicity with a lower bound $\lambda_0 > 0$

$$0 < \lambda_0 < \lambda_1 < \lambda_2 < \ldots < \lambda_l < \ldots . \tag{108}$$

Moreover, $d(\cdot)^{1/2}$ is equivalent to the norm of H, where $d(x) = (\mathcal{D}x, x)$ and H is the Sobolev space of functions $u \in L^2(I, \mathbb{R}^n)$ having weak derivatives $\dot{u} \in L^2(I, \mathbb{R}^n)$. In particular, there is a constant $c_0 > 0$ such that

$$\|x\|_\infty^2 \leq c_0 d(x), \quad x \in H. \tag{109}$$

We put $\lambda = 1/\mu$ and prove the following.

Theorem 20 *Assume:*

1. *There are a function $b(t) \in L^1(I)$ and positive constants m and $\theta < 2$ such that*

$$2V(t, x) \leq b(t)|x|^\theta, \quad |x| \leq m, \; x \in \mathbb{R}^n.$$

2. *There is a constant $M > K_0 = c_0 m^{\theta-2}\|b\|_1$ such that*

$$\liminf_{c \to 0} 2 \int_I V(t, c\varphi)/c^2 \|\varphi\|_2^2 > M\lambda_0, \tag{110}$$

where φ is an eigenfunction of $\mathcal{D}$ corresponding to the first eigenvalue λ_0 and c_0 satisfies (109).

Then the system (107) has a nontrivial solution u_λ satisfying

$$d(u_\lambda) < m^2/c_0, \quad G_\lambda(u_\lambda) < 0$$

for each $\lambda \in (K_0, M)$, where

$$G_\lambda(u) = \lambda d(u) - 2 \int_I V(t, u)\, dt, \quad 0 < \lambda < \infty.$$

Theorem 21 *Assume that hypotheses (1) and (2) of Theorem 20 are satisfied in addition to*

$$\liminf_{c \to \infty} 2 \int_I V(t, c\varphi)/c^2 \|\varphi\|_2^2 > M\lambda_0. \tag{111}$$

Then the system (107) has two nontrivial solutions u_λ, v_λ satisfying

$$d(u_\lambda) < m^2/c_0, \; G_\lambda(u_\lambda) < 0, \quad G_\lambda(v_\lambda) > 0$$

for almost all $\lambda \in (K_0, M)$.

Theorem 22 *Assume that hypotheses (1) and (2) of Theorem 20 are satisfied. Moreover,*

3. *The function* V *is such that*

$$V(t,x)/|x|^2 \to \infty, \quad \text{as } |x| \to \infty, \tag{112}$$

uniformly with respect to t.

4. *There is a function* $W(t) \in L^1(I)$ *such that*

$$\begin{aligned} 2V(t,x) - 2V(t,rx) + (r^2-1)x \cdot \nabla_x V(t,x) \\ \le W(t), \quad t \in I,\ x \in \mathbb{R}^n,\ r \in [0,1]. \end{aligned} \tag{113}$$

Then the system (107) has two nontrivial solutions u_λ, v_λ *satisfying*

$$d(u_\lambda) < m^2/c_0, \; G_\lambda(u_\lambda) < 0, \quad G_\lambda(v_\lambda) > 0$$

for each $\lambda \in (K_0, M)$.

Conclusion

The aim of this volume is to show that critical point theory is a powerful tool in the variational approach to solving nonlinear problems. In many cases, it obtains results that are better than those obtained by other methods.

Brooklyn, NY, USA Martin Schechter
TVSLB"O

Contents

Chapter 1
Linking Systems

1.1 Linking Pairs

Many problems arising in science and engineering call for the solving of the Euler equations of functionals, i.e., equations of the form

$$G'(u) = 0, \tag{1.1}$$

where $G(u)$ is a C^1 functional (usually representing the energy) arising from the given data. As an illustration, the equation

$$-\Delta u(x) = f(x, u(x))$$

is the Euler equation of the functional

$$G(u) = \frac{1}{2}\|\nabla u\|^2 - \int F(x, u(x))\, dx$$

on an appropriate space, where

$$F(x, t) = \int_0^t f(x, s)\, ds, \tag{1.2}$$

and the norm is that of L^2. The solving of the Euler equations is tantamount to finding critical points of the corresponding functional. The history of this approach goes back to the calculus of variations. Then the desire was to find extrema of certain expressions G (functionals). Following the approach of calculus, one tried to find all critical points of G, substitute them back in G, and see which one gives the required extremum. This worked fairly well in one dimension where $G'(u) = 0$ is an ordinary differential equation. However, in higher dimensions, it turned out that

M. Schechter, *Critical Point Theory*,
https://doi.org/10.1007/978-3-030-45603-0_1

it was easier to find the extrema of G than solve $G'(u) = 0$. This led to the approach of solving equations of the form $G'(u) = 0$ by finding extrema of G.

Thus the classical approach was to look for maxima or minima. If the functional is bounded from below and one is looking for a minimum, one can obtain a minimizing sequence satisfying

$$G(u_k) \to a = \inf G. \tag{1.3}$$

If such a sequence converges or has a convergent subsequence, then we indeed obtain a minimum. However, in dealing with such sequences it is difficult, in general, to establish the convergence of a subsequence because there is very little with which to work.

Luckily, there is some help. In such a case, one can show that there is a sequence, called a Palais-Smale PS sequence, satisfying

$$G(u_k) \to a, \quad G'(u_k) \to 0 \tag{1.4}$$

where $a = \inf G$. It is much easier to establish the existence of a convergent subsequence of a PS sequence than of a minimizing sequence. In fact, a minimizing sequence may not have a convergent subsequence while a PS sequence for the same functional does.

Actually, one can do better. If the functional $G(u)$ is bounded from below, then there exists a sequence (called a Cerami sequence) satisfying

$$G(u_k) \to a, \quad (1 + \|u_k\|)G'(u_k) \to 0 \tag{1.5}$$

for $a = \inf G$. As in the case of a PS sequence, if a Cerami sequence has a convergent subsequence, it will produce a minimum. The advantage of obtaining such a sequence is that the additional structure allows one to prove the convergence of a subsequence in cases where a corresponding PS sequence need not have a converging subsequence.

However, when the functional is not semi-bounded, there is no organized procedure for producing critical points. The same is true even for semibounded functionals if one wishes to obtain critical points which are not extrema. In this book we present several approaches which produce sequences similar to (1.5) when one is searching for critical points whether or not they are extrema.

An important method of detecting critical points is called **linking.** It was discovered that there are pairs of sets A, B such that whenever they separate a functional G, i.e., satisfy

$$a_0 := \sup_A G \leq b_0 := \inf_B G,$$

one obtains a Cerami sequence of the form

$$G(u_k) \to a < \infty, \ (1 + \|u_k\|)\|G'(u_k)\| \to 0, \tag{1.6}$$

provided the functional is bounded on bounded sets. If this sequence has a convergent subsequence, we obtain a critical point. The main question is to identify such pairs of sets. We now describe a method of obtaining them.

Let E be a Banach space. The set Φ of mappings $\Gamma(t) \in C(E \times [0, 1], E)$ is to have following properties:

(a) for each $t \in [0, 1)$, $\Gamma(t)$ is a homeomorphism of E onto itself and $\Gamma(t)^{-1}$ is continuous on $E \times [0, 1)$
(b) $\Gamma(0) = I$
(c) for each $\Gamma(t) \in \Phi$ there is a $u_0 \in E$ such that $\Gamma(1)u = u_0$ for all $u \in E$ and $\Gamma(t)u \to u_0$ as $t \to 1$ uniformly on bounded subsets of E.
(d) For each $t_0 \in [0, 1)$ and each bounded set $A \subset E$ we have

$$\sup_{\substack{0 \le t \le t_0 \\ u \in A}} \{\|\Gamma(t)u\| + \|\Gamma^{-1}(t)u\|\} < \infty.$$

Definition 1.1 A subset A of E links a subset B of E if $A \cap B = \phi$ and, for each $\Gamma(t) \in \Phi$, there is a $t \in (0, 1]$ such that $\Gamma(t)A \cap B \neq \phi$.

Once we have linking set, we can use the following:

Theorem 1.2 *Let G be a C^1-functional on E, and let A, B be subsets of E such that A links B and*

$$a_0 := \sup_A G \le b_0 := \inf_B G.$$

Assume that

$$a := \inf_{\Gamma \in \Phi} \sup_{\substack{0 \le s \le 1 \\ u \in A}} G(\Gamma(s)u) \tag{1.7}$$

is finite. Then there is a sequence $\{u_k\} \subset E$ such that

$$G(u_k) \to a, \ (1 + \|u_k\|)\|G'(u_k)\| \to 0. \tag{1.8}$$

If $\{u_k\}$ has a convergent subsequence, then there is a solution $u \in E$ of

$$G(u) = a, \quad G'(u) = 0.$$

Before we give the Proof of Theorem 1.2, we want to introduce some concepts.

1.2 Flows

Let $\mathcal{Q}$ be a set of positive functions $\rho(t)$ on $[0, \infty)$, which are

(a) locally Lipschitz continuous,
(b) monotone,
(c) satisfy

$$\int_0^\infty \frac{dt}{\rho(t)} = \infty. \tag{1.9}$$

Moreover, $\mathcal{Q}$ is to satisfy

$$\rho_1, \rho_2 \in \mathcal{Q} \Longrightarrow \max(\rho_1, \rho_2) \in \mathcal{Q},$$

and contain functions of the form

$$(1 + |t|)^\beta, \quad \beta \in \mathbb{R}.$$

Let $Q \neq \phi$ be a subset of a Banach space E, and let Σ_Q be the set of all continuous maps $\sigma = \sigma(t)$ from $E \times [0, 1]$ to E such that

1. $\sigma(0)$ is the identity map,
2. for each $t \in [0, 1]$, $\sigma(t)$ is a homeomorphism of E onto E,
3. $\sigma'(t)$ is piecewise continuous and satisfies

$$\|\sigma'(t)u\| \leq C\rho(d(\sigma(t)u, Q)), \quad u \in E, \tag{1.10}$$

for some $\rho \in \mathcal{Q}$. If $Q = \{0\}$, we write $\Sigma = \Sigma_Q$. The mappings in Σ_Q are called *flows*. We note the following.

Remark 1.3 If σ_1, σ_2 are in Σ_Q, define $\sigma_3 = \sigma_1 \circ \sigma_2$ by

$$\sigma_3(s) = \begin{cases} \sigma_1(2s), & 0 \leq s \leq \frac{1}{2}, \\ \sigma_2(2s-1)\sigma_1(1), & \frac{1}{2} < s \leq 1. \end{cases}$$

Then $\sigma_3 \in \Sigma_Q$, and $\sigma_3(1) = \sigma_2(1)\sigma_1(1)$.

Proof. The first two properties are obvious. To check the third, note that

$$\sigma_3'(s) = \begin{cases} 2\sigma_1'(2s), & 0 \leq s \leq (\frac{1}{2})_-, \\ 2\sigma_2'(2s-1)\sigma_1(1), & (\frac{1}{2})_+ \leq s \leq 1. \end{cases}$$

Thus, if

$$\|\sigma_i'(t)u\| \leq C_i\rho_i(d(\sigma_i(t)u, Q)), \quad u \in E,\ i = 1, 2, \tag{1.11}$$

then

$$\|\sigma_3'(s)u\| \leq \begin{cases} 2\|\sigma_1'(2s)u\|, & 0 \leq s \leq (\frac{1}{2})_-, \\ 2\|\sigma_2'(2s-1)\sigma_1(1)u\|, & (\frac{1}{2})_+ \leq s \leq 1, \end{cases}$$

or

$$\|\sigma_3'(s)u\| \leq \begin{cases} 2C_1\rho(d(\sigma_3(s)u, Q)), & 0 \leq s \leq (\frac{1}{2})_-, \\ 2C_2\rho(d(\sigma_3(s)u, Q)), & (\frac{1}{2})_+ \leq s \leq 1, \end{cases}$$

where $\rho = \max(\rho_1, \rho_2)$. We can now take $C_3 = 2\max(C_1, C_2)$. □

1.3 A General Linking Theorem

We now return to Theorem 1.2. A slight generalization is given by

Theorem 1.4 *Let G be a C^1-functional on E, and let A, B be subsets of E such that A links B and*

$$a_0 := \sup_A G \leq b_0 := \inf_B G.$$

Assume that

$$a := \inf_{\Gamma \in \Phi} \sup_{\substack{0 \leq s \leq 1 \\ u \in A}} G(\Gamma(s)u) \tag{1.12}$$

is finite. Then for each $Q \subset E$ and each $\rho \in \mathcal{Q}$ there is a sequence $\{u_k\} \subset E$ such that

$$G(u_k) \to a,\ \rho(d(u_k, Q))\|G'(u_k)\| \to 0. \tag{1.13}$$

If $\{u_k\}$ has a convergent subsequence, then there is a solution $u \in E$ of

$$G(u) = a, \quad G'(u) = 0.$$

Proof. By the definition of linking, for each $\Gamma \in \Phi$, $\Gamma(s)A$ intersects B for some $s \in (0, 1]$. Thus, $b_0 \leq a$. Assume first that $b_0 < a$. If (1.13) were false, there would exist a positive constant δ such that $3\delta < a - b_0$ and

$$\rho(d(u, Q))\|G'(u)\| \geq 3\delta \tag{1.14}$$

whenever

$$u \in U = \{u \in E : |G(u) - a| \leq 3\delta\}. \tag{1.15}$$

Since $G \in C^1(E, \mathbb{R})$, for each $\theta < 1$ there is a locally Lipschitz continuous mapping $Y(u)$ of $\hat{E} = \{u \in E : G'(u) \neq 0\}$ into E such that

$$\|Y(u)\| \leq 1, \ \ \theta\|G'(u)\| \leq (G'(u), Y(u)), \quad u \in \hat{E} \tag{1.16}$$

(cf., e.g., [184]). Let

$$\begin{aligned}
Q_0 &= \{u \in E : |G(u) - a| \leq 2\delta\} \\
Q_1 &= \{u \in E : |G(u) - a| \leq \delta\} \\
Q_2 &= E \setminus Q_0 \\
\eta(u) &= d(u, Q_2)/[d(u, Q_1) + d(u, Q_2)].
\end{aligned}$$

It is easily checked that $\eta(u)$ is locally Lipschitz continuous on E and satisfies

$$\begin{cases} \eta(u) = 1, & u \in Q_1, \\ \eta(u) = 0, & u \in \bar{Q}_2, \\ \eta(u) \in (0, 1), & \text{otherwise.} \end{cases} \tag{1.17}$$

Let

$$W(u) = -\eta(u)Y(u)\rho(d(u, Q)).$$

Then

$$\|W(u)\| \leq \rho(d(u, Q)), \quad u \in E.$$

By Corollary 6.6, for each $v \in E$ there is a unique solution $\sigma(t)v$ of

$$\sigma'(t) = W(\sigma(t)), \ \ t \in \mathbb{R}^+, \quad \sigma(0) = v. \tag{1.18}$$

We have

$$\begin{aligned}
dG(\sigma(t)v)/dt &= -\eta(\sigma(t)v)(G'(\sigma(t)v), Y(\sigma(t)v))\rho(d(\sigma(t)v, Q)) \quad (1.19) \\
&\leq -\theta\eta(\sigma)\|G'(\sigma)\|\rho(d(\sigma(t)v, Q)) \\
&\leq -3\delta\theta\eta(\sigma).
\end{aligned}$$

Let us denote the solution of (1.18) by $\sigma(t)u$. The mapping $\sigma(t)$ is in $C(E \times \mathbb{R}, E)$ and is called the flow generated by $W(u)$. Let

$$E_\alpha = \{u \in E : G(u) \leq \alpha\}. \tag{1.20}$$

I claim that there is a $T > 0$ such that

$$\sigma(T)E_{a+\delta} \subset E_{a-\delta}. \tag{1.21}$$

In fact, we take $T = 2/3\theta$. Let u be any element in $E_{a+\delta}$. If there is a $t_1 \in [0, T]$ such that $\sigma(t_1)u \notin Q_1$, then

$$G(\sigma(T)u) \leq G(\sigma(t_1)u) < a - \delta$$

by (1.19). Hence $\sigma(T)u \in E_{a-\delta}$. On the other hand, if $\sigma(t)u \in Q_1$ for all $t \in [0, T]$, then $\eta(\sigma(t)u) = 1$ for all $t \in [0, T]$, and (1.19) yields

$$G(\sigma(T)u) \leq G(u) - 3\delta\theta T \leq a - \delta. \tag{1.22}$$

Hence (1.21) holds. Now by (1.12) there is a $\Gamma \in \Phi$ such that

$$\Gamma(s)A \subset E_{a+\delta}, \quad 0 \leq s \leq 1. \tag{1.23}$$

Let

$$\Gamma_1(s) = \begin{cases} \sigma(2sT), & 0 \leq s \leq \frac{1}{2}, \\ \sigma(T)\Gamma(2s-1), & \frac{1}{2} < s \leq 1. \end{cases} \tag{1.24}$$

It is easily checked that $\Gamma_1 \in \Phi$. Moreover, by (1.19) and (1.24)

$$\Gamma_1(s)A = \sigma(2Ts)A \subset E_{a_0} \subset E_{a-\delta}, \quad 0 \leq s \leq \frac{1}{2},$$

while by (1.21), (1.23), and (1.24)

$$\Gamma_1(s)A = \sigma(T)\Gamma(2s-1)A \subset \sigma(T)E_{a+\delta} \subset E_{a-\delta}, \quad \frac{1}{2} < s \leq 1. \tag{1.25}$$

Thus

$$G(\Gamma_1(s)A) \leq a - \delta, \quad 0 \leq s \leq 1.$$

But this contradicts (1.12), proving the theorem for the case $b_0 < a$.

Now assume $b_0 = a$. If (1.13) were false, there would exist a positive constant $\delta > 0$ such that (1.14) holds whenever $u \in U$ given by (1.15). We follow the proof given above until we reach (1.24). We now note that

$$\sigma(t)A \cap B = \phi, \quad 0 \le t \le T. \tag{1.26}$$

For we have by (1.19)

$$G(\sigma(t)u) \le a_0 - 3\theta\delta \int_0^t \eta(\sigma(\tau)u)d\tau, \quad u \in A.$$

If $\sigma(t)u \in B$, we must have $G(\sigma(t)u) \ge b_0 \ge a_0$. The only way this can happen is if

$$\eta(\sigma(\tau)u) \equiv 0, \quad 0 \le \tau \le t.$$

But this implies $\sigma(\tau)u \in \bar{Q}_2$ for such τ, and this in turn implies

$$G(\sigma(\tau)u) < a - \delta, \quad 0 \le \tau \le t.$$

Since $a = b_0$, we cannot have $\sigma(\tau)u \in B$. Thus (1.26) holds. Let $\Gamma \in \Phi$ satisfy (1.23), and let Γ_1 be given by (1.24). Then $\Gamma_1 \in \Phi$. But (1.25) and (1.26) imply that $\Gamma_1(s)A \cap B = \phi$ for all $s \in [0, 1]$, contradicting the fact that A links B. This completes the proof of the theorem. □

Theorem 1.5 *Let G be a C^1-functional on E, and let A, B be subsets of E such that A links B and*

$$a_0 := \sup_A G \le b_0 := \inf_B G.$$

Assume that

$$b_0 = a := \inf_{\Gamma\in\Phi} \sup_{\substack{0\le s\le 1\\ u\in A}} G(\Gamma(s)u). \tag{1.27}$$

Then for each nondecreasing $\rho \in \mathcal{Q}$ there is a sequence $\{u_k\} \subset E$ such that

$$G(u_k) \to a, \quad d(u_k, B) \to 0, \quad \rho(d(u_k, B))\|G'(u_k)\| \to 0. \tag{1.28}$$

If $\{u_k\}$ has a convergent subsequence, then there is a solution $u \in \overline{B}$ of

$$G(u) = a, \quad G'(u) = 0.$$

Proof of Theorem 1.5 Since $G \in C^1(E, \mathbb{R})$, for each $\theta < 1$ there is a locally Lipschitz continuous mapping $Y(u)$ of $\hat{E} = \{u \in E : G'(u) \ne 0\}$ into E such that

$$\|Y(u)\| \le 1, \ \theta\|G'(u)\| \le (G'(u), Y(u)), \quad u \in \hat{E} \tag{1.29}$$

(cf., e.g., [184]). Take $\theta > 2/3$. If there did not exist a sequence satisfying the conclusion of the theorem, then there would be positive numbers ε, δ, T such that $2\delta/3\theta < \varepsilon T$ (e.g., take $\delta = \varepsilon^2 = T^2$) and

$$\rho(d(u, B))\|G'(u)\| \geq 3\varepsilon$$

holds whenever

$$u \in U = \{u \in E : d(u, B) \leq 4[u(T) - u(0)], \ |G(u) - a| \leq 3\delta\},$$

where $u(t)$ is the solution of

$$u'(t) = \rho(u(t)), \quad t \in [0, T], \ u(0) = d(u, B).$$

Let

$$\begin{aligned} Q_0 &= \{u \in E : d(u, B) \leq 3[u(T) - u(0)], \ |G(u) - a| \leq 2\delta\} \\ Q_1 &= \{u \in E : d(u, B) \leq 2[u(T) - u(0)], \ |G(u) - a| \leq \delta\}. \\ Q_2 &= E \setminus Q_0, \\ \eta(u) &= d(u, Q_2)/[d(u, Q_1) + d(u, Q_2)]. \end{aligned}$$

It is easily checked that $\eta(u)$ is locally Lipschitz continuous on E and satisfies

$$\begin{cases} \eta(u) = 1, & u \in Q_1, \\ \eta(u) = 0, & u \in \bar{Q}_2, \\ \eta(u) \in (0, 1), & \text{otherwise}. \end{cases} \tag{1.30}$$

Let

$$W(u) = -\eta(u)Y(u)\rho(d(u, B)).$$

Then

$$\|W(u)\| \leq \rho(d(u, B)), \quad u \in E.$$

By (1.27), we see that $Q_1 \neq \phi$. By Corollary 6.6, for each $v \in E$ there is a unique solution $\sigma(t)v$ of

$$\sigma'(t) = W(\sigma(t)), \ t \in \mathbb{R}^+, \quad \sigma(0) = v. \tag{1.31}$$

We have

$$dG(\sigma(t)v)/dx = -\eta(\sigma(t)v)(G'(\sigma(t)v), Y(\sigma(t)v))\rho(d(\sigma(t)v, B)) \tag{1.32}$$

$$\leq -\theta\eta(\sigma)\|G'(\sigma)\|\rho(d(\sigma, B))$$
$$\leq -3\varepsilon\theta\eta(\sigma)$$

since $\eta(\sigma) = 0$ when $\sigma \notin Q_0$. Let u satisfy $G(u) < a + \delta$. If there is a $t_1 \leq T$ such that $\sigma(t_1)u \notin Q_1$, then either

$$G(\sigma(T)u) \leq G(\sigma(t_1)u) < a - \delta \tag{1.33}$$

or

$$d(\sigma(t_1)u, B) > 2[\tilde{u}(T) - \tilde{u}(0)], \tag{1.34}$$

where

$$\tilde{u}'(t) = \rho(\tilde{u}(t)), \quad \tilde{u}(0) = h(t_1),$$

and $h(t) := d(\sigma(t)u, B)$. By Corollary 6.6,

$$h(t + t_1) \leq \tilde{u}(t), \quad t \in [0, T - t_1].$$

Thus,

$$\begin{aligned}
\|\sigma(t + t_1)u - \sigma(t_1)u\| &\leq \int_{t_1}^{t+t_1} \rho(d(\sigma(s), B))\, ds \\
&= \int_{t_1}^{t+t_1} \rho(h(s))\, ds \\
&\leq \int_0^t \rho(\tilde{u}(s))\, ds \\
&= \tilde{u}(t) - \tilde{u}(0).
\end{aligned}$$

Consequently,

$$|h(t + t_1) - h(t_1)| \leq \tilde{u}(t) - \tilde{u}(0), \quad t \in [0, T - t_1]$$

and, in particular,

$$h(t_1) - h(t + t_1) \leq \tilde{u}(t) - \tilde{u}(0).$$

This implies

$$h(t + t_1) > h(t_1) - [\tilde{u}(t) - \tilde{u}(0)] > 0, \quad when \ \ t \in [0, T - t_1].$$

Thus

$$d(\sigma(T)u, B) > 0. \tag{1.35}$$

On the other hand, if $\sigma(t)u \in Q_1$ for all $t \in [0, T]$, then

$$G(\sigma(T)u) \leq G(u) - 3\theta\varepsilon T < a + \delta - 2\delta = a - \delta. \tag{1.36}$$

Thus we have either

$$G(\sigma(T)u) < a - \delta \tag{1.37}$$

or

$$d(\sigma(T)u, B) > 0. \tag{1.38}$$

In either case we cannot have $\sigma(T)u \in B$. Let

$$E_\alpha = \{u \in E : G(u) < \alpha\}. \tag{1.39}$$

Then

$$\sigma(T)u \cap B = \phi, \quad u \in E_{a+\delta}. \tag{1.40}$$

We also note that

$$\sigma(t)A \cap B = \phi, \quad 0 \leq t \leq T. \tag{1.41}$$

For we have by (1.32)

$$G(\sigma(t)u) \leq a_0 - 3\theta\varepsilon \int_0^t \eta(\sigma(\tau)u)d\tau, \quad u \in A.$$

If $\sigma(t)u \in B$, we must have $G(\sigma(t)u) \geq b_0 \geq a_0$. The only way this can happen is if

$$\eta(\sigma(\tau)u) \equiv 0, \quad 0 \leq \tau \leq t.$$

But this implies $\sigma(\tau)u \in \bar{Q}_2$ for such τ, and this in turn implies either

$$G(\sigma(\tau)u) < a - 2\delta, \quad 0 \leq \tau \leq t$$

or

$$d(\sigma(\tau)u, B) > 0, \quad 0 \leq \tau \leq t.$$

In either case we cannot have $\sigma(\tau)u \in B$. Thus (1.41) holds. Now by (1.27) there is a $\Gamma \in \Phi$ such that

$$\Gamma(s)A \subset E_{a+\delta}, \quad 0 \leq s \leq 1. \tag{1.42}$$

Let

$$\Gamma_1(s) = \begin{cases} \sigma(2sT), & 0 \leq s \leq \frac{1}{2}, \\ \sigma(T)\Gamma(2s-1), & \frac{1}{2} < s \leq 1. \end{cases} \tag{1.43}$$

It is easily checked that $\Gamma_1 \in \Phi$.

Let $\Gamma \in \Phi$ satisfy (1.42), and let Γ_1 be given by (1.43). Then $\Gamma_1 \in \Phi$. But (1.40) and (1.41) imply that $\Gamma_1(s)A \cap B = \phi$ for all $s \in [0, 1]$, contradicting the fact that A links B. This completes the proof of the theorem. □

Proof of Theorem 1.2 We use Theorem 1.4 with $Q = \{0\}$ and $\rho(t) = 1 + |t|$. If $G(u)$ maps bounded sets into bounded sets, then a given by (1.12) is finite. □

Theorem 1.6 *Let G be a C^1-functional on E and let A be a subset of E such that the quantity a given by (1.12) is finite. Assume that, for each $\Gamma \in \Phi$, the set*

$$g_\Gamma := \{v = \Gamma(s)u : s \in (0, 1], u \in A, v \notin A, G(v) \geq a_0\} \tag{1.44}$$

is not empty, where

$$a_0 := \sup_A G.$$

Then for each $Q \subset E$ and each $\rho \in \mathcal{Q}$ there is a sequence $\{u_k\} \subset E$ such that

$$G(u_k) \to a, \quad \rho(d(u_k, Q))\|G'(u_k)\| \to 0. \tag{1.45}$$

If $\{u_k\}$ has a convergent subsequence, then there is a solution $u \in E$ of

$$G(u) = a, \quad G'(u) = 0.$$

Proof. Let

$$B = \bigcup_{\Gamma \in \Phi} g_\Gamma. \tag{1.46}$$

Then $A \cap B = \phi$, and for each $\Gamma \in \Phi$, there are a $v \in B$, an $s \in (0, 1]$, and a $u \in A$ such that $v = \Gamma(s)u$. Thus $\Gamma(s)A \cap B \neq \phi$. This means that A links B. Since $a_0 \leq G(v)$ for all $v \in B$, we have $a_0 \leq b_0$. We can now apply Theorem 1.4 to conclude that a sequence satisfying (1.45) exists. □

Corollary 1.7 *If $a < \infty$ and $a_0 \neq a$, then a sequence satisfying (1.45) exists.*

Proof. If $a_0 < a$, then for each $\Gamma \in \Phi$ there are a $u \in A$, $s \in (0, 1]$ such that $G(\Gamma(s)u) > a_0$. Clearly $v = \Gamma(s)u \notin A$. Thus the set g_Γ given by (1.44) is not empty. We can now apply Theorem 1.4. □

Corollary 1.8 *There is a $B \subset E$ such that A links B and $a_0 \leq b_0$ if and only if g_Γ defined by (1.44) is not empty for each $\Gamma \in \Phi$.*

Proof of Corollary 1.8 The "if" part was already proved in the Proof of Theorem 1.6. Conversely, suppose A links B and $a_0 \leq b_0$. If $g_\Gamma = \phi$ for some $\Gamma \in \Phi$, then $\Gamma(s)A \cap B = \phi$ for each $s \in [0, 1]$. This says that A does not link B, a contradiction. □

Corollary 1.9 *If $a < \infty$ and for each $\Gamma \in \Phi$*

$$\max_{0 \leq s \leq 1, u \in A} G(\Gamma(s)u) \tag{1.47}$$

is attained at a point not in A, then there is a sequence satisfying (1.45).

Proof. If the maximum (1.47) is attained at a point outside A, then this point is in g_Γ. Hence g_Γ is not empty for any $\Gamma \in \Phi$. Apply Theorem 1.6. □

Theorem 1.10 *Let N be a finite dimensional subspace of a Banach space E, and let Ω be a bounded open subset of N containing a point p. Let F be a continuous map of E onto N such that F is bijective on Ω. Then $\partial\Omega$ links $F^{-1}(p)$.*

Proof. We may take $p = 0$. Assume that $\partial\Omega$ does not link $F^{-1}(0)$. Then there is a $\Gamma \in \Phi$ such that

$$\Gamma(t)\partial\Omega \cap F^{-1}(0) = \phi, \quad 0 \leq t \leq 1$$

or, equivalently,

$$F(\Gamma(t)\partial\Omega) \cap \{0\} = \phi, \quad 0 \leq t \leq 1. \tag{1.48}$$

Let

$$\gamma(t) = F \circ \Gamma(t) \circ F_0^{-1},$$

where $F_0 = F|_N$. Then $\gamma(t) \in C(\bar{\Omega}, N)$ for each $t \in [0, 1]$. Also

$$\gamma(0) = I. \tag{1.49}$$

If $\Gamma(1)E = \{u_0\}$, then

$$\gamma(1)x = F(u_0) \neq 0, \quad x \in \overline{\Omega}, \tag{1.50}$$

since

$$F(\Gamma(1)\partial\Omega) \cap \{0\} = \phi$$

by (1.48). In view of (1.48) and (1.49), the Brouwer degree satisfies

$$i(\gamma(t), \Omega, 0) = i(\gamma(0), \Omega, 0) = 1$$

for all $t \in [0, 1]$. But this contradicts (1.50). Hence ∂Q links $F^{-1}(0)$. □

1.4 Linking Sets

In order to apply Theorem 1.2, we must find sets that link in the sense of Definition 1.1. We give a partial list below.

Example 1 Let B be an open set in E, and let A consist of two points e_1, e_2 with $e_1 \in B$ and $e_2 \notin \bar{B}$. Then A links ∂B. ∂B links A as well if ∂B is bounded.

Example 2 Let M, N be closed subspaces such that $\dim N < \infty$ and $E = M \oplus N$. Let

$$\mathcal{B}_R = \{u \in E : \|u\| < R\} \tag{1.51}$$

and take $A = \partial\mathcal{B}_R \cap N, \; B = M$. Then A links B.

Example 3 We take M, N as in Example 2. Let $w_0 \neq 0$ be an element of M, and take

$$A = \{v \in N : \|v\| \leq R\} \cup \{sw_0 + v : v \in N, s \geq 0, \|sw_0 + v\| = R\},$$
$$B = \partial\mathcal{B}_\delta \cap M, \; 0 < \delta < R.$$

Then A and B link each other.

Example 4 Take M, N as before and let $v_0 \neq 0$ be an element of N. We write $N = \{v_0\} \oplus N'$. We take

$$A = \{v' \in N' : \|v'\| \leq R\} \cup \{sv_0 + v' : v' \in N', s \geq 0, \|sv_0 + v'\| = R\},$$
$$B = \{w \in M : \|w\| \geq \delta\} \cup \{sv_0 + w : w \in M, s \geq 0, \|sv_0 + w\| = \delta\},$$

where $0 < \delta < R$. Then A links B.

Example 5 This is the same as Example 4 with A replaced by $A = \partial\mathcal{B}_R \cap N$.

Example 6 Let M, N be as in Example 2. Take $A = \partial\mathcal{B}_\delta \cap N$, and let v_0 be any element in $\partial\mathcal{B}_1 \cap N$. Take B to be the set of all u of the form

$$u = w + sv_0, \ w \in M,$$

satisfying any of the following:

(a) $\|w\| \leq R, \ s = 0$
(b) $\|w\| \leq R, \ s = 2R_0$
(c) $\|w\| = R, \ 0 \leq s \leq 2R_0$,

where $0 < \delta < \min(R, R_0)$. Then A and B link each other.

Example 7 Let M, N be as in Example 2. Let v_0 be in $\partial\mathcal{B}_1 \cap N$ and write $N = \{v_0\} \oplus N'$. Let

$$A = \partial\mathcal{B}_\delta \cap N, \ Q = \bar{\mathcal{B}}_\delta \cap N,$$

and

$$B = \{w \in M : \|w\| \leq R\} \cup \{w + sv_0 : w \in M, s \geq 0, \|w + sv_0\| = R\},$$

where $0 < \delta < R$. Then A and B link each other.

Example 8 Let M, N be closed subspaces of E, one of which is finite-dimensional, and such that

$$E = M \oplus N.$$

If

$$\mathcal{B}_R := \{u \in E : \|u\| < R\},$$

then $M \cap \partial\mathcal{B}_R$ links N for each $R > 0$.

Example 9 Let M, N be closed subspaces of E such that

$$E = M \oplus N,$$

with one of them being finite-dimensional. Let w_0 be an element of $M \setminus \{0\}$, and let $0 < \delta < r < R$. Take

$$\begin{aligned} A = \{v \in N : \delta \leq \|v\| \leq R\} &\cup \{sw_0 + v : v \in N, s \geq 0, \|sw_0 + v\| = \delta\} \\ &\cup \{sw_0 + v : v \in N, s \geq 0, \|sw_0 + v\| = R\}, \\ B = \partial\mathcal{B}_r \cap M, \ 0 < \delta < r < R. \end{aligned}$$

Then A and B link each other.

Example 10 Let M, N be closed subspaces of E such that

$$E = M \oplus N,$$

with one of them being finite-dimensional. Let w_0 be an element of $M \setminus \{0\}$, and let $0 < r < R$,

$$A = \{w \in M : \|w\| = R\},$$

$$B = \{v \in N : \|v\| \geq r\} \cup \{u = v + sw_0 : v \in N, s \geq 0, \|u\| = r\}.$$

Then A links B.

Example 11 Let M, N be as in Example 2. Take $A = \partial\mathcal{B}_\delta \cap N$, and let v_0 be any element in $\partial\mathcal{B}_1 \cap N$. Take B to be the set of all u of the form

$$u = w + sv_0, \quad w \in M,$$

satisfying any of the following:

(a) $s = 0$
(b) $s = 2R_0$

where $0 < \delta < R_0$. Then A links B.

Example 12 Let M, N be as in Example 2. Take $A = \partial\mathcal{B}_\delta \cap N$, and let v_0 be any element in $\partial\mathcal{B}_1 \cap N$. Take B to be the set of all u of the form

$$u = w + sv_0, \quad w \in M,$$

satisfying any of the following:

(a) $\|w\| \leq R, \ s = 0,$
(b) $\|w\| = R, \ s > 0,$

where $0 < \delta < \infty$. Then A links B.

Example 13 Let M be a closed subspace of a Hilbert space E with complement $N \oplus \{v_0\}$, where v_0 is an element in E having unit norm, and let δ be any positive number. Let $\varphi(t) \in C^1(\mathbb{R})$ be such that

$$0 \leq \varphi(t) \leq 1, \quad \varphi(0) = 1,$$

and

$$\varphi(t) = 0, \quad |t| \geq 1.$$

Let

$$F(v+w+sv_0)=w+[s+\delta-\delta\varphi(\|v\|^2/\delta^2)]v_0,\quad v\in N,\ w\in M,\ s\in\mathbb{R}.$$

Assume that one of the subspaces M, N is finite-dimensional. Take

$$A=[M\oplus\{v_0\}]\cap\partial\mathcal{B}_R$$

and

$$B=\{v+rv_0: v\in N, r=\delta\varphi(\|v\|^2/\delta^2)\}.$$

Then A links B provided $0<\delta<R$.

1.5 Various Geometries

We now apply the theorems of the preceding sections to various geometries in Banach space. As before, we assume that $G\in C^1(E,\mathbb{R})$ and that $\rho\in\mathcal{Q}$ with $\psi=1/\rho$.

Theorem 1.11 *Assume that there is a* $\delta>0$ *such that*

$$G(0)\le\alpha\le G(u),\quad u\in\partial\mathcal{B}_\delta,\tag{1.52}$$

and that there are a $R_0<\infty$ *and a* $\varphi_0\in\partial\mathcal{B}_1$ *such that*

$$G(R\varphi_0)\le\gamma,\quad R>R_0.\tag{1.53}$$

Then, for each function $\rho(t)\in\mathcal{Q}$*, there is a sequence* $\{u_k\}\subset E$ *such that*

$$G(u_k)\to c,\quad \alpha\le c\le\gamma,\quad \rho(\|u_k\|)G'(u_k)\to 0.\tag{1.54}$$

Proof. We take $A=\{0,R\varphi_0\}$, $B=\partial\mathcal{B}_\delta$. Note that a given by (1.12) is finite for each R since

$$a_R\le\max_{0\le r\le R}G(r\varphi_0).$$

We note that in each case

$$a_R\le\gamma,\quad R>0,\tag{1.55}$$

since the mapping

$$\Gamma(s)u = (1-s)u \tag{1.56}$$

(which is in Φ) satisfies

$$G(\Gamma(s)u) \leq \gamma, \quad 0 \leq s \leq 1, \ u \in A. \tag{1.57}$$

This implies (1.55). Moreover, $a_R \geq \alpha$ for $R > \delta$ since

$$\Gamma(t)u \cap \mathcal{B}_\delta \neq \phi$$

for some $t \in [0, 1]$ whenever $u \in A$. We replace $\rho(t)$ with $\tilde{\rho}(t) = \rho(t+\delta)$, which is also in $\mathcal{Q}$. By Theorem 1.4, we can find a sequence satisfying

$$\alpha - (1/k) \leq G(u_k) \leq \gamma + (1/k), \quad \tilde{\rho}(d(u_k, B))G'(u_k) \to 0. \tag{1.58}$$

This implies (1.54), since $\|u\| \leq d(u, B) + \delta$. □

Theorem 1.12 *Let M, N be closed subspaces of E such that*

$$E = M \oplus N, \quad M \neq E, \quad N \neq E, \tag{1.59}$$

with

$$\dim M < \infty \ \ or \ \ \dim N < \infty. \tag{1.60}$$

Let $G \in C^1(E, \mathbb{R})$ be such that

$$G(v) \leq \gamma, \quad v \in \partial\mathcal{B}_R \cap N, \ R > R_0, \tag{1.61}$$

and

$$G(w) \geq \alpha, \quad w \in M. \tag{1.62}$$

Then, for each function $\rho(t) \in \mathcal{Q}$, there is a sequence $\{u_k\} \subset E$ such that

$$G(u_k) \to c, \quad \alpha \leq c \leq \gamma, \quad \rho(d(u_k, M))G'(u_k) \to 0. \tag{1.63}$$

Proof. This time we take A and B as in Example 2 above. Thus, A links B. Again, a given by (1.12) is finite for each R since

$$a_R \leq \max_{u \in \bar{B}_R \cap N} G(u).$$

Again we see that we can apply Theorem 1.4 to conclude that the desired sequence exists. □

Theorem 1.13 *Let M, N be as in Theorem 1.12, and let $G \in C^1(E, \mathbb{R})$ satisfy*

$$G(v) \leq \alpha, \quad v \in N, \tag{1.64}$$

$$G(w) \geq \alpha, \quad w \in \partial\mathcal{B}_\delta \cap M, \tag{1.65}$$

$$G(sw_0 + v) \leq \gamma, \quad s \geq 0, \ v \in N, \ \|sw_0 + v\| = R > R_0, \tag{1.66}$$

for some $w_0 \in \partial\mathcal{B}_1 \cap M$, where $0 < \delta < R_0$. Then, for each function $\rho \in \mathcal{Q}$, there is a sequence $\{u_k\} \subset E$ such that (1.54) holds.

Proof. Here we take A, B as in Example 3 above. Thus, A and B link each other. Here

$$A'' = \{sw_0 + v : s \geq 0, v \in N, \|sw_0 + v\| = R\}.$$

Again, for each R, the quantity a given by (1.12) is finite since

$$a_R \leq \max_Q G,$$

where

$$Q = \{sw_0 + v : s \geq 0, v \in N, \|sw_0 + v\| \leq R\}.$$

We now apply Theorem 1.4 to conclude that the desired sequence exists. □

Theorem 1.14 *Let M, N be as in Theorem 1.12, and let $v_0 \in \partial\mathcal{B}_1 \cap N$. Take $N = \{v_0\} \oplus N'$. Let $G \in C^1(E, \mathbb{R})$ be such that*

$$G(v) \leq \gamma, \quad v \in \partial\mathcal{B}_R \cap N, \ R > R_0, \tag{1.67}$$

$$G(w) \geq \alpha, \quad w \in M, \ \|w\| \geq \delta, \tag{1.68}$$

$$G(sv_0 + w) \geq \alpha, \quad s \geq 0, \ w \in M, \ \|sv_0 + w\| = \delta, \tag{1.69}$$

where $0 < \delta < R_0$. Then, for each function $\rho(t) \in \mathcal{Q}$, there is a sequence $\{u_k\} \subset E$ such that (1.54) holds.

Proof. We take A, B as in Example 5 above. Thus, A links B. As before, we note that $a_R < \infty$ for each R. Hence, (1.54) holds by Theorem 1.4. □

1.6 Notes and Remarks

The approach of this chapter follows [226]. The history of this approach can be traced back to the calculus of variations. There the problem was to find maxima or minima of G. They did this by solving (1.1) and then showing that some of the solutions are extrema. This approach worked well for one-dimensional problems. However, in higher dimensions it was realized quite early that it is easier to find maxima and minima of G than it is to solve (1.1). Consequently, the tables were turned, and critical point theory was devoted to finding extrema of G.

Previous results can be found in [4, 180]. More details can be found in [185, 193, 197, 275].

Chapter 2
Sandwich Systems

2.1 Sandwich Pairs

As we saw in Chap. 1, an important requirement for a linking pair is that they must separate the functional, i.e., they must satisfy

$$a_0 := \sup_A G \leq b_0 := \inf_B G.$$

If a linking pair does not separate the functional, nothing can be said concerning a potential critical point. This raises the questions, "Is there anything one can do if one cannot find linking sets that separate the functional?" "Are there sets that can lead to critical sequences even though they do not separate the functional?" Fortunately, there are.

One approach is to search for sets A, B such that

$$a_0 := \sup_A G < \infty, \qquad b_0 := \inf_B G > -\infty \tag{2.1}$$

implies the existence of a sequence satisfying either

$$G(u_k) \to a, \quad G'(u_k) \to 0, \tag{2.2}$$

or

$$G(u_k) \to a, \quad (1 + \|u_k\|)G'(u_k) \to 0. \tag{2.3}$$

If it implies (2.2), we call A, B a **sandwich pair**; if it implies (2.3), we call A, B a **strong sandwich pair.**

M. Schechter, *Critical Point Theory*,
https://doi.org/10.1007/978-3-030-45603-0_2

How can one find such pairs? Since so much more is demanded of them, one would expect that they are harder to find. This is indeed true, and we shall have to work quite a bit harder to find them than we did in the case of linking pairs. Moreover, we shall not be able to produce a neat list of such pairs as we did in the case of linking pairs. We shall not have a workable test to check if two sets form a sandwich, as we do for linking sets. Sometimes the criteria will be difficult to verify. Lastly, we shall need to involve linking sets in order to find sandwich sets. We shall follow many different paths. We outline one procedure here; others will be given in subsequent sections.

2.2 Some Criteria

Let $Q \neq \phi$ be a subset of a Banach space E, and let Σ_Q be the set of all continuous flows $\sigma = \sigma(t)$ from $E \times [0, 1]$ to E such that $\sigma'(t)$ is piecewise continuous and satisfies

$$\|\sigma'(t)u\| \leq C(1 + d(\sigma(t)u, Q)), \quad u \in E. \tag{2.4}$$

If $Q = \{0\}$, we write Σ for Σ_Q. We shall show that a sufficient condition for A, B to be a sandwich pair is that

$$B \cap \sigma(1)A \neq \phi, \quad \sigma \in \Sigma_E, \tag{2.5}$$

and a sufficient condition for A, B to be a strong sandwich pair is that

$$B \cap \sigma(1)A \neq \phi, \quad \sigma \in \Sigma. \tag{2.6}$$

In general, if

$$B \cap \sigma(1)A \neq \phi, \quad \sigma \in \Sigma_Q, \tag{2.7}$$

then (2.1) implies

$$G(u_k) \to a, \quad (1 + d(u_k, Q))G'(u_k) \to 0, \tag{2.8}$$

where $b_0 \leq a \leq a_0$.

How does one find sets A, B that satisfy (2.1) and (2.7)? First, one needs a set A that satisfies

$$a_0 := \sup_A G < \infty.$$

Then one needs a $b_0 > -\infty$ such that

$$g_\sigma = \{v = \sigma(1)u : u \in A, \ \sigma \in \Sigma_Q, \ G(v) \geq b_0\} \neq \phi. \tag{2.9}$$

If these are true, then we can take

$$B = \bigcup_{\sigma \in \Sigma_Q} g_\sigma.$$

It is then easy to verify that

$$\inf_B G > -\infty$$

and (2.7) holds.

Our first result is the following.

Theorem 2.1 *Let G be a C^1-functional on E, and let A be a nonempty subset of E. Let*

$$c_Q := \inf_{\sigma \in \Sigma_Q} \sup_{u \in A} G(\sigma(1)\, u). \tag{2.10}$$

Then for every Q such that c_Q is finite, and every $\rho \in \mathcal{Q}$, there is a sequence $\{u_k\} \subset E$ such that

$$G(u_k) \to c_Q, \quad \rho(d(u_k, Q))\|G'(u_k)\| \to 0. \tag{2.11}$$

If Q is bounded, then

$$(G'(u_k), u_k) \to 0. \tag{2.12}$$

We shall prove Theorem 2.1 later. We now state some consequences.

Corollary 2.2 *Under the hypotheses of Theorem 2.1, for each sequence $\nu_k \to \infty$ there is a sequence $\{u_k\} \subset E$ such that*

$$G(u_k) \to c_Q, \quad (\nu_k + d(u_k, Q))\|G'(u_k)\| \to 0. \tag{2.13}$$

Proof. For each positive n, let $\rho_n = n(\nu_n + t) \in \mathcal{Q}$. By Theorem 2.1, for each n there is a sequence $\{u_k\}$ such that

$$G(u_k) \to c_Q, \quad \rho_n(d(u_k, Q))\|G'(u_k)\| \to 0.$$

Choose one of them to satisfy

$$|G(u_n) - c_Q| < 1/n, \quad (\nu_n + d(u_n, Q))\|G'(u_n)\| \leq 1/n.$$

Let $n \to \infty$. □

Theorem 2.3 *Let A, B be subsets of E such that*

$$\sigma(1)A \cap B \neq \phi, \quad \forall \sigma \in \Sigma_Q \tag{2.14}$$

and

$$a_0 := \sup_A G < \infty, \qquad b_0 := \inf_B G > -\infty. \tag{2.15}$$

Then c_Q given by (2.10) satisfies $b_0 \leq c_Q \leq a_0$, and there is a sequence $\{u_k\} \subset E$ such that

$$G(u_k) \to c_Q, \quad (1 + d(u_k, Q))\|G'(u_k)\| \to 0. \tag{2.16}$$

If $Q = E$, then there is a sequence $\{u_k\} \subset E$ such that

$$G(u_k) \to c_Q, \quad \|G'(u_k)\| \to 0. \tag{2.17}$$

If $Q = \{0\}$, then there is a sequence $\{u_k\} \subset E$ such that

$$G(u_k) \to c_Q, \quad (1 + \|u_k\|)\|G'(u_k)\| \to 0. \tag{2.18}$$

Moreover, for each sequence $\nu_k \to \infty$ there is a sequence $\{u_k\} \subset E$ such that

$$G(u_k) \to c_Q, \quad (\nu_k + d(u_k, Q))\|G'(u_k)\| \to 0. \tag{2.19}$$

Proof. Apply Theorem 2.1. The quantity c_Q satisfies $b_0 \leq c_Q \leq a_0$ by (2.14), (2.15), and the fact that $\sigma(t)u \equiv u$ is in Σ_Q. If $Q = E$, we take $\rho(t) = 1$. If $Q = \{0\}$, we take $\rho(t) = 1 + t$. For the last statement, apply Corollary 2.2. □

Corollary 2.4 *If*

$$b_0 := \inf_E G > -\infty,$$

then there is a sequence such that

$$G(u_k) \to b_0, \quad (1 + \|u_k\|)G'(u_k) \to 0. \tag{2.20}$$

Proof. Let $\varepsilon > 0$, and let $p \in E$ be such that $G(p) < b_0 + \varepsilon$. We take $A = \{p\},\ B = E,\ Q = \{0\}$ in Theorem 2.1. Then

$$c_Q := \inf_{\sigma \in \Sigma} \sup G(\sigma(1)p) < b_0 + \varepsilon. \tag{2.21}$$

By Theorem 2.1, there is a $u \in E$ such that

$$|G(u) - c_Q| < \varepsilon, \quad (1 + \|u\|)\|G'(u)\| < \varepsilon. \tag{2.22}$$

If we take a sequence $\varepsilon_k \to 0$, then the corresponding functions satisfy

$$|G(u_k) - b_0| < 2\varepsilon_k, \quad (1 + \|u_k\|)\|G'(u_k)\| < \varepsilon_k. \tag{2.23}$$

□

2.3 Sandwich Systems

Let $Q \neq \phi$ be a subset of a Banach space E. We define a nonempty collection $\mathcal{K}_Q$ of nonempty subsets $K \subset E$ to be a **sandwich system** if $\mathcal{K}_Q$ has the following property:

For each $\sigma \in \Sigma_Q, \ K \in \mathcal{K}_Q$, there is a $\tilde{K} \in \mathcal{K}_Q$ such that $\tilde{K} \subset \sigma(1)K$.

If $Q = \{0\}$, we write $\mathcal{K} = \mathcal{K}_Q$. We have

Theorem 2.5 *Let $\mathcal{K}_Q$ be a sandwich system, and let $G(u)$ be a C^1 functional on E. Define*

$$a_Q := \inf_{K \in \mathcal{K}_Q} \sup_K G, \tag{2.24}$$

and assume that a_Q is finite. Then for each $\rho \in \mathcal{Q}$ there is a sequence $\{u_k\} \subset E$ such that

$$G(u_k) \to a_Q, \quad \rho(d(u_k, Q))\|G'(u_k)\| \to 0. \tag{2.25}$$

Theorem 2.5 will be proved at the end of this section. The following are consequences of Theorem 2.5.

Corollary 2.6 *Under the hypotheses of Theorem 2.5, for any $\nu > 0$ there is a sequence $\{u_k\} \subset E$ such that*

$$G(u_k) \to a_Q, \quad (\nu + d(u_k, Q))\|G'(u_k)\| \to 0. \tag{2.26}$$

Theorem 2.7 *Let $\mathcal{K}_Q$ be a sandwich system, and let G(u) be a C^1 functional on E. Assume that there are subsets A,B of E such that*

$$a_0 := \sup_A G < \infty, \qquad b_0 := \inf_B G > -\infty, \tag{2.27}$$

$A \in \mathcal{K}_Q$ and

$$B \cap K \neq \phi, \quad K \in \mathcal{K}_Q. \tag{2.28}$$

Then for each $\rho \in \mathcal{Q}$ *there is a constant* a_Q *and a sequence* $\{u_k\} \subset E$ *such that*

$$G(u_k) \to a_Q, \quad \rho(d(u_k, Q))\|G'(u_k)\| \to 0. \tag{2.29}$$

2.4 Q Sandwich Pairs

Definition 2.8 We shall say that the sets A,B form a Q sandwich pair if for each $G \in C^1(E, \mathbb{R})$ satisfying

$$a_0 := \sup_A G < \infty, \qquad b_0 := \inf_B G > -\infty \tag{2.30}$$

and each $\rho \in \mathcal{Q}$ there is a constant c_Q and a sequence $\{u_k\} \subset E$ such that

$$G(u_k) \to c_Q, \quad \rho(d(u_k, Q))\|G'(u_k)\| \to 0. \tag{2.31}$$

Corollary 2.9 *Let* $\mathcal{K}_Q$ *be a sandwich system, and let G(u) be a* C^1 *functional on E. Assume that there are subsets A,B of E such that*

$$a_0 := \sup_A G < \infty, \qquad b_0 := \inf_B G > -\infty, \tag{2.32}$$

$A \in \mathcal{K}_Q$ *and*

$$B \cap K \neq \phi, \quad K \in \mathcal{K}_Q. \tag{2.33}$$

Then the sets A,B form a Q sandwich pair.

Theorem 2.10 *Let* $\mathcal{K}_Q$ *be a sandwich system containing a nonempty set A. Let G(u) be a* C^1 *functional on E such that there is a constant* $b_0 > -\infty$ *satisfying*

$$g_K = \{v \in K : G(v) \geq b_0\} \neq \phi, \quad K \in \mathcal{K}_Q, \tag{2.34}$$

and

$$a_0 = \sup_A G < \infty. \tag{2.35}$$

Then for every $\rho \in \mathcal{Q}$ *there is a sequence* $\{u_k\} \subset E$ *such that*

$$G(u_k) \to c, \ b_0 \leq c \leq a_0, \quad \rho(d(u_k, Q))\|G'(u_k)\| \to 0. \tag{2.36}$$

Proof of Theorem 2.5 First we note that if the theorem was false, then there would be a $\rho \in \mathcal{Q}$ and a $\delta > 0$ such that

$$\rho(d(u,Q))\|G'(u)\| \geq 3\delta \tag{2.37}$$

when

$$u \in \{u \in E : |G(u) - a_Q| \leq 3\delta\}. \tag{2.38}$$

Since $G \in C^1(E, \mathbb{R})$, for each $\theta < 1$ there is a locally Lipschitz continuous mapping $Y(u)$ of $\hat{E} = \{u \in E : G'(u) \neq 0\}$ into E such that

$$\|Y(u)\| \leq 1, \ \theta\|G'(u)\| \leq (G'(u), Y(u)), \quad u \in \hat{E} \tag{2.39}$$

(cf., e.g., [184]). Let

$$\begin{aligned} Q_0 &= \{u \in E : |G(u) - a_Q| \leq 2\delta\}, \\ Q_1 &= \{u \in E : |G(u) - a_Q| \leq \delta\}, \\ Q_2 &= E \setminus Q_0, \\ \eta(u) &= d(u, Q_2)/[d(u, Q_1) + d(u, Q_2)]. \end{aligned}$$

It is easily checked that $\eta(u)$ is locally Lipschitz continuous on E and satisfies

$$\begin{cases} \eta(u) = 1, & u \in Q_1, \\ \eta(u) = 0, & u \in \bar{Q}_2, \\ \eta(u) \in (0,1), & \text{otherwise.} \end{cases} \tag{2.40}$$

Let

$$W(u) = -\eta(u)Y(u)\rho(d(u,Q)).$$

Then

$$\|W(u)\| \leq \rho(d(u,Q)), \quad u \in E.$$

By Theorem 6.10, for each $v \in E$ there is a unique solution $\sigma(t)v$ of

$$\sigma'(t) = W(\sigma(t)), \ t \in \mathbb{R}^+, \quad \sigma(0) = v. \tag{2.41}$$

We have

$$\begin{aligned} dG(\sigma(t)v)/dt &= -\eta(\sigma(t)v)(G'(\sigma(t)v), Y(\sigma(t)v))\rho(d(\sigma(t)v, Q)) \qquad (2.42) \\ &\leq -\theta\eta(\sigma)\|G'(\sigma)\|\rho(d(\sigma(t)v, Q)) \\ &\leq -3\delta\theta\eta(\sigma). \end{aligned}$$

By the definition (2.24) of a_Q, there is a $K \in \mathcal{K}_Q$ such that

$$G(v) < a_Q + \delta, \quad v \in K. \tag{2.43}$$

Let v be any element of K. Take $T \geq 2/3\theta$. If there is a $t_1 \leq T$ such that $\sigma(t_1)v \notin Q_1$, then

$$G(\sigma(T)v) < a_Q - \delta, \tag{2.44}$$

since

$$G(\sigma(T)v) \leq G(\sigma(t_1)v) \leq G(v) \leq a_Q + \delta$$

by (2.42) and (2.43). On the other hand, if $\sigma(t)v \in Q_1$ for all $t \in [0, T]$, then we have by (2.42)

$$G(\sigma(T)v) \leq a_Q + \delta - 3\delta\theta \int_0^T dt < a_Q - \delta.$$

Hence

$$G(\sigma(T)v) < a_Q - \delta, \quad v \in K. \tag{2.45}$$

Let

$$\sigma_1(s) = \sigma(sT). \tag{2.46}$$

Then $\sigma_1 \in \Sigma_Q$ by (2.41) and

$$G(\sigma_1(1)v) < a_Q - \delta, \quad v \in K. \tag{2.47}$$

By hypothesis, there is a $\tilde{K} \in \mathcal{K}_Q$ such that $\tilde{K} \subset \sigma_1(1)K$. This means that

$$G(w) < a_Q - \delta, \quad w \in \tilde{K}. \tag{2.48}$$

But this contradicts the definition (2.24) of a_Q. Hence (2.37) cannot hold for u satisfying (2.38). □

Proof of Corollary 2.6 Take $\rho(r) = \nu + r$ and apply Theorem 2.5. □

Proof of Theorem 2.7 We take $c = a_Q$ given by (2.24). We note that

$$b_0 = \inf_B G \leq \inf_{B \cap K} G \leq \sup_{B \cap K} G \leq \sup_K G, \quad K \in \mathcal{K}_Q.$$

Thus $a_Q \geq b_0$. Since $A \in \mathcal{K}_Q$, we see that $a_Q \leq a_0$. We can now apply Theorem 2.5 to obtain the desired result. □

Proof of Theorem 2.10 Let

$$g_K = \{u \in K : G(u) \geq b_0\}, \quad K \in \mathcal{K}_Q,$$

and

$$B = \bigcup_{K \in \mathcal{K}_Q} g_K.$$

Then

$$\inf_B G \geq b_0 > -\infty$$

and

$$B \cap K \neq \phi, \quad K \in \mathcal{K}_Q. \tag{2.49}$$

By Theorem 2.7 for each $\rho \in \mathcal{Q}$ there is a sequence $\{u_k\} \subset E$ such that

$$G(u_k) \to c, \quad \rho(d(u_k, Q))\|G'(u_k)\| \to 0, \tag{2.50}$$

where c satisfies $b_0 \leq c \leq a_0$. □

Proof of Theorem 2.1 Define

$$\mathcal{K}_Q = \{\sigma(1)A : \sigma \in \Sigma_Q\}.$$

Then $\mathcal{K}_Q$ is a sandwich system. To see this, let $K = \tilde{\sigma}(1)A$ be a set in $\mathcal{K}_Q$. If $\sigma \in \Sigma_Q$, then $\sigma_1 = \sigma \circ \tilde{\sigma}$ is also in Σ_Q. But $\sigma_1(1) = \sigma(1)\tilde{\sigma}(1)$. Thus, $\mathcal{K}_Q$ is a sandwich system and c_Q given by (2.10) equals the quantity a_Q given by (2.24). We can now apply Theorem 2.5. If $Q \subset \mathcal{B}_\nu$, we have

$$\nu + d(u, Q) \geq \|u\|.$$

Take $\rho(r) = \nu + r$. □

2.5 Notes and Remarks

The chapter follows the presentation in [196]. Sandwich pairs originated in [89, 186, 187]. The first discovered sandwich pair is described in Corollary 3.5 given in the next chapter. It consists of a closed subspace M and its orthogonal complement $M^\perp$ provided one of the subspaces is finite dimensional.

Chapter 3
Linking Sandwich Sets

3.1 Using Linking

The theory presented in Chap. 2 gives precise criteria for the existence of sandwich pairs, but it does not provide a list of such pairs. There is good reason for this. The criterion

$$\sigma(1)\, A \cap B \neq \phi, \quad \forall \sigma \in \Sigma_Q \tag{3.1}$$

is very difficult to verify in practice, while the corresponding statement for linking pairs is easier. We were able to provide a reasonable list of linking sets at the end of Chap. 1, but we have not yet been able to do so for sandwich sets. In this chapter we shall focus our attention on this matter. It turns out that we can obtain a lot of help from the theory of linking sets. For instance, we have the following.

Theorem 3.1 *Let $\{A, B\}$ be subsets of E such that A links B. For $G \in C^1(E, \mathbb{R})$, assume*

$$a_0 = \sup_A G < \infty, \quad b_0 = \inf_B G > -\infty, \tag{3.2}$$

with

$$a := \inf_{\Gamma \in \Phi} \sup_{\substack{0 \leq s \leq 1 \\ u \in A}} G(\Gamma(s)u) \tag{3.3}$$

finite. Let $A' = \{u \in A : G(u) > b_0\}$ and assume that $d = d(A', B) > 0$, where we take $d(\phi, B) = \infty$. Assume also that $\rho \in \mathcal{Q}$ satisfies

$$\beta \int_0^d \frac{dt}{\rho(t)} > a_0 - b_0 \tag{3.4}$$

M. Schechter, *Critical Point Theory*,
https://doi.org/10.1007/978-3-030-45603-0_3

for some $\beta > 0$. Then there is a sequence $\{u_k\} \subset E$ such that

$$G(u_k) \to c, \; b_0 \le c \le a, \quad \rho(d(u_k, B))\|G'(u_k)\| \le \beta. \tag{3.5}$$

Proof. The case $a_0 \le b_0 \le a$ was considered in Theorem 1.2. The case $a_0 < a$ was considered in Corollary 1.7. Therefore, we may assume that $b_0 < a_0 = a$. We note that if the theorem were false, then there would be a $\delta > 0$ such that

$$\rho(d(u, B))\|G'(u)\| \ge \beta \tag{3.6}$$

when

$$u \in U = \{u \in E : b_0 - 3\delta \le G(u) \le a_0 + 3\delta\}. \tag{3.7}$$

Since $G \in C^1(E, \mathbb{R})$, for each $\theta < 1$ there is a locally Lipschitz continuous mapping $Y(u)$ of $\hat{E} = \{u \in E : G'(u) \ne 0\}$ into E such that

$$\|Y(u)\| \le 1, \; \theta\|G'(u)\| \le (G'(u), Y(u)), \quad u \in \hat{E} \tag{3.8}$$

(cf., e.g., [193]). If necessary, we reduce δ to satisfy

$$\int_\delta^d \frac{dt}{\rho(t)} \ge \frac{a_0 - b_0 + \delta}{\beta\theta}.$$

Let

$$\begin{aligned} Q_0 &= \{u \in E : b_0 - 2\delta \le G(u) \le a_0 + 2\delta\}, \\ Q_1 &= \{u \in E : b_0 - \delta \le G(u) \le a_0 + \delta\}, \\ Q_2 &= E \setminus Q_0, \\ \eta(u) &= d(u, Q_2)/[d(u, Q_1) + d(u, Q_2)]. \end{aligned}$$

It is easily checked that $\eta(u)$ is locally Lipschitz continuous on E and satisfies

$$\begin{cases} \eta(u) = 1, & u \in Q_1, \\ \eta(u) = 0, & u \in \bar{Q}_2, \\ \eta(u) \in (0, 1), & \text{otherwise.} \end{cases} \tag{3.9}$$

Let

$$W(u) = -\eta(u)Y(u)\rho(d(u, B)).$$

Then

$$\|W(u)\| \leq \rho(d(u, B)), \quad u \in E.$$

By Corollary 6.6, for each $v \in E$ there is a unique solution $\sigma(t)v$ of

$$\sigma'(t) = W(\sigma(t)), \ t \in \mathbb{R}^+, \quad \sigma(0) = v. \tag{3.10}$$

We have

$$\begin{aligned} dG(\sigma(t)v)/dt &= -\eta(\sigma(t)v)(G'(\sigma(t)v), Y(\sigma(t)v))\rho(d(\sigma(t)v, B)) \\ &\leq -\theta\eta(\sigma)\|G'(\sigma)\|\rho(d(\sigma(t)v, B)) \\ &\leq -\beta\theta\eta(\sigma). \end{aligned} \tag{3.11}$$

Let $v \in A$. Take

$$T = \int_\delta^d \frac{dt}{\rho(t)} \geq \frac{a_0 - b_0 + \delta}{\beta\theta}. \tag{3.12}$$

If there is a $t_1 \leq T$ such that $\sigma(t_1)v \notin Q_1$, then

$$G(\sigma(T)v) \leq G(\sigma(t_1)v) < b_0 - \delta. \tag{3.13}$$

On the other hand, if $\sigma(t)v \in Q_1$ for all $t \in [0, T]$, then we have by (3.12)

$$G(\sigma(T)v) \leq a_0 - \beta\theta T < b_0 - \delta.$$

Hence

$$G(\sigma(T)v) < b_0 - \delta, \quad v \in A. \tag{3.14}$$

Let $u(t)$ be the solution of

$$u'(t) = -\rho(u(t)), \ t \in [0, T], \quad u(0) = d \leq d(v, B), \quad v \in A'.$$

By Corollary 6.9,

$$d(\sigma(t)v, B) \geq u(t), \ t \in [0, T].$$

But

$$\int_{u(t)}^d \frac{d\tau}{\rho(\tau)} = t, \quad t \in [0, T].$$

Consequently,

$$u(t) \geq u(T) = \delta, \quad t \in [0, T]$$

since

$$T = \int_\delta^d \frac{dt}{\rho(t)} > (a_0 - b_0 + \delta)/\beta\theta.$$

Thus,

$$d(\sigma(t)v, B) \geq \delta, \quad t \in [0, T], \ v \in A'.$$

If $v \in A \backslash A'$, then $v \notin B$ and $G(\sigma(t)v) < b_0$ for $t \in (0, T]$. Thus $\sigma(t)v \cap B = \phi, \ t \in [0, T]$. This means that

$$\sigma(t)v \cap B = \phi, \quad v \in A, \ t \in [0, T].$$

Hence,

$$\sigma(t)A \cap B = \phi, \ t \in [0, T],$$

and

$$\sup_{\sigma(T)A} G \leq b_0 - \delta.$$

If we can now show that $\tilde{A} = \sigma(T)A$ links B, then we can use Theorem 1.2 to conclude that there is a sequence $\{u_k\} \subset E$ such that

$$G(u_k) \to \hat{a}, \quad \rho(d(u_k, B))\|G'(u_k)\| \to 0, \tag{3.15}$$

where

$$\hat{a} = \inf_{\Gamma \in \Phi} \sup_{0 \leq s \leq 1; \ u \in A} G(\Gamma(s)\sigma(T)u).$$

This follows from

$$\tilde{a}_0 = \sup_{\tilde{A}} G \leq b_0 \leq \hat{a}.$$

Since $b_0 \leq \hat{a} \leq a$, we can take $c = \hat{a}$.

It remains only to show that $\sigma(T)A$ links B. Let Γ be any map in Φ, and let

$$\begin{aligned} \Gamma_1(s) &= \sigma(2sT), & 0 \leq s \leq \frac{1}{2}. \\ &= \Gamma(2s-1)\sigma(T), & \frac{1}{2} < s \leq 1. \end{aligned}$$

Then $\Gamma_1 \in \Phi$. Since A links B, there is an $s_1 \in [0, 1]$ such that $\Gamma_1(s_1)A \cap B \neq \phi$. Since $\sigma(2sT)A \cap B = \phi$ for $0 \leq s \leq \frac{1}{2}$, we must have $s_1 > \frac{1}{2}$. Hence

$$\Gamma(2s_1 - 1)\sigma(T)A \cap B \neq \phi.$$

This shows that $\sigma(T)A$ links B. □

Theorem 3.2 *Let $\{A, B\}$ be subsets of E such that there is a sequence $A_n \subset A$ such that A_n links B for each n. For $G \in C^1(E, \mathbb{R})$, assume*

$$a_0 = \sup_A G < \infty, \quad b_0 = \inf_B G > -\infty, \tag{3.16}$$

with

$$a := \inf_{\Gamma \in \Phi} \sup_{\substack{0 \leq s \leq 1 \\ u \in A}} G(\Gamma(s)u) \tag{3.17}$$

finite. Assume that $d_n = d(A_n, B) \to \infty$ as $n \to \infty$. Then for each $\rho \in \mathcal{Q}$ there is a sequence $\{u_k\} \subset E$ such that

$$G(u_k) \to c, \ b_0 \leq c \leq a, \quad \rho(d(u_k, B))\|G'(u_k)\| \to 0.$$

Proof. Let $\varepsilon > 0$ and take n so large that

$$\varepsilon \int_0^{d_n} \frac{dt}{\rho(t)} > a_0 - b_0.$$

By Theorem 3.1, for each n there is a sequence $\{u_k\} \subset E$ such that

$$G(u_k) \to c_n, \ b_0 \leq c_n \leq a_n \leq a, \ \rho(d(u_k, B))\|G'(u_k)\| \leq \varepsilon, \tag{3.18}$$

where

$$a_n := \inf_{\Gamma \in \Phi} \sup_{\substack{0 \leq s \leq 1 \\ u \in A_n}} G(\Gamma(s)u). \tag{3.19}$$

In particular, there is a u_n in this sequence such that

$$b_0 - \varepsilon < G(u_n) < a + \varepsilon, \quad \rho(d(u_n, B))\|G'(u_n)\| \leq \varepsilon. \tag{3.20}$$

Letting $\varepsilon \to 0$ produces a sequence satisfying (3.5). □

Corollary 3.3 *Let N be a finite dimensional subspace of a Hilbert space E, and assume that $A_R = \partial\mathcal{B}_R \cap N$ links a subset B of E for each $R \geq R_0 > 0$. Let G be a C^1-functional on E such that*

$$-\infty < b_0 = \inf_B G, \quad \sup_{A^R} G \leq a_0 < \infty, \tag{3.21}$$

for $R > R_0$, where $A^R = N\backslash\mathcal{B}_R$. Assume also that

$$d(A_R, B) \to \infty, \quad R \to \infty.$$

Then for each $\rho \in \mathcal{Q}$ there is a sequence $\{u_k\} \subset E$ such that

$$G(u_k) \to c, \; b_0 \leq c \leq a_0, \quad \rho(d(u_k, B))\|G'(u_k)\| \to 0. \tag{3.22}$$

Proof. Let $R_0 \leq R_k \to \infty$. Then $A_k = A_{R_k}$ links B for each k. Note that

$$\inf_{\Gamma\in\Phi} \sup_{\substack{0\leq s\leq 1\\ u\in A_k}} G(\Gamma(s)u) \tag{3.23}$$

is finite for each $R_k > R_0$ since N is finite dimensional and $\sup_{\mathcal{B}_{R_k}} G < \infty$. We can now apply Theorem 3.2 to any $A^R = N/\mathcal{B}_R$, when $R > R_0$. □

Theorem 3.4 *Let N be a finite dimensional subspace of a Banach space E, and for each $R > 0$ let $\Omega_R(p)$ be an open bounded set in N containing a point p such that $d(\partial\Omega_R(p), p) \to \infty$ as $R \to \infty$. Let F be a continuous map of E onto N such that $F = I$ on N. Assume also that*

$$d(A_R, F^{-1}(p)) \to \infty, \quad R \to \infty,$$

where $A_R = \partial\Omega_R(p)$. Then for each $R > 0$ and each $\rho \in \mathcal{Q}$, $A^R = N\backslash\Omega_R(p)$ and $B = F^{-1}(p)$ form a B sandwich pair.

Proof. Note that $A_R = \partial\Omega_R(p)$ links B (cf., e.g., Theorem 1.10) for each $R > 0$. Also note that

$$\inf_{\Gamma\in\Phi} \sup_{\substack{0\leq s\leq 1\\ u\in A_R}} G(\Gamma(s)u) \tag{3.24}$$

is finite for $R > R_0$ since N is finite dimensional and $\sup_{\Omega_R(p)} G < \infty$. Apply Corollary 6.6. □

Corollary 3.5 *Let N be a finite dimensional subspace of a Hilbert space E and let $M = N^{\perp}$. Then M,N form an M sandwich pair.*

Proof. We let F be the (orthogonal) projection of E onto N. Then F satisfies the hypotheses of Theorem 3.4, with $M = F^{-1}(0)$. Thus for each $R > 0$ and each $\rho \in \mathcal{Q}$, $A^R = N \backslash \Omega_R(0)$ and $M = F^{-1}(0)$ form an M sandwich pair. □

Theorem 3.6 *Let N be a finite dimensional subspace of a Hilbert space E with complement $M \oplus \{v_0\}$, where v_0 is an element in E having unit norm, and let δ be any positive number. Let $\varphi(t) \in C^1(\mathbb{R})$ be such that*

$$0 \leq \varphi(t) \leq 1, \ \varphi(0) = 1,$$

and

$$\varphi(t) = 0, \quad |t| \geq 1.$$

Let

$$F(v + w + sv_0) = v + [s + \delta - \delta\varphi(\|w\|^2/\delta^2)]v_0, \quad w \in M, \ v \in N, \ s \in \mathbb{R}. \tag{3.25}$$

Let G be a C^1-functional on E such that 3.21 holds with $A = [N \oplus \{v_0\}] \backslash \mathcal{B}_{R_0}$ and $B = F^{-1}(\delta v_0) = \{w + rv_0 : w \in M, \ r = \delta\varphi(\|w\|^2/\delta^2)\}$. Then for any sequence $\{R_k\} \subset \mathbb{R}_0^+$, such that $R_k \to \infty$, there are a constant $c \in \mathbb{R}$ and a sequence $\{u_k\} \subset E$ satisfying (3.22).

Proof. Let $A_R = \partial\mathcal{B}_R \cap [N \oplus \{v_0\}]$ and $B = F^{-1}(\delta v_0)$. Then A_R links B for $R > \delta$ (Theorem 1.10). Now,

$$A_R = \{v + sv_0 : \|v\|^2 + s^2 = R^2, \ v \in N\}$$

and

$$B = \{w + rv_0 : w \in M, \ r = \delta\varphi(\|w\|^2/\delta^2)\}.$$

Thus,

$$\begin{aligned} d(A_R, B)^2 &= \inf \|v + sv_0 - w - rv_0\|^2 \\ &= \|v\|^2 + \|w\|^2 + (s - r)^2 \\ &= R^2 - s^2 + \|w\|^2 + s^2 - 2s\delta\varphi(\|w\|^2/\delta^2) + \delta^2\varphi(\|w\|^2/\delta^2)^2 \\ &\geq R^2 - 2R\delta\varphi(\|w\|^2/\delta^2) + \delta^2\varphi(\|w\|^2/\delta^2)^2 \\ &= [R - \delta\varphi(\|w\|^2/\delta^2)]^2 \\ &\geq (R - \delta)^2 \to \infty \ \ as \ \ R \to \infty. \end{aligned}$$

This implies (3.22). □

Corollary 3.7 *Let N be a finite dimensional subspace of a Hilbert space E with complement $M \oplus \{v_0\}$, where v_0 is an element in E having unit norm, and let δ be any positive number. Let φ and F be defined as in Theorem 3.6. Then $A = [N \oplus \{v_0\}]\backslash \mathcal{B}_{R_0}$ and $B = F^{-1}(\delta v_0)$ form a strong sandwich pair for $R_0 > \delta$.*

Theorem 3.8 *Let N be a finite dimensional subspace of a Banach space E, and let p be any point of N. Let F be a continuous map of E onto N such that $F=I$ on N and*

$$\|Fv - Fw\| \leq C\|v - w\|, \quad v, w \in E. \tag{3.26}$$

Let G be a C^1-functional on E such that

$$-\infty < b_0 = \inf_B G, \quad \sup_{A^R} G \leq a_0 < \infty, \tag{3.27}$$

for $R > R_0$, where $A^R = N\backslash \mathcal{B}_R$ and $B = F^{-1}(p)$. Then for each $R > 0$ and each $\rho \in \mathcal{Q}$, A^R and B form a B sandwich pair.

Proof. Let A_R be the set of those $u \in N$ whose distance from p is R, and let $B = F^{-1}(p)$. Then A_R links B for each $R > 0$ (Theorem 1.10). If $u \in A_R$ and $v \in B$, then $Fv = p$. Consequently,

$$\|u - p\| = \|Fu - Fv\| \leq C\|u - v\|.$$

Thus

$$R \leq Cd(u, B).$$

This implies

$$R \leq Cd(A_R, B).$$

Then $d(A_R, B) \to \infty$ as $R \to \infty$. We can now apply Theorem 3.4. □

Corollary 3.9 *Let N be a finite dimensional subspace of a Banach space E, and let p be any point of N. Let F be a continuous map of E onto N such that $F=I$ on N and*

$$\|Fv - Fw\| \leq C\|v - w\|, \quad v, w \in E. \tag{3.28}$$

Let G be a C^1-functional on E such that

$$-\infty < b_0 = \inf_B G, \quad \sup_A G \leq a_0 < \infty, \tag{3.29}$$

where $A = N$ and $B = F^{-1}(p)$. Then for each $\rho \in \mathcal{Q}$ there is a sequence $\{u_k\} \subset E$ such that

$$G(u_k) \to c,\ b_0 \le c \le a_0, \quad \rho(d(u_k, B))\|G'(u_k)\| \to 0. \tag{3.30}$$

Proof. By Theorem 3.8, for R sufficiently large and each $p \in \mathcal{Q}$,$A^R = N/\mathcal{B}_R$ and $B = F^{-1}(p)$ form a B sandwich pair. The same is true of N and B. □

Corollary 3.10 *Let N be a finite dimensional subspace of a Hilbert space E with complement M, let v_0 be an element in N having unit norm, and let $\delta < R$ be positive numbers. Let $A = N \backslash \mathcal{B}_R$ and*

$$B = \{w \in M : \|w\| \ge \delta\} \cup \{sv_0 + w : w \in M, s \ge 0, \|sv_0 + w\| = \delta\}.$$

Then A, B form a B sandwich pair.

Proof. We write $N = \{v_0\} \oplus N'$. If

$$u = v' + w + sv_0,\ v' \in N',\ w \in M,\ s \in \mathbb{R},$$

we take $A_R = \partial\mathcal{B}_R \cap N$ and

$$\begin{aligned} F(u) &= v' + \left(s + \delta - \sqrt{\delta^2 - \|w\|^2}\right) v_0, & \|w\| \le \delta \\ &= v' + (s + \delta)v_0, & \|w\| > \delta. \end{aligned}$$

Note that $F|_N = I$ while $F^{-1}(\delta v_0)$ is precisely the set B. Hence we can conclude via Theorem 1.10 that A_R links B. Since, $d(A_R, B) \ge R - \delta \to \infty$, we can apply Theorem 3.4 to conclude that the theorem follows. □

3.2 Examples

The following examples obey the previous theorems.

Example 1 Let B be the boundary of an open set in E containing $p \in E$, and let $A = \{p + r\varphi_0 : r \ge 0\}$, where $\varphi_0 \ne 0$ is a given element of E. If

$$a_0 := \sup_A G < \infty, \quad G(p) \le b_0 := \inf_B G, \tag{3.31}$$

then for each sequence $v_k \to \infty$, there are constants $c, C \in \mathbb{R}$ and a sequence $\{u_k\} \subset E$ such that

$$G(u_k) \to c, \quad b_0 \le c \le a_0, \quad (v_k + \|u_k\|)\|G'(u_k)\| \le C. \tag{3.32}$$

Example 1a Let B be the boundary of a bounded open set in E containing $p \in E$, and let $A = \{p + r\varphi_0 : r \ge 0\}$, where $\varphi_0 \ne 0$ is a given element of B. If $G(p) \le$

$b_0 = \inf_B G > 0$, and

$$G(p + R\varphi_0) \le m_R, \tag{3.33}$$

where

$$m_R/R^{\beta+1} \to 0 \; as \; R \to \infty \; for \; some \; \beta \in \mathbb{R}, \tag{3.34}$$

then there are constants $c, C \in \mathbb{R}$ and a sequence $\{u_k\} \subset E$ such that

$$G(u_k) \to c, \quad b_0 \le c \le C, \quad \|G'(u_k)\|/(1 + d(u_k, B))^\beta \to 0. \tag{3.35}$$

Proof. For each $R > 0$, let

$$K_R = \{p, (p + R\varphi_0)\},$$

For R sufficiently large, the set K_R links B. Let

$$A'_k = \{p\}, \; A''_k = \{p + k\varphi_0\}.$$

Then

$$\sup_{A''_k} G \le m_k, \; m_k/k^{\beta+1} \to 0.$$

(If $\beta = -1$, then $m_k \to 0$, and eventually $m_k < b_0$.) Let $\rho(t) = 1/(1 + t)^\beta$. Then (3.4) holds, and we can apply Theorem 3.1. □

Lemma 3.11 *Let $\varphi \in E$, $\|\varphi\| = 1$, and let $N = \{\varphi\}$, $M = N^\perp$. If $G \in C^1(E)$ satisfies*

$$G(s\varphi) \le \alpha, \quad |s| \le \rho,$$

and

$$G(s\varphi + w) \ge \alpha, \quad (\rho - s)^2 + \|w\|^2 = \rho^2, w \in M, \|w\| < \rho, |s| \le \rho,$$

where $\rho > 0$, then $G'(0) = 0$.

Proof. To see this, let

$$G_1(s\varphi + w) = G([s + \rho - (\rho^2 - \|w\|^2)^{1/2}]\varphi + w), \quad s \in \mathbb{R}, \; w \in M, \|w\| < \rho.$$

Then $G_1 \in C^1$ for $s \in \mathbb{R}$, $\|w\| < \rho$. Moreover,

$$G_1(w) \ge \alpha, \quad w \in M, \|w\| < \rho.$$

Since $G_1(0) = G(0) \leq \alpha$, we have $D_w G_1(0) = 0$. Also,

$$D_w G_1(s\varphi + w) = D_w G + D_s G \cdot (\rho^2 - \|w\|^2)^{-1/2}) w.$$

Since $D_s G(0) = 0$, this shows that $D_w G(0) = 0$. Thus $G'(0) = D_s G(0) + D_w G(0) = 0$. □

Example 1b Assume $G(u) \in C(E, \mathbb{R})$ and

$$\sup_{r_1 \leq r \leq r_2} G(r\varphi) \leq \alpha,$$

and

$$G(u) \geq \alpha, \quad u \in \partial \mathcal{B}_\rho,$$

where

$$\mathcal{B}_R = \{u \in E : \|u\| < R\}, \tag{3.36}$$

$\|\varphi\| = 1$ and $r_1 < \rho < r_2$. Then $u_0 = \rho\varphi$ satisfies

$$G(u_0) = \alpha, \quad G'(u_0) = 0.$$

Proof. Let $A = \{r_1\varphi, r_2\varphi\}, \quad B = \partial\mathcal{B}_\rho$. Then A links B. Let $\Omega = [r_1\varphi, r_2\varphi]$. Then Ω is the convex hull of A. By Lemma 3.11, the point $u_0 \in \Omega \cap B$ satisfies

$$G(u_0) = \min_B G = \max_\Omega G. \tag{3.37}$$

This implies

$$G(u_0) = \alpha, \quad G'(u_0) = 0.$$

□

Example 2 Let M, N be closed subspaces such that $\dim N < \infty$ and $E = M \oplus N$. Take $A = \partial B_R \cap N, \ B = M$. Then A links B.

Example 2a Let M, N be closed subspaces such that $\dim N < \infty$ and $E = M \oplus N$. Take $A_R = \partial\mathcal{B}_R \cap N, \ B = M$. Then A_R links B for each $R > 0$. Assume

$$\sup_{A_R} G \leq m_R, \quad R > 0. \tag{3.38}$$

If (3.34) holds, then there is a sequence satisfying (3.35).

Proof. In this case we take $A_k' = \phi, \quad A_k'' = A_k$. Then (3.4) holds, and we can apply Theorem 3.1. □

Example 3 We take M, N as in Example 2. Let $w_0 \neq 0$ be an element of M, and take

$$A = \{v \in N : \|v\| \leq R\} \cup \{sw_0 + v : v \in N, s \geq 0, \|sw_0 + v\| = R\}$$

$$B = \partial B_\delta \cap M, \ 0 < \delta < R.$$

Then A and B link each other.

Example 3a Let M, N be closed subspaces of E such that

$$E = M \oplus N, \ M \neq E, \ N \neq E$$

with one of the subspaces finite dimensional. Let $G \in C^1(E, \mathbb{R})$ be such that

$$G(v) \leq \alpha, \quad v \in N,$$

$$G(w) \geq \alpha, \quad w \in \partial B_\delta \cap M,$$

$$G(sw_0 + v) \leq m_R, \quad s \geq 0, \ v \in N, \ \|sw_0 + v\| = R > R_0,$$

for some $w_0 \in \partial B_1 \cap M$, where $0 < \delta < R_0$. If (3.34) holds, then there is a sequence satisfying (3.35).

Example 3b Let M, N be as in Example 3.2, and let $v_0 \in \partial B_1 \cap N$. Take $N = \{v_0\} \oplus N'$. Let $G \in C^1(E, \mathbf{R})$ be such that

$$G(v) \leq m_R, \quad v \in \partial B_R \cap N,$$

$$G(w) \geq \alpha, \quad w \in M, \|w\| \geq \delta,$$

$$G(sv_0 + w) \geq \alpha, \quad s \geq 0, \ w \in M, \ \|sv_0 + w\| = \delta,$$

where $0 < \delta < R$ and $\alpha > -\infty$. If (3.34) holds, then there is a sequence satisfying (3.35).

Example 3c Let M, N be closed subspaces of E such that

$$E = M \oplus N, \ M \neq E, \ N \neq E$$

with one of the subspaces finite dimensional. Let $G \in C^1(E, \mathbb{R})$ be such that

$$G(v) \leq \alpha, \quad v \in N \cap \mathcal{B}_R$$

$$G(w) \geq \alpha, \quad w \in \partial\mathcal{B}_\delta \cap M,$$

$$G(sw_0 + v) \leq \alpha, \quad s \geq 0, \ v \in N, \ \|sw_0 + v\| \leq R > R_0,$$

for some $w_0 \in \partial\mathcal{B}_1 \cap M$, where $0 < \delta < R_0$. Then there is a

$$u_0 \in \partial\mathcal{B}_1 \cap M \cap \{sw_0 + v : s \geq 0, \ v \in N, \ \|sw_0 + v\| \leq R\}$$

such that $G'(u_0) = 0$.

Proof. Let

$$A = N \cap \mathcal{B}_R \cup \{sw_0 + v : s \geq 0, \ v \in N, \ \|sw_0 + v\| = R\}$$

and $B = \partial\mathcal{B}_\delta \cap M$. Then A and B link each other. Moreover

$$\Omega = \{sw_0 + v : s \geq 0, \ v \in N, \ \|sw_0 + v\| \leq R\}.$$

By hypothesis

$$\sup_\Omega G \leq \alpha \leq b_0 := \inf_B G.$$

In view of Corollary 3.11 there is a point $u_0 \in \Omega \cap B$ such that

$$G(u_0) = \min_B G = \max_\Omega G. \tag{3.39}$$

Since u_0 is an interior point of Ω and is a local maximum, $G'_N(u_0) = 0$. Since B is a smooth surface in M and u_0 is a minimum on B, we also have $G'_M(u_0) = 0$. Thus, $G'(u_0) = 0$. □

Example 4 Take M, N as before and let $v_0 \neq 0$ be an element of N. We write $N = \{v_0\} \oplus N'$. We take

$$A = \{v' \in N' : \|v'\| \leq R\} \cup \{sv_0 + v' : v' \in N', s \geq 0, \|sv_0 + v'\| = R\}$$
$$B = \{w \in M : \|w\| \geq \delta\} \cup \{sv_0 + w : w \in M, s \geq 0, \|sv_0 + w\| = \delta\},$$

where $0 < \delta < R$. Then A links B.

Example 4a Take M, N as before and let $v_0 \neq 0$ be an element of N. We write $N = \{v_0\} \oplus N'$. We take

$$A_R = \{v' \in N' : \|v'\| \leq R\} \cup \{sv_0 + v' : v' \in N', s \geq 0, \|sv_0 + v'\| = R\}$$
$$B = \{w \in M : \|w\| \geq \delta\} \cup \{sv_0 + w : w \in M, s \geq 0, \|sv_0 + w\| = \delta\},$$

where $0 < \delta < R$. Then A_R links B for each $R > 0$. Assume

$$b_0 = \inf_B G > -\infty$$

and

$$\sup_{A_R} G \le m_R, \quad R > 0.$$

If (3.34) holds, then there is a sequence satisfying (3.35).

Example 4b Take M, N as before and let $v_0 \neq 0$ be an element of N. We write $N = \{v_0\} \oplus N'$. We take

$$A = \{v' \in N' : \|v'\| \le R\} \cup \{sv_0 + v' : v' \in N', s \ge 0, \|sv_0 + v'\| = R\}$$
$$B = \{w \in M : \|w\| \ge \delta\} \cup \{sv_0 + w : w \in M, s \ge 0, \|sv_0 + w\| = \delta\},$$

where $0 < \delta < R$. Then A links B. Let $G \in C^1(E, \mathbb{R})$ be such that

$$G(sv_0 + v') \le \alpha, \quad v' \in N', s \ge 0, \|sv_0 + v'\| \le R$$

and

$$G(u) \ge \alpha, \quad u \in B.$$

Then there is a

$$u_0 \in \{sv_0 + v' : v' \in N', s \ge 0, \|sv_0 + v'\| \le R\} \cap B$$

such that $G'(u_0) = 0$.

Proof. Note that

$$\Omega = \{sv_0 + v' : v' \in N', s \ge 0, \|sv_0 + v'\| \le R\}.$$

By hypothesis

$$\sup_\Omega G \le \alpha \le b_0 := \inf_B G.$$

In view of Corollary 3.11 there is a point $u_0 \in \Omega \cap B$ such that

$$G(u_0) = \min_B G = \max_\Omega G. \tag{3.40}$$

Since u_0 is an interior point of Ω and is a local maximum on it, $G'_\Omega(u_0) = 0$. Since B is a smooth surface in M near u_0 and u_0 is a minimum on B, we also have $G'_M(u_0) = 0$. Thus, $G'(u_0) = 0$. □

Example 5 This is the same as Example 4 with A replaced by $A = \partial\mathcal{B}_R \cap N$.

Example 5a Take M, N as before and let $v_0 \neq 0$ be an element of N. We write $N = \{v_0\} \oplus N'$. We take

$$A_R = \partial\mathcal{B}_R \cap N$$
$$B = \{w \in M : \|w\| \geq \delta\} \cup \{sv_0 + w : w \in M, s \geq 0, \|sv_0 + w\| = \delta\},$$

where $0 < \delta < R$. Then A_R links B for each $R > 0$. Assume

$$b_0 = \inf_B G > -\infty$$

and

$$\sup_{A_R} G \leq m_R, \qquad R > 0.$$

If (3.34) holds, then there is a sequence satisfying (3.35).

Example 6 Let M, N be as in Example 2. Take $A = \partial B_\delta \cap N$, and let v_0 be any element in $\partial B_1 \cap N$. Take B to be the set of all u of the form

$$u = w + sv_0, \; w \in M$$

satisfying the following:

(a) $\|w\| \leq R, \; s = 0$
(b) $\|w\| \leq R, \; s = 2R_0$
(c) $\|w\| = R, \; 0 \leq s \leq 2R_0$,

where $0 < \delta < \min(R, R_0)$. Then A and B link each other.

Example 6a Let M, N be as in Example 2. Take $A = \partial\mathcal{B}_\delta \cap N$, and let v_0 be any element in $\partial\mathcal{B}_1 \cap N$. Take B_R to be the set of all u of the form

$$u = w + sv_0, \; w \in M$$

satisfying the following:

(a) $\|w\| \leq R, \; s = 0$
(b) $\|w\| \leq R, \; s = 2R_0$
(c) $\|w\| = R, \; 0 \leq s \leq 2R_0$,

where $0 < \delta < \min(R, R_0)$. Then A and B_R link each other. Assume

$$b_R = \inf_{B_R} G > -m_R, \quad R > 0$$

and

$$a_0 = \sup_A G < \infty.$$

If (3.34) holds, then there is a sequence satisfying (3.35).

3.3 Notes and Remarks

Sandwich pairs are discussed in [186, 187, 196]. Strong sandwich pairs were introduced in [203] and infinite dimensional sandwich pairs in [204]. Additional discussions can be found in [205, 206, 217].

Chapter 4
The Monotonicity Trick

4.1 The Basic Problem

The use of linking or sandwich pairs cannot produce critical points by themselves. The most they can produce are sequences satisfying

$$G(u_k) \to a, \quad (1 + \|u_k\|)G'(u_k) \to 0. \tag{4.1}$$

If such a sequence has a convergent subsequence, we obtain a critical point. Lacking such information, we cannot eliminate the possibility that

$$\|u_k\| \to \infty, \tag{4.2}$$

which destroys any hope of obtaining a critical point from this sequence. On the other hand, knowing that the sequence is bounded does not guarantee a critical point either. But there is a difference. In many applications, knowing that a sequence satisfying (4.1) is bounded allows one to obtain a convergent subsequence. This is just what is needed. For such applications it would be very helpful if we could obtain a bounded sequence satisfying (4.1). This leads to the question: Is there anything we can do to obtain such a sequence?

Fortunately, there is, but as usual, it comes at a price. In this chapter we explore one way of obtaining this goal. It involves embedding the functional in a family of functionals depending on a parameter and then showing that for almost all values of the parameter, members of this family produce bounded sequences satisfying (4.1). One is then left with the choice of either accepting the fact that for almost all values of the parameter the functionals have critical points, or trying to show that the particular value of the parameter that one is interested in happens to be one of the "almost all" values.

The method we use involves sandwich and linking systems.

M. Schechter, *Critical Point Theory*,
https://doi.org/10.1007/978-3-030-45603-0_4

4.2 Sandwich Systems

Let E be a Banach space and let $\tilde{\Sigma} \neq \phi$ be a subset of Σ. We define a nonempty collection $\mathcal{K}$ of nonempty subsets $K \subset E$ to be a $\tilde{\Sigma}$ **sandwich system** if

$$\sigma \in \tilde{\Sigma},\ K \in \mathcal{K} \Longrightarrow \exists \tilde{K} \in \mathcal{K} : \tilde{K} \subset \sigma(1)K.$$

We have

Theorem 4.1 *Let $G(u)$ be a C^1 functional on E and let $\mathcal{K}$ be a $\tilde{\Sigma}$ sandwich system, where*

$$\tilde{\Sigma} \supset \{\sigma \in \Sigma : G(\sigma(t)u) \leq G(u),\ u \in E,\ t \in [0,1]\}. \tag{4.3}$$

Define

$$a := \inf_{K \in \mathcal{K}} \sup_K G, \tag{4.4}$$

and assume that a is finite. Assume, in addition, that there is a constant C_0 such that for each $\delta > 0$ there is a $K \in \mathcal{K}$ satisfying

$$\sup_K G \leq a + \delta, \tag{4.5}$$

such that

$$u \in K,\ G(u) \geq a - \delta \Longrightarrow \|u\| \leq C_0. \tag{4.6}$$

Then there is a sequence $\{u_k\} \subset E$ such that

$$\|u_k\| \leq C_0 + 3, \quad G(u_k) \to a, \quad \|G'(u_k)\| \to 0. \tag{4.7}$$

Theorem 4.2 *Let $G(u)$ be a C^1 functional on E and let $\mathcal{K}$ be a $\tilde{\Sigma}$ sandwich system satisfying (4.3). Assume that there are subsets A,B of E such that*

$$a_0 := \sup_A G < \infty, \qquad b_0 := \inf_B G > -\infty, \tag{4.8}$$

$A \in \mathcal{K}$ and

$$B \cap K \neq \phi, \quad K \in \mathcal{K}. \tag{4.9}$$

Then the quantity a given by (4.4) satisfies $b_0 \leq a \leq a_0$. Assume, in addition, that there is a constant C_0 such that for each $\delta > 0$ there is a $K \in \mathcal{K}$ satisfying (4.5) and (4.6). Then there is a sequence $\{u_k\} \subset E$ such that

$$\|u_k\| \le C_0 + 3, \quad G(u_k) \to a, \quad \|G'(u_k)\| \to 0. \tag{4.10}$$

Theorem 4.3 *Let A be a subset of E. For $G \in C^1(E, \mathbb{R})$, let*

$$\tilde{\Sigma} = \{\sigma \in \Sigma : G(\sigma(t)u) \le G(u), \quad t \in [0,1], \ u \in E\},$$

and let $\mathcal{K} = \{K = \sigma(1)A : \sigma \in \tilde{\Sigma}\}$. Define

$$a := \inf_{K \in \mathcal{K}} \sup_K G, \tag{4.11}$$

and assume a is finite. Assume, in addition, that there is a constant C_0 such that for each $\delta > 0$ there is a $K \in \mathcal{K}$ satisfying (4.5) and (4.6). Then there is a bounded sequence $\{u_k\} \subset E$ such that

$$\|u_k\| \le C_0 + 3, \quad G(u_k) \to a, \quad \|G'(u_k)\| \to 0. \tag{4.12}$$

Theorem 4.4 *Let $G(u)$ be a C^1 functional on E. Let*

$$\tilde{\Sigma} = \{\sigma \in \Sigma : G(\sigma(t)u) \le G(u), \quad t \in [0,1], \ u \in E\},$$

and let $\mathcal{K} = \{K = \sigma(1)A : \sigma \in \tilde{\Sigma}\}$. Assume that there are subsets A,B of E such that

$$a_0 := \sup_A G < \infty, \qquad b_0 := \inf_B G > -\infty, \tag{4.13}$$

and

$$B \cap K \ne \phi, \quad K \in \mathcal{K}. \tag{4.14}$$

Then the quantity a given by (4.11) satisfies $b_0 \le a \le a_0$. Assume, in addition, that there is a constant C_0 such that for each $\delta > 0$ there is a $K \in \mathcal{K}$ satisfying (4.5) and (4.6). Then there is a sequence $\{u_k\} \subset E$ such that

$$\|u_k\| \le C_0 + 3, \quad G(u_k) \to a, \quad \|G'(u_k)\| \to 0. \tag{4.15}$$

4.3 Linking Systems

Let E be a Banach space and let $\tilde{\Sigma} \ne \phi$ be a subset of Σ. We define a nonempty collection $\mathcal{K}$ of nonempty subsets $K \subset E$ each containing the set A to be a $\tilde{\Sigma}$ **linking system for** A if $\mathcal{K}$ has the following property:

$$\sigma \in \tilde{\Sigma},\ K \in \mathcal{K} \Longrightarrow \exists \tilde{K} \in \mathcal{K} : \tilde{K} \subset \bigcup_{t\in[0,1]} \sigma(t)A \cup \sigma(1)K.$$

We have the following.

Theorem 4.5 *Let $G(u)$ be a C^1 functional on E and let $\mathcal{K}$ be a $\tilde{\Sigma}$ linking system for a set A, where*

$$\tilde{\Sigma} \supset \{\sigma \in \Sigma : G(\sigma(t)u) \leq G(u),\ u \in E,\ t \in [0,1]\}. \tag{4.16}$$

Define

$$a := \inf_{K\in\mathcal{K}} \sup_K G, \tag{4.17}$$

and assume that $a_0 = \sup_A G < a < \infty$. Assume, in addition, that there is a constant C_0 such that for each $\delta > 0$ there is a $K \in \mathcal{K}$ satisfying (4.5) and (4.6). Then there is a bounded sequence $\{u_k\} \subset E$ such that

$$\|u_k\| \leq C_0 + 3, \quad G(u_k) \to a, \quad \|G'(u_k)\| \to 0. \tag{4.18}$$

Theorem 4.6 *Let $G(u)$ be a C^1 functional on E and let $\mathcal{K}$ be a $\tilde{\Sigma}$ linking system for a set A, where $\tilde{\Sigma}$ satisfies (4.16). Assume that the quantity a given by (4.17) is $< \infty$ and there is a subset B of E such that*

$$a_0 := \sup_A G \leq b_0 := \inf_B G \tag{4.19}$$

and

$$B \cap K \neq \phi, \quad K \in \mathcal{K}. \tag{4.20}$$

Assume, in addition, that there is a constant C_0 such that for each $\delta > 0$ there is a $K \in \mathcal{K}$ satisfying (4.5) and (4.6). Then the quantity a given by (4.17) satisfies $a_0 \leq b_0 \leq a$ and there is a bounded sequence $\{u_k\} \subset E$ such that

$$\|u_k\| \leq C_0 + 3, \quad G(u_k) \to a, \quad \|G'(u_k)\| \to 0. \tag{4.21}$$

Theorem 4.7 *Let $G(u)$ be a C^1 functional on E and let $\mathcal{K}$ be a $\tilde{\Sigma}$ linking system for a set A, where*

$$\tilde{\Sigma} \supset \{\sigma \in \Sigma : G(\sigma(t)u) \leq G(u),\ u \in E,\ t \in [0,1]\}. \tag{4.22}$$

Define

$$a := \inf_{K\in\mathcal{K}} \sup_K G, \tag{4.23}$$

and assume that a is finite and

$$(K\backslash A) \cap E^{a_0} \neq \phi, \quad K \in \mathcal{K},$$

where $a_0 = \sup_A G$, *and* $E^\beta = \{v \in E : G(v) \geq \beta\}$. *Assume, in addition, that there is a constant* C_0 *such that for each* $\delta > 0$ *there is a* $K \in \mathcal{K}$ *satisfying (4.5) and (4.6). Then there is a bounded sequence* $\{u_k\} \subset E$ *such that*

$$\|u_k\| \leq C_0 + 3, \quad G(u_k) \to a, \quad \|G'(u_k)\| \to 0. \tag{4.24}$$

4.4 The Parameter Problem

Let E be a reflexive Banach space with norm $\|\cdot\|$. Suppose that $G \in C^1(E, \mathbb{R})$ is of the form: $G(u) := I(u) - J(u)$, $u \in E$, where $I, J \in C^1(E, \mathbb{R})$ map bounded sets to bounded sets. Define

$$G_\lambda(u) = \lambda I(u) - J(u), \quad \lambda \in \Lambda,$$

where Λ is an open interval contained in $(0, +\infty)$. Assume one of the following alternatives holds.

(H$_1$) $I(u) \geq 0$ for all $u \in E$ and $I(u) + |J(u)| \to \infty$ as $\|u\| \to \infty$.

(H$_2$) $I(u) \leq 0$ for all $u \in E$ and $|I(u)| + |J(u)| \to \infty$ as $\|u\| \to \infty$.

Furthermore, we suppose that

(H$_3$) There is a $\tilde{\Sigma}$ sandwich system $\mathcal{K}$ such that $a(\lambda) := \inf_{K \in \mathcal{K}} \sup_K G_\lambda$ is finite for each $\lambda \in \Lambda$.

Theorem 4.8 *Assume that* **(H$_1$)** *(or* **(H$_2$)***) and* **(H$_3$)** *hold. Then we have*

(1) For almost all $\lambda \in \Lambda$ *there exists a constant* $k_0(\lambda) := k_0$ *(depending only on* λ*) such that for each* $\delta > 0$ *there exists a* $K \in \mathcal{K}$ *such that*

$$\sup_K G_\lambda \leq a(\lambda) + \delta$$

and

$$\|u\| \leq k_0 \quad \text{whenever} \quad u \in K \text{ and } G_\lambda(u) \geq a(\lambda) - \delta. \tag{4.25}$$

(2) For almost all $\lambda \in \Lambda$ *there exists a bounded sequence* $u_k(\lambda) \in E$ *such that*

$$\|G_\lambda'(u_k)\| \to 0, \quad G_\lambda(u_k) \to a(\lambda) := \inf_{K \in \mathcal{K}} \sup_K G_\lambda, \quad \text{as } k \to \infty.$$

Theorem 4.9 *The conclusions of Theorem 4.8 hold if we replace hypothesis* **(H₃)** *with*

(H′₃) *There is a $\tilde{\Sigma}$ sandwich system $\mathcal{K}$ and sets A,B such that $A \in \mathcal{K}$,*

$$B \cap K \neq \phi, \quad K \in \mathcal{K} \tag{4.26}$$

and

$$a_0(\lambda) := \sup_A G_\lambda < \infty, \qquad b_0(\lambda) := \inf_B G_\lambda > -\infty \tag{4.27}$$

for each $\lambda \in \Lambda$.

Theorem 4.10 *The conclusions of Theorem 4.8 hold if we replace hypothesis* **(H₃)** *with*

(H″₃) *There is a $\tilde{\Sigma}$ linking system $\mathcal{K}$ for a set A such that*

$$a_0(\lambda) := \sup_A G_\lambda < a(\lambda) := \inf_{K \in \mathcal{K}} \sup_K G_\lambda < \infty$$

for each $\lambda \in \Lambda$.

Theorem 4.11 *The conclusions of Theorem 4.8 hold if we replace hypothesis* **(H₃)** *with*

(H‴₃) *There is a $\tilde{\Sigma}$ linking system $\mathcal{K}$ for a set A, and a set B such that*

$$B \cap K \neq \phi, \quad K \in \mathcal{K} \tag{4.28}$$

and

$$a_0(\lambda) := \sup_A G_\lambda \leq b_0(\lambda) := \inf_B G_\lambda \tag{4.29}$$

for each $\lambda \in \Lambda$.

Corollary 4.12 *The conclusions of Theorem 4.8 hold if we replace hypothesis* **(H₃)** *with*

(H⁗₃) *There is a sandwich pair A,B such that*

$$a_0 := \sup_A G_\lambda < \infty, \qquad b_0 := \inf_B G_\lambda > -\infty \tag{4.30}$$

for each $\lambda \in \Lambda$.

Corollary 4.13 *The conclusions of Theorem 4.8 hold if we replace hypothesis* **(H₃)** *with*

$(\mathbf{H}_3''''')$ *There are sets A,B such that A links B and*

$$a_0 := \sup_A G_\lambda \leq b_0 := \inf_B G_\lambda, \quad a(\lambda) := \inf_{\Gamma\in\Phi} \sup_{\substack{0\leq s\leq 1\\ u\in A}} G_\lambda(\Gamma(s)u) < \infty \tag{4.31}$$

for each $\lambda \in \Lambda$.

4.5 Finding the Sequences

Proof of Theorem 4.1. Let $M = C_0 + 1$. Then

$$\|\sigma(1)v\| \leq M$$

whenever $\sigma \in \Sigma$ satisfies $\|\sigma'(t)\| \leq 1$ and $v \in E$ satisfies $\|v\| \leq C_0$. If the theorem were false, then there would be a $\delta > 0$ such that

$$\|G'(u)\| \geq 3\delta \tag{4.32}$$

when

$$u \in \{u \in E : \|u\| \leq M + 2, |G(u) - a| \leq 3\delta\}. \tag{4.33}$$

Note that this set is not empty since for every $K \in \mathcal{K}$

$$\sup_K G \geq a, \tag{4.34}$$

and there is a $K \in \mathcal{K}$ such that

$$\sup_K G \leq a + \delta. \tag{4.35}$$

Since $G \in C^1(E, \mathbb{R})$, for each $\theta < 1$ there is a locally Lipschitz continuous mapping $Y(u)$ of $\hat{E} = \{u \in E : G'(u) \neq 0\}$ into E such that

$$\|Y(u)\| \leq 1, \ \theta\|G'(u)\| \leq (G'(u), Y(u)), \quad u \in \hat{E} \tag{4.36}$$

(cf., e.g., [193]). Take $\theta > 2/3$. Let

$$Q_0 = \{u \in E : \|u\| \leq M + 1, |G(u) - a| \leq 2\delta\},$$

$$Q_1 = \{u \in E : \|u\| \leq M, |G(u) - a| \leq \delta\},$$

$$Q_2 = E \setminus Q_0,$$

$$\eta(u) = a(u, Q_2)/[a(u, Q_1) + a(u, Q_2)].$$

It is easily checked that $\eta(u)$ is locally Lipschitz continuous on E and satisfies

$$\begin{cases} \eta(u) = 1, & u \in Q_1, \\ \eta(u) = 0, & u \in \bar{Q}_2, \\ \eta(u) \in (0, 1), & \text{otherwise.} \end{cases} \tag{4.37}$$

Let

$$W(u) = -\eta(u)Y(u).$$

Then

$$\|W(u)\| \leq 1, \quad u \in E.$$

By Theorem 6.10, for each $v \in E$ there is a unique solution $\sigma(t)v$ of

$$\sigma'(t) = W(\sigma(t)), \ t \in \mathbb{R}^+, \quad \sigma(0) = v. \tag{4.38}$$

We have

$$\begin{aligned} dG(\sigma(t)v)/dt &= -\eta(\sigma(t)v)(G'(\sigma(t)v), Y(\sigma(t)v)) \\ &\leq -\theta\eta(\sigma)\|G'(\sigma)\| \\ &\leq -3\theta\delta\eta(\sigma). \end{aligned} \tag{4.39}$$

Let $K \in \mathcal{K}$ satisfy (4.5) and (4.6), and let $u = \sigma(1)v$, where v is an element of $K \cap Q_1$. Then $\|v\| \leq C_0$ since $G(v) \geq a - \delta$. If there is a $t_1 \leq 1$ such that $\sigma(t_1)v \notin Q_1$, then

$$G(\sigma(1)v) < a - \delta, \tag{4.40}$$

since $\|\sigma(1)v\| \leq M$,

$$G(\sigma(1)v) \leq G(\sigma(t_1)v) \leq G(v) \leq a + \delta.$$

On the other hand, if $\sigma(t)v \in Q_1$ for all $t \in [0, 1]$, then we have by (4.39)

$$G(\sigma(1)v) \leq a + \delta - 3\delta\theta < a - \delta.$$

If $v \in K \backslash Q_1$, then we must have $G(v) < a - \delta$, since $G(v) \geq a - \delta$ would imply $\|v\| \leq C_0$ and put v into Q_1. Hence

$$G(\sigma(1)v) < a - \delta, \quad v \in K. \tag{4.41}$$

Since $G(\sigma(t)u) \leq G(u)$ for $u \in E$ and $t > 0$, we have $\sigma \in \tilde{\Sigma}$. Hence, there is a $\tilde{K} \in \mathcal{K}$ such that $\tilde{K} \subset \sigma(1)K$. But

$$\sup_{\tilde{K}} G \leq a - \delta.$$

This contradicts the definition of a and gives the theorem. □

Proof of Theorem 4.2. Since $A \in \mathcal{K}$, clearly $a \leq a_0$. Moreover, for any $K \in \mathcal{K}$, we have

$$b_0 = \inf_B G \leq \inf_{B \cap K} G \leq \sup_{B \cap K} G \leq \sup_K G.$$

Hence, $b_0 \leq a$. Apply Theorem 4.1. □

Proof of Theorems 4.3 and 4.4. All we need to show is that $\mathcal{K}$ is a $\tilde{\Sigma}$ sandwich system. I claim that

$$\sigma, \tilde{\sigma} \in \tilde{\Sigma} \Longrightarrow \sigma \circ \tilde{\sigma} \in \tilde{\Sigma}.$$

To see this, note that

$$G(\sigma(1)\tilde{\sigma}(1)w) \leq G(\tilde{\sigma}(1)w) \leq G(w), \quad w \in A.$$

Hence, $\sigma \circ \tilde{\sigma} \in \tilde{\Sigma}$. But

$$\sup_{\tilde{K}} G \leq a - \delta,$$

where $\tilde{K} = (\sigma \circ \tilde{\sigma})A \in \mathcal{K}$. This contradicts the definition of a and gives the theorems. □

Proof of Theorem 4.5. Let $M = C_0 + 1$. Then

$$\|\sigma(1)v\| \leq M$$

whenever $\sigma \in \Sigma$ satisfies $\|\sigma'(t)\| \leq 1$ and $v \in E$ satisfies $\|v\| \leq C_0$. If the theorem were false, then there would be a $\delta > 0$ such that

$$\|G'(u)\| \geq 3\delta \tag{4.42}$$

when

$$u \in \{u \in E : \|u\| \leq M + 2, |G(u) - a| \leq 3\delta\}. \tag{4.43}$$

Take $\delta < 1/3$ and $\delta < (a - a_0)/2$. Since $G \in C^1(E, \mathbb{R})$, for each $\theta < 1$ there is a locally Lipschitz continuous mapping $Y(u)$ of $\hat{E} = \{u \in E : G'(u) \neq 0\}$ into E such that

$$\|Y(u)\| \leq 1, \ \theta\|G'(u)\| \leq (G'(u), Y(u)), \quad u \in \hat{E} \tag{4.44}$$

(cf., e.g., [193]). Take $\theta > 2/3$. Let

$$\begin{aligned} Q_0 &= \{u \in E : \|u\| \leq M + 1, |G(u) - a| \leq 2\delta\}, \\ Q_1 &= \{u \in E : \|u\| \leq M, |G(u) - a| \leq \delta\}, \\ Q_2 &= E \setminus Q_0, \\ \eta(u) &= a(u, Q_2)/[a(u, Q_1) + a(u, Q_2)]. \end{aligned}$$

It is easily checked that $\eta(u)$ is locally Lipschitz continuous on E and satisfies

$$\begin{cases} \eta(u) = 1, & u \in Q_1, \\ \eta(u) = 0, & u \in \bar{Q}_2, \\ \eta(u) \in (0, 1), & \text{otherwise}. \end{cases} \tag{4.45}$$

Let

$$W(u) = -\eta(u)Y(u).$$

Then

$$\|W(u)\| \leq 1, \quad u \in E.$$

By Theorem 6.10, for each $v \in E$ there is a unique solution $\sigma(t)v$ of

$$\sigma'(t) = W(\sigma(t)), \ t \in \mathbb{R}^+, \quad \sigma(0) = v. \tag{4.46}$$

We have

$$\begin{aligned} dG(\sigma(t)v)/dt &= -\eta(\sigma(t)v)(G'(\sigma(t)v), Y(\sigma(t)v)) \\ &\leq -\theta\eta(\sigma)\|G'(\sigma)\| \\ &\leq -3\theta\delta\eta(\sigma). \end{aligned} \tag{4.47}$$

Let $K \in \mathcal{K}$ satisfy the hypotheses of the theorem and let v be any element of $K \cap Q_1$. Then $\|v\| \leq C_0$ since $G(v) \geq a - \delta$. If there is a $t_1 \leq 1$ such that $\sigma(t_1)v \notin Q_1$, then

$$G(\sigma(1)v) < a - \delta, \tag{4.48}$$

since $\|\sigma(1)v\| \leq M$, and

$$G(\sigma(1)v) \leq G(\sigma(t_1)v) \leq G(v) \leq a + \delta.$$

On the other hand, if $\sigma(t)v \in Q_1$ for all $t \in [0, 1]$, then we have by (4.47)

$$G(\sigma(1)v) \leq a + \delta - 3\delta\theta < a - \delta.$$

If $v \in K \backslash Q_1$, then we must have

$$G(\sigma(1)v) \leq G(v) < a - \delta,$$

since $G(v) \geq a - \delta$ would put v into Q_1. Hence

$$G(\sigma(1)v) < a - \delta, \quad v \in K. \tag{4.49}$$

Moreover,

$$G(\sigma(t)u) \leq G(u) \leq a_0 < a - \delta, \quad u \in A, \ t \in [0, 1].$$

By hypothesis, $\exists \tilde{K} \in \mathcal{K}$:

$$\tilde{K} \subset \bigcup_{t \in [0,1]} \sigma(t)A \cup \sigma(1)K.$$

This means that

$$G(w) < a - \delta, \quad w \in \tilde{K}. \tag{4.50}$$

But this contradicts the definition (4.17) of a. Hence (4.42) cannot hold for u satisfying (4.43). This proves the theorem. □

Proof of Theorem 4.6. By assumption

$$B \cap K \neq \phi, \quad K \in \mathcal{K}. \tag{4.51}$$

Thus, $b_0 \leq a$. Assume first that $b_0 < a$. Then $a_0 < a$, and the result follows by Theorem 4.5. Finally, assume $b_0 = a$. If the theorem were false, there would exist a positive constant $\delta > 0$ such that (4.42) holds whenever $u \in U$ given by (4.43). We follow the proof given above until we reach (4.49). We now note that

$$\sigma(t)A \cap B = \phi, \quad 0 \leq t \leq T. \tag{4.52}$$

For we have by (4.47)

$$G(\sigma(t)u) \leq a_0 - 3\theta\delta \int_0^t \eta(\sigma(\tau)u)d\tau, \quad u \in A.$$

If $\sigma(t)u \in B$, we must have $G(\sigma(t)u) \geq b_0 \geq a_0$. The only way this can happen is if

$$\eta(\sigma(\tau)u) \equiv 0, \quad 0 \leq \tau \leq t.$$

But this implies $\sigma(\tau)u \in \bar{Q}_2$ for such τ, and this in turn implies

$$G(\sigma(\tau)u) < a - \delta, \quad 0 \leq \tau \leq t.$$

Since $a = b_0$, we cannot have $\sigma(\tau)u \in B$. Thus (4.52) holds.

By hypothesis, $\exists \tilde{K} \in \mathcal{K}$:

$$\tilde{K} \subset \bigcup_{t\in[0,1]} \sigma(t)A \cup \sigma(1)K.$$

But $\sigma(1)K \cap B = \phi$ and

$$\sigma(t)A \cap B = \phi, \quad 0 \leq t \leq T. \tag{4.53}$$

Thus $\tilde{K} \cap B = \phi$, contrary to hypothesis. This proves the theorem. □

Proof of Theorem 4.7. Let

$$B = \bigcup_{K} [(K\backslash A) \cap E^{a_0}].$$

Then

$$a_0 \leq G(v), \quad v \in B.$$

Thus,

$$a_0 \leq b_0 = \inf_B G.$$

Since

$$\inf_{(K\backslash A)\cap E^{a_0}} G \leq \sup_K G, \quad K \in \mathcal{K},$$

we have

$$b_0 \leq \inf_{\mathcal{K}} \inf_{(K\backslash A)\cap E^{a_0}} G \leq \inf_{\mathcal{K}} \sup_K G = a.$$

Moreover,

$$B \cap K \neq \phi, \quad K \in \mathcal{K}.$$

The theorem now follows from Theorem 4.6. □

4.6 The Monotonicity Trick

We now give the proof of Theorem 4.8.

Proof. We prove conclusion (1) assuming the first alternative hypothesis ($\mathbf{H}_1$). By ($\mathbf{H}_1$), the map $\lambda \mapsto a(\lambda)$ is nondecreasing. Hence, $a'(\lambda) := da(\lambda)/d\lambda$ exists for almost every $\lambda \in \Lambda$. From this point on, we consider those λ where $a'(\lambda)$ exists. For fixed $\lambda \in \Lambda$, let $\lambda_n \in (\lambda, 2\lambda) \cap \Lambda$, $\lambda_n \to \lambda$ as $n \to \infty$. Then there exists $\bar{n}(\lambda)$ such that

$$a'(\lambda) - 1 \leq \frac{a(\lambda_n) - a(\lambda)}{\lambda_n - \lambda} \leq a'(\lambda) + 1 \quad \text{for } n \geq \bar{n}(\lambda). \tag{4.54}$$

Next, we note that there exist $K_n \in \mathcal{K}$, $k_0 := k_0(\lambda) > 0$ such that

$$\|u\| \leq k_0 \quad \text{whenever} \quad u \in K_n \ \ and \ \ G_\lambda(u) \geq a(\lambda) - (\lambda_n - \lambda). \tag{4.55}$$

In fact, by the definition of $a(\lambda_n)$, there exists a K_n such that

$$\sup_{K_n} G_\lambda(u) \leq \sup_{K_n} G_{\lambda_n}(u) \leq a(\lambda_n) + (\lambda_n - \lambda). \tag{4.56}$$

If $G_\lambda(u) \geq a(\lambda) - (\lambda_n - \lambda)$ for some $u \in K_n$, then, by (4.54) and (4.56), we have that

$$\begin{aligned} I(u) &= \frac{G_{\lambda_n}(u) - G_\lambda(u)}{\lambda_n - \lambda} \\ &\leq \frac{a(\lambda_n) + (\lambda_n - \lambda) - a(\lambda) + (\lambda_n - \lambda)}{\lambda_n - \lambda} \\ &\leq a'(\lambda) + 3, \end{aligned} \tag{4.57}$$

and it follows that

$$\begin{aligned} J(u) &= \lambda_n I(u) - G_{\lambda_n}(u) \\ &\leq \lambda_n (a'(\lambda) + 3) - G_\lambda(u) \\ &\leq \lambda_n (a'(\lambda) + 3) - a(\lambda) + (\lambda_n - \lambda) \end{aligned} \tag{4.58}$$

$$\leq 2\lambda(a'(\lambda)+3)-a(\lambda)+\lambda.$$

On the other hand, by $(\mathbf{H_1})$, (4.54), and (4.56),

$$\begin{aligned} J(u) &= \lambda_n I(u) - G_{\lambda_n}(u) \\ &\geq -G_{\lambda_n}(u) \\ &\geq -(a(\lambda_n)+(\lambda_n-\lambda)) \\ &\geq -(a(\lambda)+(\lambda_n-\lambda)(a'(\lambda)+2)) \\ &\geq -a(\lambda)-\lambda|a'(\lambda)+2|. \end{aligned} \tag{4.59}$$

Combining (4.57)–(4.59) and $(\mathbf{H_1})$, we see that there exists a $k_0(\lambda) := k_0$ (depending only on λ) such that (4.55) holds.

By the choice of K_n and (4.54), we see that

$$\begin{aligned} G_\lambda(u) &\leq G_{\lambda_n}(u) \\ &\leq \sup_{K_n} G_{\lambda_n}(u) \\ &\leq a(\lambda_n)+(\lambda_n-\lambda) \\ &\leq (a'(\lambda)+1)(\lambda_n-\lambda)+a(\lambda)+(\lambda_n-\lambda) \\ &\leq a(\lambda)+(a'(\lambda)+2)(\lambda_n-\lambda) \end{aligned}$$

for all $u \in K_n$. We take n sufficiently large to ensure that $|a'(\lambda)+2|(\lambda_n-\lambda) < \delta$. This proves conclusion (1). Conclusion (2) now follows from Theorem 4.1. The proof under hypothesis $(\mathbf{H_2})$ is similar and is omitted. □

Proof of Theorem 4.9 Use Theorem 4.2. □

Proof of Theorem 4.10 Use Theorem 4.5. □

Proof of Theorem 4.11 Use Theorem 4.6. □

4.7 Notes and Remarks

The monotonicity trick was introduced by Struwe in [239, 240] for minimization problems. This trick was also used by others to solve Landesman-Lazer type problems [118], bifurcation problems [119], superlinear problems [230], Hamiltonian systems [208, 218, 219, 276], Schrödinger equations [212, 213, 215, 220, 223, 230, 299], infinite dimensional sandwich systems [207], elliptic systems [211], etc.

Chapter 5
Infinite Dimensional Linking

5.1 Infinite Dimensional Splitting

Let N be a closed, separable subspace of a Hilbert space E. We can define a new norm $|v|_w$ satisfying $|v|_w \leq \|v\|, \forall v \in N$ and such that the topology induced by this norm is equivalent to the weak topology of N on bounded subsets of N. This can be done as follows: Let $\{e_k\}$ be an orthonormal basis for N. Define

$$(u, v)_w = \sum_{k=1}^{\infty} \frac{(u, e_k)(v, e_k)}{2^k}, \quad u, v \in N.$$

This is a scalar product. The corresponding norm squared is

$$|v|_w^2 = \sum_{k=1}^{\infty} \frac{|(v, e_k)|^2}{2^k}, \quad v \in N.$$

Then $|v|_w$ satisfies $|v|_w \leq \|v\|, \ v \in N$. If $v_j \to v$ weakly in N, then there is a $C > 0$ such that

$$\|v_j\|, \|v\| \leq C, \quad \forall j > 0.$$

For any $\varepsilon > 0$, there exist $K > 0, M > 0$, such that $1/2^K < \varepsilon^2/(8C^2)$ and $|(v_j - v, e_k)| < \varepsilon/2$ for $1 \leq k \leq K, j > M$. Therefore,

$$|v_j - v|_w^2 = \sum_{k=1}^{\infty} \frac{|(v_j - v, e_k)|^2}{2^k}$$

M. Schechter, *Critical Point Theory*,
https://doi.org/10.1007/978-3-030-45603-0_5

$$\leq \sum_{k=1}^{K} \frac{\varepsilon^2/4}{2^k} + \sum_{k=K+1}^{\infty} \frac{4C^2}{2^k}$$

$$\leq \frac{\varepsilon^2}{4} \sum_{k=1}^{\infty} \frac{1}{2^k} + \frac{4C^2}{2^K} \sum_{k=1}^{\infty} \frac{1}{2^k}$$

$$\leq \frac{\varepsilon^2}{2} + \frac{\varepsilon^2}{2}.$$

Therefore, $v_j \to v$ weakly in N implies $|v_j - v|_w \to 0$.

Conversely, let $\|v_j\|, \|v\| \leq C$ for all $j > 0$ and $|v_j - v|_w \to 0$. Let $\varepsilon > 0$ be given. If $h = \sum_{k=1}^{\infty} \alpha_k e_k \in N$, take K so large that $\|h_K\| < \varepsilon/(4C)$, where $h_K = \sum_{k=K+1}^{\infty} \alpha_k e_k$. Take M so large that $|v_j - v|_w^2 < \varepsilon^2/(4 \sum_{k=1}^{K} 2^k |\alpha_k|^2)$ for all $j > M$. Then

$$|(v_j - v, h - h_K)|^2 = |\sum_{k=1}^{K} \alpha_k (v_j - v, e_k)|^2$$

$$\leq \sum_{k=1}^{K} 2^k |\alpha_k|^2 \sum_{k=1}^{\infty} \frac{|(v_j - v, e_k)|^2}{2^k}$$

$$< \varepsilon^2/4$$

for $j > M$. Also, $|(v_j - v, h_K)| \leq 2C\|h_K\| < \varepsilon/2$. Therefore,

$$|(v_j - v, h)| < \varepsilon, \quad \forall j > M,$$

that is, $v_j \to v$ weakly in N.

For $u = v + h,\ u_1 = v_1 + h_1 \in E = N \oplus N^\perp$ with $v, v_1 \in N, h, h_1 \in N^\perp$, we define the scalar product $(u, u_1)_w = (v, v_1)_w + (h, h_1)$. Thus, the corresponding norm satisfies $|u|_w \leq \|u\|, \forall u \in E$.

We denote E equipped with this scalar product and norm by E_w. It is a scalar product space with the same elements as E. In particular, if $(u_n = v_n + w_n)$ is $|\cdot|_w$-bounded and $u_n \overset{|\cdot|_w}{\to} u$, then $v_n \rightharpoonup v$ weakly in N, $w_n \to w$ strongly in $N^\perp$, $u_n \rightharpoonup v + w$ weakly in E.

Let L be a bounded, convex, closed subset of N. Then L is $|\cdot|_w$-compact. In fact, since L is bounded with respect to both norms $|\cdot|_w$ and $\|\cdot\|$, for any $v_n \in L$, there is a renamed subsequence such that $v_n \rightharpoonup v_0$ weakly in E. Then $v_0 \in L$ since L is convex, and on the bounded set L the $|\cdot|_w$-topology is equivalent to the weak topology. Thus, $v_n \overset{|\cdot|_w}{\to} v_0$ and L is $|\cdot|_w$-compact.

We define Σ_w to be the set of all $\sigma(t) \in \Sigma : [0, 1] \times E \mapsto E$ such that

1. $\sigma(t)$ is $|\cdot|_w$-continuous.
2. $\sigma(0)u = u, \ u \in E$.
3. For each compact subset L of $I \times E_w$ there is a finite dimensional subspace E_f of E such that $\dim E_f > 0$ and $\sigma(t)u - u \in E_f, \ (t, u) \in L$.

Here we use E_f to denote various finite-dimensional subspaces of E when exact dimensions are irrelevant.

We have

Lemma 5.1 *If L is compact in E_w and $\sigma \in \Sigma_w$, then*

$$\tilde{L} = \{\sigma(t)L : t \in I\}$$

is compact in E_w.

Proof. Suppose $\{t_k\} \subset I, \ \{u_k\} \subset L$ are sequences. Then there are renamed subsequences such that

$$t_j \to t_0, \quad |u_k - u_0|_w \to 0.$$

Thus $I \times L$ is a compact subset of $I \times E_w$. Consequently, there is a finite dimensional subspace E_f containing the set $\{\sigma(t)u - u, \ t \in I, \ u \in L\}$. Since it is bounded, every sequence has a convergent subsequence. Since every sequence in L has a convergent subsequence, the same must be true of $\tilde{L}$. □

Lemma 5.2 *If $\sigma_1, \ \sigma_2 \in \Sigma_w$, then $\sigma_3 = \sigma_1 \circ \sigma_2 \in \Sigma_w$.*

Proof. Let L be a compact subset of $I \times E_w$. By the definition of Σ_w, for any $(s_0, u_0) \in L$, there is a $|\cdot|_w$-neighborhood $U_{(s_0,u_0)}$ such that $\{u - \sigma_1(t)u : (t, u) \in U_{(s_0,u_0)} \cap L\} \subset E_f$. Note that, $L \subset \bigcup_{(s,u)\in L} U_{(s,u)}$. Since L is $|\cdot|_w$-compact, $L \subset \bigcup_{i=1}^{j_0} U_{(s_i,u_i)}$ where $(s_i, u_i) \in L$. Consequently, $\{u - \sigma_1(t)u : (t, u) \in L\} \subset E_f$. The same is true of σ_2. Since

$$\sigma_3(s) = \begin{cases} \sigma_1(2s), & 0 \le s \le \frac{1}{2}, \\ \sigma_2(2s-1)\sigma_1(1), & \frac{1}{2} < s \le 1, \end{cases}$$

$u - \sigma_3(t)u \in E_f$ as well. □

5.2 Weak Sandwich Systems

Let $\tilde{\Sigma}$ be a subset of Σ_w containing $\sigma(t) \equiv 1$. We define a nonempty collection $\mathcal{K}$ of nonempty subsets $K \subset E$ to be a $\tilde{\Sigma}$ sandwich system if $\mathcal{K}$ has the following property:

$$\sigma \in \tilde{\Sigma},\ K \in \mathcal{K} \Longrightarrow \exists \tilde{K} \in \mathcal{K} : \tilde{K} \subset \sigma(1)K.$$

Theorem 5.3 *Let N be a closed separable subspace of a Hilbert space E. Let G be a continuously differentiable functional on E such that*

$$v_n = Pu_n \to v \text{ weakly in } E, \quad w_n = (I - P)u_n \to w \text{ strongly in } E \tag{5.1}$$

implies

$$G'(v_n + w_n) \to G'(v + w) \text{ weakly in } E, \tag{5.2}$$

where P is the projection of E onto N. Let $\mathcal{K}$ be a $\tilde{\Sigma}$ sandwich system. Assume

$$a := \inf_{K \in \mathcal{K}} \sup_K G \tag{5.3}$$

is finite. Assume, in addition, that there is a constant C_0 such that for each $\delta > 0$ there is a $K \in \mathcal{K}$ satisfying

$$\sup_K G \le a + \delta, \tag{5.4}$$

such that the inequality

$$G(u) \ge a - \delta, \quad u \in K, \tag{5.5}$$

implies $\|u\| \le C_0$. Then there is a sequence $\{u_k\} \subset E$ such that

$$\|u_k\| \le C_0 + 3, \quad G(u_k) \to a, \quad \|G'(u_k)\| \to 0. \tag{5.6}$$

Theorem 5.4 *Let $\mathcal{K}$ be a $\tilde{\Sigma}$ sandwich system, and let G(u) be a C^1 functional on E_w satisfying (5.2) when (5.1) holds. Assume that there are subsets A,B of E such that $A \in \mathcal{K}$,*

$$a_0 := \sup_A G < \infty, \quad b_0 := \inf_B G > -\infty, \tag{5.7}$$

$$B \cap K \neq \phi, \quad K \in \mathcal{K}. \tag{5.8}$$

Assume, in addition, that there is a constant C_0 such that for each $\delta > 0$ there is a $K \in \mathcal{K}$ satisfying (5.4) such that the inequality (5.5) implies $\|u\| \leq C_0$, where a is given by (5.3). Then there is a sequence $\{u_k\} \subset E$ such that

$$\|u_k\| \leq C_0 + 3, \quad G(u_k) \to a, \quad \|G'(u_k)\| \to 0. \tag{5.9}$$

Theorem 5.5 *Let N be a closed separable subspace of a Hilbert space E. Let Ω be a convex, open, bounded subset of N, and let p be any point of Ω. Let G be a continuously differentiable functional on E such that*

$$v_n = Pu_n \to v \text{ weakly in } E, \quad w_n = (I - P)u_n \to w \text{ strongly in } E \tag{5.10}$$

implies

$$G'(v_n + w_n) \to G'(v + w) \text{ weakly in } E, \tag{5.11}$$

where P is the projection of E onto N. Let F be a continuous map of E_w onto N such that $F|_\Omega = I$. Assume that for each finite dimensional subspace S of E containing p such that $FS \neq \{0\}$, there is a finite dimensional subspace $S_0 \neq \{0\}$ of N containing p such that

$$v \in S_0, \ w \in S \Longrightarrow F(v + w) \in S_0. \tag{5.12}$$

Let $\mathcal{K}$ be the collection of sets

$$K = \{\sigma(1)A : \sigma \in \tilde{\Sigma}\},$$

where

$$\tilde{\Sigma} = \{\sigma \in \Sigma_w : G(\sigma(t)u) \leq G(u), \quad t \in I\},$$

and assume that

$$a_0 := \sup_A G < \quad b_0 := \inf_B G, \tag{5.13}$$

where $A = \partial\Omega$ and $B = F^{-1}(p)$. Then $\mathcal{K}$ is a $\tilde{\Sigma}$ sandwich system containing A; also B satisfies (5.8).

Theorem 5.6 *Assume, in addition, that there is a constant C_0 such that for each $\delta > 0$ there is a $K \in \mathcal{K}$ satisfying (5.4) such that the inequality (5.5) implies $\|u\| \leq C_0$, where a is given by (5.3). Then there is a sequence $\{u_k\} \subset E$ such that*

$$\|u_k\| \leq C_0 + 3, \quad G(u_k) \to a, \quad \|G'(u_k)\| \to 0. \tag{5.14}$$

Theorem 5.7 *Let F be a $|\cdot|_w$-continuous map from E onto N satisfying*

- $F|_\Omega = I$; *F maps any finite dimensional subspace of N containing p into itself; it maps bounded sets into bounded sets;*
- *there exists a fixed finite-dimensional subspace E_0 of E such that* $F(u-v)-(F(u)-F(v)) \in E_0, \forall v, u \in E$;
- *F maps finite-dimensional subspaces of E to finite-dimensional subspaces of E;*

Then F satisfies (5.12).

Proof. We show that F satisfies the hypotheses of Theorem 5.5. Clearly $E_0 \subset N$. Let S be a finite dimensional subspace of E containing p such that $FS \neq \{0\}$. Let S_0 be a finite dimensional subspace of N containing $FS + E_0$. If $v \in S_0,\ w \in S$, then

$$F(v+w)+F(-w)-F(v) \in E_0.$$

Thus, $F(v+w) \in S_0$. □

Remark 5.8 Let $E = E^- \oplus E^+$, $z_0 \in E^+$ with $\|z_0\| = 1$. For any $u \in E$, we write $u = u^- \oplus sz_0 \oplus w^+$ with $u^- \in E^-, s \in \mathbb{R}, w^+ \in (E^- \oplus \mathbb{R}z_0)^\perp := E_1^+$. Let $N := E^- \oplus \mathbb{R}z_0$. For $R > 0$, let $Q := \{u := u^- + sz_0 : s \in \mathbb{R}^+, u^- \in E^-, \|u\| < R\}$, $p_0 = s_0 z_0 \in Q,\quad s_0 > 0$. Let $F : E \mapsto N$ be defined by $Fu := u^- + \|sz_0 + w^+\| z_0$, then F, Q, p_0 satisfy the above conditions with

$$A = \partial Q = \{u := u^- + sz_0 : s \in \mathbb{R}^+, u^- \in E^-, \|u\| = R\} \ \cup\ (\mathcal{B}_R \cap E^-).$$

$$B = F^{-1}(s_0 z_0) = \{u := sz_0 + w^+ : s \in \mathbb{R},\ w^+ \in E_1^+, \|sz_0 + w^+\| = s_0\}.$$

In fact, according to the definition, $F|_Q = I$ and F maps bounded sets to bounded sets. On the other hand, for any $u, v \in E$, we can write $u = u^- + sz_0 + w^+$, $v = v^- + tz_0 + w_1^+$. Then

$$F(u) = u^- + \|sz_0 + w^+\| z_0, \quad F(v) = v^- + \|tz_0 + w_1^+\| z_0,$$

$$F(u-v) = u^- - v^- + (\|(s-t)z_0 + w^+ - w_1^+\| z_0.$$

Therefore,

$$\begin{aligned}
&F(u-v)-(F(u)-F(v)) \\
&= \Big(\|(s-t)z_0 + w^+ - w_1^+\| - \|sz_0 + w^+\| + \|tz_0 + w_1^+\|\Big) z_0 \\
&\subset \mathbb{R}z_0 := E_0 \quad \text{(a one-dimensional subspace).}
\end{aligned}$$

Proof of Theorem 5.3 Take $\delta < 1/3$. Let $M = C_0 + 1$. Then

$$\|\sigma(1)v\| \leq M$$

whenever $\sigma \in \Sigma_w$ satisfies $\|\sigma'(t)\| \leq 1$ and $v \in E$ satisfies $\|v\| \leq C_0$. If the theorem were false, then there would be a $\delta > 0$ such that

$$\|G'(u)\| \geq 3\delta \tag{5.15}$$

when

$$u \in \tilde{E} = \{u \in E : \|u\| \leq M + 2, |G(u) - a| \leq 3\delta\}. \tag{5.16}$$

For $u \in \tilde{E}$, let $q(u) = G'(u)/\|G'(u)\|$. Then by (5.15)

$$(G'(u), q(u)) \geq 3\delta, \quad u \in \tilde{E}. \tag{5.17}$$

For each $u \in \tilde{E}$ there is a E_w neighborhood $\mathcal{N}(u)$ of u such that

$$(G'(h), q(u)) > 2\delta, \quad h \in \mathcal{N}(u) \cap \tilde{E}. \tag{5.18}$$

For otherwise there would be a sequence $\{h_k\} \subset \mathcal{N}(u) \cap \tilde{E}$ such that

$$|h_k - u|_w \to 0 \text{ and } (G'(h_k), q(u)) \leq 2\delta. \tag{5.19}$$

Since $\tilde{E}$ is bounded in E, $Ph_k \to Pu$ weakly in N and $(I - P)h_k \to (I - P)u$ strongly in M. Hence, by hypothesis,

$$(G'(h_k), q(u)) \to (G'(u), q(u)) \geq 3\delta$$

in view of (5.17). This contradicts (5.19). Hence, (5.18) would hold. Let $\tilde{E}_w$ be the set $\tilde{E}$ with the inherited topology of E_w. It is a metric space and $\mathcal{N}(u) \cap \tilde{E}$ is an open set in this space. Thus $\{\mathcal{N}(u) \cap \tilde{E}\}$, $u \in \tilde{E}$, is an open covering of the paracompact space $\tilde{E}_w$. Consequently, there is a locally finite refinement $\{\mathcal{N}_\tau\}$ of this cover. For each τ there is an element u_τ such that $\mathcal{N}_\tau \subset \mathcal{N}(u_\tau)$. Let $\{\psi_\tau\}$ be a partition of unity subordinate to this covering. Each ψ_τ is locally Lipschitz continuous with respect to the norm $|u|_w$ and consequently with respect to the norm of E. Let

$$Y(u) = \sum \psi_\tau(u) q(u_\tau), \quad u \in E. \tag{5.20}$$

Then $Y(u)$ is locally Lipschitz continuous with respect to both norms. Moreover,

$$\|Y(u)\| \leq \sum \psi_\tau(u) \|q(u_\tau)\| \leq 1. \tag{5.21}$$

Let

$$
\begin{aligned}
Q_0 &= \{u \in E : \|u\| \le M+1, |G(u)-a| \le 2\delta\},\\
Q_1 &= \{u \in E : \|u\| \le M, |G(u)-a| \le \delta\},\\
Q_2 &= E \setminus Q_0,\\
\eta(u) &= d_w(u, Q_2)/[d_w(u, Q_1) + d_w(u, Q_2)],
\end{aligned}
$$

where

$$d_w(u, Q) = \inf_{v \in Q} |u - v|_w .$$

It is easily checked that $\eta(u)$ is locally Lipschitz continuous on E (in both norms) and satisfies

$$
\begin{cases}
\eta(u) = 1, & u \in Q_1,\\
\eta(u) = 0, & u \in \bar{Q}_2 \ in \ E_w,\\
\eta(u) \in (0, 1), & \text{otherwise.}
\end{cases}
\tag{5.22}
$$

Let

$$W(u) = -\eta(u)Y(u).$$

Then

$$\|W(u)\| \le 1, \quad u \in E.$$

By Theorem 6.10, for each $v \in E$ there is a unique solution $\sigma(t)v$ of

$$\sigma'(t) = W(\sigma(t)), \ t \in \mathbb{R}^+, \quad \sigma(0) = v. \tag{5.23}$$

We have

$$
\begin{aligned}
dG(\sigma(t)v)/dt &= -\eta(\sigma(t)v)(G'(\sigma(t)v), Y(\sigma(t)v)) \\
&\le -\theta\eta(\sigma)\|G'(\sigma)\| \\
&\le -3\theta\delta\eta(\sigma).
\end{aligned}
\tag{5.24}
$$

We note that $\sigma(t)$ is locally Lipschitz continuous with respect to the weak norm. Since

$$\sigma(t)u - \sigma(t)u_0 = u - u_0 + \int_0^t [W(\sigma(s)u) - W(\sigma(s)u_0)]\, ds,$$

we have

$$|\sigma(t)u - \sigma(t)u_0|_w \leq |u - u_0|_w + \int_0^t |W(\sigma(s)u) - W(\sigma(s)u_0)|_w \, ds,$$

which implies

$$|\sigma(t)u - \sigma(t)u_0|_w \leq |u - u_0|_w + C \int_0^t |\sigma(s)u - \sigma(s)u_0|_w \, ds$$

and

$$|\sigma(t)u - \sigma(t)u_0|_w \leq C|u - u_0|_w$$

for $|u - u_0|_w$ sufficiently small.

Note that $\sigma(t) \in \Sigma_w$. To see this, we first observe that for each $u_0 \in E$, there is a neighborhood $\mathcal{N}(u_0)$ such that

$$W(u) \in E_f, \quad u \in \mathcal{N}(u_0).$$

Thus,

$$\int_t^{t'} W(\sigma(s)u) \, dt \in E_f, \quad u \in \mathcal{N}(u_0),$$

for $t' - t$ sufficiently small. If L is compact in E_w, this implies

$$\int_0^t W(\sigma(s)L) \, ds \subset E_f, \quad t \in I.$$

Thus,

$$\sigma(t)u - u = \int_0^t W(\sigma(s)u) \, ds \in E_f, \quad u \in L, \ t \in I.$$

The other properties are easily checked. Let $K \in \mathcal{K}$ satisfy the hypotheses of the theorem for δ satisfying (5.15) and (5.16). Let v be any element of $K \cap Q_1$. Then $\|v\| \leq C_0$. If there is a $t_1 \leq 1$ such that $\sigma(t_1)v \notin Q_1$, then

$$G(\sigma(1)v) < a - \delta, \tag{5.25}$$

since $\|\sigma(1)v\| \leq M$,

$$G(\sigma(1)v) \leq G(\sigma(t_1)v)$$

and the right-hand side cannot be greater than $a+\delta$ by (5.24). On the other hand, if $\sigma(t)v \in Q_1$ for all $t \in [0, 1]$, then we have by (5.24)

$$G(\sigma(1)v) \leq a+\delta-3\delta\theta < a-\delta.$$

If $v \in K\backslash Q_1$, then we must have

$$G(\sigma(1)v) \leq G(v) < a-\delta,$$

since $G(v) \geq a-\delta$ would put v into Q_1. Hence

$$G(\sigma(1)v) < a-\delta, \quad v \in K. \tag{5.26}$$

By hypothesis, there is a $\tilde{K} \in \mathcal{K}$ such that

$$\tilde{K} \subset \sigma(1)K.$$

Consequently,

$$G(v) \leq a-\delta, \quad v \in \tilde{K},$$

contrary to the definition (5.3) of the quantity a.

Thus, (5.15) cannot hold for u satisfying (5.16). This proves the theorem. □

Proof of Theorem 5.4 We must show that $-\infty < a < \infty$. Since,

$$B \cap K \neq \phi, \quad K \in \mathcal{K}, \tag{5.27}$$

we have

$$b_0 = \inf_B G \leq \inf_{B\cap K} G \leq \sup_{B\cap K} G \leq \sup_K G, \quad K \in \mathcal{K}.$$

Thus $-\infty < b_0 \leq a \leq \sup_A G < \infty$ since $A \in \mathcal{K}$. □

Proof of Theorem 5.5 First we note that if σ_1, σ_2 are in $\tilde{\Sigma}$, then $\sigma_3 = \sigma_1 \circ \sigma_2 \in \tilde{\Sigma}$. This follows from the fact that

$$\sigma_3(s) = \begin{cases} \sigma_1(2s), & 0 \leq s \leq \frac{1}{2}, \\ \sigma_2(2s-1)\sigma_1(1), & \frac{1}{2} < s \leq 1. \end{cases}$$

Consequently,

$$G(\sigma_3(t)u) \leq \begin{cases} G(\sigma_1(2t)u) \leq G(u), & 0 \leq t \leq \frac{1}{2}, \\ G(\sigma_2(2t)\sigma_1(1)u) \leq G(\sigma_1(1)u) \leq G(u), & \frac{1}{2} < t \leq 1. \end{cases}$$

Thus, $\mathcal{K}$ is a $\tilde{\Sigma}$ sandwich system containing $A = \partial\Omega$. Now (5.8) is equivalent to

$$F^{-1}(p) \cap \sigma(1)A \neq \phi, \quad \sigma \in \tilde{\Sigma}, \quad G(\sigma(t)u) \leq G(u), \quad t \in I.$$

Let $\sigma(t)$ be any flow in $\tilde{\Sigma}$. Since $\overline{\Omega}$ is compact in E_w, there is a finite dimensional subspace $S \in E$ such that

$$\sigma(t)u - u \in S, \quad u \in \overline{\Omega}.$$

Enlarge S to contain p and satisfy $FS \neq \{0\}$. By hypothesis, there is a finite dimensional subspace $S_0 \neq \{0\}$ of N containing p such that

$$v \in S_0, \ w \in S \Longrightarrow F(v+w) \in S_0. \tag{5.28}$$

Take $w = \sigma(t)u - u \in S$. Then

$$F\sigma(t)u \in S_0, \quad u \in \overline{\Omega} \cap S_0.$$

Let

$$\varphi_t(v) = \varphi(v,t) = F\sigma(t)v, \quad v \in \overline{\Omega}, \ 0 \leq t \leq 1. \tag{5.29}$$

Then $\varphi(v,t)$ maps $\overline{\Omega} \cap S_0 \times [0,1]$ into S_0. We claim that

$$F\sigma(t)u \neq p \text{ for all } u \in \partial(\Omega \cap S_0) = \partial\Omega \cap S_0, \quad t \in [0,1]. \tag{5.30}$$

To see this, assume that there exist $t_0 \in [0,1]$ and $u_0 \in \partial\Omega \cap S_0$ such that $F\sigma(t_0)u_0 = p$, i.e., $\sigma(t_0)u_0 \in B$. It follows that $G(u_0) \geq G(\sigma(t_0)u_0) \geq \inf_B G > \sup_{\partial\Omega} G$, in view of (5.13). This is impossible. Thus, (5.30) is true. Hence,

$$\varphi_t(v) \neq p, \quad v \in \partial\Omega \cap S_0, \ 0 \leq t \leq 1. \tag{5.31}$$

Consequently the Brouwer degree $d(\varphi_t, \overline{\Omega} \cap S_0, p)$ can be defined. Since $\varphi_t(v)$ is continuous, we have

$$d(\varphi_1, \overline{\Omega} \cap S_0, p) = d(\varphi_0, \overline{\Omega} \cap S_0, p) = d(I, \overline{\Omega} \cap S_0, p) = 1.$$

Hence there is a $v \in \overline{\Omega}$ such that $F\sigma(1)v = p$. Consequently, $\sigma(1)v \in B$. Thus (5.8) holds. This completes the proof.

□

Definition 5.9 We shall say that a pair of subsets A, B of a Banach space E forms a **weak sandwich pair** for a continuously differentiable $G \in C^1(E_w, \mathbb{R})$ if A, B satisfy (5.7) and (5.8).

We have the following.

Theorem 5.10 *Let N be a separable subspace of a Banach space E, and let p be any point of N. Let G, Ω, F satisfy the hypotheses of Theorem 5.5. Then $A = \partial(\Omega)$ and $B = F^{-1}(p)$ form a weak sandwich pair for G.*

Proof. Apply Theorem 5.5. □

Corollary 5.11 *Let N be a closed subspace of a Hilbert space E and let $M = N^{\perp}$. Assume that at least one of the subspaces M, N is separable. Then M,N form a weak sandwich pair for G.*

Proof. Apply Theorem 5.5. □

Corollary 5.12 *Let N be a closed, separable subspace of a Hilbert space E with complement $M' = M \oplus \{v_0\}$, where v_0 is an element in E having unit norm. Let δ be any positive number, and let $\varphi(t) \in C^1(\mathbb{R})$ be such that*

$$0 \le \varphi(t) \le 1, \ \varphi(0) = 1,$$

and

$$\varphi(t) = 0, \quad |t| \ge 1.$$

Let

$$F(v + w + sv_0) = v + [s + \delta - \delta\varphi(\|w\|^2/\delta^2)]v_0, \quad v \in N, \ w \in M, \ s \in \mathbb{R}. \tag{5.32}$$

Then $A = N' = N \oplus \{v_0\}$, $B = F^{-1}(\delta v_0)$ form a weak sandwich pair for G.

Proof. Define

$$J(v + w + sv_0) = v + w + [s - \delta + \delta\varphi(\|w\|^2/\delta^2)]v_0, \quad v \in N, \ w \in M, \ s \in \mathbb{R}.$$

Then J is a diffeomorphism on E. Moreover, $A = JN'$ and $B = J[M + \delta v_0]$. Since N' and $M + \delta v_0$ form a weak sandwich pair by Corollary 5.11, we see that A, B also form a weak sandwich pair. □

5.3 Weak Linking Systems

Let E be a Banach space and let $\tilde{\Sigma} \neq \phi$ be a subset of Σ_w. We define a nonempty collection $\mathcal{K}$ of nonempty subsets $K \subset E$ each containing the set A to be a $\tilde{\Sigma}$ **linking system for** A if $\mathcal{K}$ has the following property:

$$\sigma \in \tilde{\Sigma}, \ K \in \mathcal{K} \Longrightarrow \exists \tilde{K} \in \mathcal{K} : \tilde{K} \subset \bigcup_{t \in [0,1]} \sigma(t)A \cup \sigma(1)K.$$

We have the following.

Theorem 5.13 *Let $G(u)$ be a C^1 functional on E_w and let $\mathcal{K}$ be a $\tilde{\Sigma}$ linking system for a set A, where*

$$\tilde{\Sigma} \supset \{\sigma \in \Sigma_w : G(\sigma(t)u) \leq G(u),\ u \in E,\ t \in [0, 1]\}. \tag{5.33}$$

Define

$$a := \inf_{K \in \mathcal{K}} \sup_K G, \tag{5.34}$$

and assume that $a_0 = \sup_A G < a < \infty$. Assume, in addition, that there is a constant C_0 such that for each $\delta > 0$ there is a $K \in \mathcal{K}$ satisfying (5.4) and (5.5). Then there is a bounded sequence $\{u_k\} \subset E$ such that

$$\|u_k\| \leq C_0 + 3, \quad G(u_k) \to a, \quad \|G'(u_k)\| \to 0. \tag{5.35}$$

Theorem 5.14 *Let $G(u)$ be a C^1 functional on E and let $\mathcal{K}$ be a $\tilde{\Sigma}$ linking system for a set A, where $\tilde{\Sigma}$ satisfies (5.33). Assume that the quantity a given by (5.34) is $< \infty$ and there is a subset B of E such that*

$$a_0 := \sup_A G \leq b_0 := \inf_B G \tag{5.36}$$

and

$$B \cap K \neq \phi, \quad K \in \mathcal{K}. \tag{5.37}$$

Assume, in addition, that there is a constant C_0 such that for each $\delta > 0$ there is a $K \in \mathcal{K}$ satisfying (5.4) and (5.5). Then the quantity a given by (5.34) satisfies $a_0 \leq b_0 \leq a$ and there is a bounded sequence $\{u_k\} \subset E$ such that

$$\|u_k\| \leq C_0 + 3, \quad G(u_k) \to a, \quad \|G'(u_k)\| \to 0. \tag{5.38}$$

Theorem 5.15 *Let $G(u)$ be a C^1 functional on E and let $\mathcal{K}$ be a $\tilde{\Sigma}$ linking system for a set A, where*

$$\tilde{\Sigma} \supset \{\sigma \in \Sigma : G(\sigma(t)u) \leq G(u),\ u \in E,\ t \in [0, 1]\}. \tag{5.39}$$

Define

$$a := \inf_{K \in \mathcal{K}} \sup_K G, \tag{5.40}$$

and assume that a is finite and

$$(K \backslash A) \cap E^{a_0} \neq \phi, \quad K \in \mathcal{K},$$

where $a_0 = \sup_A G$, and $E^\beta = \{v \in E : G(v) \geq \beta\}$. Assume, in addition, that there is a constant C_0 such that for each $\delta > 0$ there is a $K \in \mathcal{K}$ satisfying (5.4) and (5.5). Then there is a bounded sequence $\{u_k\} \subset E$ such that

$$\|u_k\| \leq C_0 + 3, \quad G(u_k) \to a, \quad \|G'(u_k)\| \to 0. \tag{5.41}$$

5.4 The Parameter Problem

Let E be a reflexive Banach space with norm $\|\cdot\|$, and let A, B be two closed subsets of E. Let G be a continuously differentiable functional on E such that

$$v_n = Pu_n \to v \text{ weakly in } E, \quad w_n = (I - P)u_n \to w \text{ strongly in } E \tag{5.42}$$

implies

$$G'(v_n + w_n) \to G'(v + w) \text{ weakly in } E, \tag{5.43}$$

where P is the projection of E onto N. Suppose that G is of the form: $G(u) := I(u) - J(u), u \in E$, where $I, J \in C^1(E, \mathbb{R})$ map bounded sets to bounded sets. Define

$$G_\lambda(u) = \lambda I(u) - J(u), \quad \lambda \in \Lambda,$$

where Λ is an open interval contained in $(0, +\infty)$. Assume one of the following alternatives holds.

(H$_1$) $I(u) \geq 0$ for all $u \in E$ and $I(u) + |J(u)| \to \infty$ as $\|u\| \to \infty$.

(H$_2$) $I(u) \leq 0$ for all $u \in E$ and $|I(u)| + |J(u)| \to \infty$ as $\|u\| \to \infty$.

Furthermore, we suppose that

(H$_3$) There is a weak $\tilde{\Sigma}$ sandwich system $\mathcal{K}$ such that $a(\lambda) := \inf_{K \in \mathcal{K}} \sup_K G_\lambda$ is finite for any $\lambda \in \Lambda$.

Theorem 5.16 *Assume that* (**H$_1$**) *(or* (**H$_2$**)*) and* (**H$_3$**) *hold. Then we have*

(1) For almost all $\lambda \in \Lambda$ there exists a constant $k_0(\lambda) := k_0$ (depending only on λ) such that for each $\delta > 0$ there exists a $K \in \mathcal{K}$ such that

$$\sup_K G_\lambda \leq a(\lambda) + \delta$$

and

$$\|u\| \leq k_0 \quad \text{whenever} \quad u \in K \text{ and } G_\lambda(u) \geq a(\lambda) - \delta. \tag{5.44}$$

(2) For almost all $\lambda \in \Lambda$ *there exists a bounded sequence* $u_k(\lambda) \in E$ *such that*

$$\|G'_\lambda(u_k)\| \to 0, \quad G_\lambda(u_k) \to a(\lambda) := \inf_{K \in \mathcal{K}} \sup_K G_\lambda, \qquad \text{as } k \to \infty.$$

Theorem 5.17 *The conclusions of Theorem 5.16 hold if we replace hypothesis* (**H**$_3$) *with*

(**H**$'_3$) *There is a weak* $\tilde{\Sigma}$ *sandwich system* $\mathcal{K}$ *and sets A,B such that* $A \in \mathcal{K}$,

$$B \cap K \neq \phi, \quad K \in \mathcal{K} \tag{5.45}$$

and

$$a_0(\lambda) := \sup_A G_\lambda < \infty, \qquad b_0(\lambda) := \inf_B G_\lambda > -\infty \tag{5.46}$$

for each $\lambda \in \Lambda$.

Theorem 5.18 *The conclusions of Theorem 5.16 hold if we replace hypothesis* (**H**$_3$) *with*

(**H**$''_3$) *Let* Ω *be a convex, open, bounded subset of* N, *and let p be any point of* Ω. *Let F be a continuous map of* E_w *onto N such that* $F|_\Omega = I$. *Assume that for each finite dimensional subspace S of E containing p such that* $FS \neq \{0\}$, *there is a finite dimensional subspace* $S_0 \neq \{0\}$ *of N containing p such that*

$$v \in S_0, \ w \in S \Longrightarrow F(v+w) \in S_0. \tag{5.47}$$

Let $\mathcal{K}$ *be the collection of sets*

$$K = \{\sigma(1)A : \sigma \in \tilde{\Sigma}\},$$

where

$$\tilde{\Sigma} = \{\sigma \in \Sigma_w : G(\sigma(t)u) \leq G(u), \ \ t \in I\},$$

and assume that

$$a_0(\lambda) := \sup_A G_\lambda < \ b_0(\lambda) := \inf_B G_\lambda, \quad \lambda \in \Lambda, \tag{5.48}$$

where $A = \partial\Omega$ *and* $B = F^{-1}(p)$.

Theorem 5.19 *The conclusions of Theorem 5.16 hold if we replace hypothesis* (**H**$_3$) *with*

(**H**$'''_3$) *There is a weak* $\tilde{\Sigma}$ *linking system* $\mathcal{K}$ *for a set A such that*

$$a_0(\lambda) := \sup_A G_\lambda < a(\lambda) := \inf_{K \in \mathcal{K}} \sup_K G_\lambda < \infty$$

for each $\lambda \in \Lambda$.

Theorem 5.20 *The conclusions of Theorem 5.16 hold if we replace hypothesis* ($\mathbf{H_3}$) *with*

($\mathbf{H_3''''}$) *There is a weak* $\tilde{\Sigma}$ *linking system* $\mathcal{K}$ *for a set A, and a set B such that*

$$B \cap K \neq \phi, \quad K \in \mathcal{K} \tag{5.49}$$

and

$$a_0(\lambda) := \sup_A G_\lambda \le b_0(\lambda) := \inf_B G_\lambda \tag{5.50}$$

for each $\lambda \in \Lambda$.

Corollary 5.21 *The conclusions of Theorem 5.16 hold if we replace hypothesis* ($\mathbf{H_3}$) *with*

($\mathbf{H_3'''''}$) *There is a weak sandwich pair A,B such that*

$$a_\lambda := \sup_A G_\lambda < \infty, \qquad b_\lambda := \inf_B G_\lambda > -\infty \tag{5.51}$$

for each $\lambda \in \Lambda$. *Thus, for a.e.* $\lambda \in \Lambda$ *there is a bounded sequence* $\{u_k\} \subset E$ *such that*

$$G_\lambda(u_k) \to c_\lambda, \quad \|G'_\lambda(u_k)\| \to 0, \tag{5.52}$$

where c_λ *satisfies* $b_\lambda \le c_\lambda \le a_\lambda$.

5.5 Finding the Sequences

Proof of Theorem 5.13 Let $M = C_0 + 1$. Then

$$\|\sigma(1)v\| \le M$$

whenever $\sigma \in \Sigma$ satisfies $\|\sigma'(t)\| \le 1$ and $v \in E$ satisfies $\|v\| \le C_0$. If the theorem were false, then there would be a $\delta > 0$ such that

$$\|G'(u)\| \ge 3\delta \tag{5.53}$$

when

$$u \in \tilde{E} := \{u \in E : \|u\| \leq M + 2, |G(u) - a| \leq 3\delta\}. \tag{5.54}$$

Take $\delta < 1/3$ and $\delta < (a - a_0)/2$.

For $u \in \tilde{E}$, let $q(u) = G'(u)/\|G'(u)\|$. Then by (5.53)

$$(G'(u), q(u)) \geq 2\delta, \quad u \in \tilde{E}. \tag{5.55}$$

For each $u \in \tilde{E}$ there is a E_w neighborhood $\mathcal{N}(u)$ of u such that

$$(G'(h), q(u)) > \delta, \quad h \in \mathcal{N}(u) \cap \tilde{E}. \tag{5.56}$$

For otherwise there would be a sequence $\{h_k\} \subset \tilde{E}$ such that

$$|h_k - u|_w \to 0 \text{ and } (G'(h_k), q(u)) \leq \delta. \tag{5.57}$$

Since $\tilde{E}$ is bounded in E, $Ph_k \to Pu$ weakly in N and $(I - P)h_k \to (I - P)u$ strongly in M. Hence, by hypothesis,

$$(G'(h_k), q(u)) \to (G'(u), q(u)) \geq 2\delta$$

in view of (5.55). This contradicts (5.57). Let $\tilde{E}_w$ be the set $\tilde{E}$ with the inherited topology of E_w. It is a metric space and $\mathcal{N}(u) \cap \tilde{E}$ is an open set in this space. Thus $\{\mathcal{N}(u) \cap \tilde{E}\}, u \in \tilde{E}$, is an open covering of the paracompact space $\tilde{E}_w$. Consequently, there is a locally finite refinement $\{\mathcal{N}\tau\}$ of this cover. For each τ there is an element u_τ such that $\mathcal{N}_\tau \subset \mathcal{N}(u_\tau)$. Let $\{\psi_\tau\}$ be a partition of unity subordinate to this covering. Each ψ_τ is locally Lipschitz continuous with respect to the norm $|u|_w$ and consequently with respect to the norm of E. Let

$$Y(u) = \sum \psi_\tau(u) q(u_\tau), \quad u \in E. \tag{5.58}$$

Then $Y(u)$ is locally Lipschitz continuous with respect to both norms. Moreover,

$$\|Y(u)\| \leq \sum \psi_\tau(u) \|q(u_\tau)\| \leq 1. \tag{5.59}$$

Since $G \in C^1(E, \mathbb{R})$, for each $\theta < 1$ there is a locally Lipschitz continuous mapping $Y(u)$ of $\hat{E} = \{u \in E : G'(u) \neq 0\}$ into E such that

$$\|Y(u)\| \leq 1, \ \theta \|G'(u)\| \leq (G'(u), Y(u)), \quad u \in \hat{E} \tag{5.60}$$

(cf., e.g., [193]). Take $\theta > 2/3$. Let

$$Q_0 = \{u \in E : \|u\| \leq M + 1, |G(u) - a| \leq 2\delta\},$$

$$Q_1 = \{u \in E : \|u\| \leq M, |G(u) - a| \leq \delta\},$$

$$Q_2 = E \setminus Q_0,$$
$$\eta(u) = d_w(u, Q_2)/[d_w(u, Q_1) + d_w(u, Q_2)],$$

where

$$d_w(u, Q) = \inf_{v \in Q} |u - v|_w.$$

It is easily checked that $\eta(u)$ is locally Lipschitz continuous on E and satisfies

$$\begin{cases} \eta(u) = 1, & u \in Q_1, \\ \eta(u) = 0, & u \in \bar{Q}_2 \ in \ E_w, \\ \eta(u) \in (0, 1), & \text{otherwise}. \end{cases} \tag{5.61}$$

Let

$$W(u) = -\eta(u)Y(u).$$

Then

$$\|W(u)\| \leq 1, \quad u \in E.$$

By Theorem 6.10, for each $v \in E$ there is a unique solution $\sigma(t)v$ of

$$\sigma'(t) = W(\sigma(t)), \ t \in \mathbb{R}^+, \quad \sigma(0) = v. \tag{5.62}$$

We have

$$\begin{aligned} dG(\sigma(t)v)/dt &= -\eta(\sigma(t)v)(G'(\sigma(t)v), Y(\sigma(t)v)) \\ &\leq -\theta\eta(\sigma)\|G'(\sigma)\| \\ &\leq -3\theta\delta\eta(\sigma). \end{aligned} \tag{5.63}$$

Let $K \in \mathcal{K}$ satisfy the hypotheses of the theorem and let v be any element of $K \cap Q_1$. Then $\|v\| \leq C_0$ since $G(v) \geq a - \delta$. If there is a $t_1 \leq 1$ such that $\sigma(t_1)v \notin Q_1$, then

$$G(\sigma(1)v) < a - \delta, \tag{5.64}$$

since $\|\sigma(1)v\| \leq M$, and

$$G(\sigma(1)v) \leq G(\sigma(t_1)v) \leq G(v) \leq a + \delta.$$

On the other hand, if $\sigma(t)v \in Q_1$ for all $t \in [0, 1]$, then we have by (5.63)

$$G(\sigma(1)v) \leq a + \delta - 3\delta\theta < a - \delta.$$

If $v \in K \backslash Q_1$, then we must have

$$G(\sigma(1)v) \leq G(v) < a - \delta,$$

since $G(v) \geq a - \delta$ would put v into Q_1. Hence

$$G(\sigma(1)v) < a - \delta, \quad v \in K. \tag{5.65}$$

Moreover,

$$G(\sigma(t)u) \leq G(u) \leq a_0 < a - \delta, \quad u \in A, \quad t \in [0, 1].$$

By hypothesis, $\exists \tilde{K} \in \mathcal{K}$:

$$\tilde{K} \subset \bigcup_{t \in [0,1]} \sigma(t)A \cup \sigma(1)K.$$

This means that

$$G(w) < a - \delta, \quad w \in \tilde{K}. \tag{5.66}$$

But this contradicts the definition (5.34) of a. Hence (5.53) cannot hold for u satisfying (5.54). This proves the theorem. □

Proof of Theorem 5.14 By assumption

$$B \cap K \neq \phi, \quad K \in \mathcal{K}. \tag{5.67}$$

Thus, $b_0 \leq a$. Assume first that $b_0 < a$. Then $a_0 < a$, and the result follows by Theorem 5.13. Finally, assume $b_0 = a$. If the theorem were false, there would exist a positive constant $\delta > 0$ such that (5.53) holds whenever $u \in U$ given by (5.54). We follow the proof given above until we reach (5.65). We now note that

$$\sigma(t)A \cap B = \phi, \quad 0 \leq t \leq T. \tag{5.68}$$

For we have by (5.63)

$$G(\sigma(t)u) \leq a_0 - 3\theta\delta \int_0^t \eta(\sigma(\tau)u)d\tau, \quad u \in A.$$

If $\sigma(t)u \in B$, we must have $G(\sigma(t)u) \geq b_0 \geq a_0$. The only way this can happen is if

$$\eta(\sigma(\tau)u) \equiv 0, \quad 0 \leq \tau \leq t.$$

But this implies $\sigma(\tau)u \in \bar{Q}_2$ for such τ, and this in turn implies

$$G(\sigma(\tau)u) < a - \delta, \quad 0 \leq \tau \leq t.$$

Since $a = b_0$, we cannot have $\sigma(\tau)u \in B$. Thus (5.68) holds.

By hypothesis, $\exists \tilde{K} \in \mathcal{K}$:

$$\tilde{K} \subset \bigcup_{t\in[0,1]} \sigma(t)A \cup \sigma(1)K.$$

But $\sigma(1)K \cap B = \phi$ and

$$\sigma(t)A \cap B = \phi, \quad 0 \leq t \leq T. \tag{5.69}$$

Thus $\tilde{K} \cap B = \phi$, contrary to hypothesis. This proves the theorem. □

Proof of Theorem 5.15 Let

$$B = \bigcup_K [(K\backslash A) \cap E^{a_0}].$$

Then

$$a_0 \leq G(v), \quad v \in B.$$

Thus,

$$a_0 \leq b_0 = \inf_B G.$$

Since

$$\inf_{(K\backslash A)\cap E^{a_0}} G \leq \sup_K G, \quad K \in \mathcal{K},$$

we have

$$b_0 \leq \inf_{\mathcal{K}} \inf_{(K\backslash A)\cap E^{a_0}} G \leq \inf_{\mathcal{K}} \sup_K G = a.$$

Moreover,

$$B \cap K \neq \phi, \quad K \in \mathcal{K}.$$

The theorem now follows from Theorem 5.14. □

Proof of Theorem 5.16 Use Theorem 5.3 together with Theorem 4.8. □

Proof of Theorem 5.17 Use Theorem 5.4. □

Proof of Theorem 5.18 Use Theorems 5.5 and 5.7 and Remark 5.8. □

Proof of Theorem 5.19 Use Theorem 5.13. □

Proof of Theorem 5.20 Use Theorem 5.14. □

5.6 Notes and Remarks

We follow [189, 191, 204, 207, 229].

Chapter 6
Differential Equations

6.1 Extensions of Picard's Theorem

We shall make use of various extensions of Picard's theorem in a Banach space. Some are well known.

Theorem 6.1 *Let X be a Banach space, and let*

$$B_0 = \{x \in X : \|x - x_0\| \leq R_0\}$$

and

$$I_0 = \{t \in \mathbb{R} : |t - t_0| \leq T_0\}.$$

Assume that $g(t, x)$ is a continuous map of $I_0 \times B_0$ into X such that

$$\|g(t, x) - g(t, y)\| \leq K_0\|x - y\|, \quad x, y \in B_0, \ t \in I_0 \tag{6.1}$$

and

$$\|g(t, x)\| \leq M_0, \quad x \in B_0, \ t \in I_0. \tag{6.2}$$

Let T_1 be such that

$$T_1 \leq \min(T_0, \ R_0/M_0), \quad K_0T_1 < 1. \tag{6.3}$$

Then there is a unique solution $x(t)$ of

$$\frac{dx(t)}{dt} = g(t, x(t)), \quad |t - t_0| \leq T_1, \quad x(t_0) = x_0. \tag{6.4}$$

M. Schechter, *Critical Point Theory*,
https://doi.org/10.1007/978-3-030-45603-0_6

For a Proof of Theorem 6.1 c.f., e.g., [183].

Lemma 6.2 *Let $\gamma(t)$ and $\rho(t)$ be continuous functions on $[0, \infty)$, with $\gamma(t)$ nonnegative and $\rho(t)$ positive. Assume that*

$$\int_{u_0}^{\infty} \frac{d\tau}{\rho(\tau)} > \int_{t_0}^{T} \gamma(s)\, ds, \tag{6.5}$$

where $t_0 < T$ and $u_0 \geq 0$ are given numbers. Then there is a unique solution of

$$u'(t) = \gamma(t)\rho(u(t)), \quad t \in [t_0, T), \quad u(t_0) = u_0 \tag{6.6}$$

which is positive in (t_0, T) and depends continuously on u_0.

Proof. One can separate variables to obtain

$$W(u) = \int_{u_0}^{u} \frac{d\tau}{\rho(\tau)} = \int_{t_0}^{t} \gamma(s)\, ds.$$

The function $W(u)$ is differentiable and increasing in $\mathbb{R}$, positive in $[u_0, \infty)$, depends continuously on u_0 and satisfies

$$W(u) \to L = \int_{u_0}^{\infty} \frac{d\tau}{\rho(\tau)} > \int_{t_0}^{T} \gamma(s)\, ds, \quad \text{as } u \to \infty.$$

Thus, for each $t \in [t_0, T)$ there is a unique $u \in [u_0, \infty)$ such that

$$u = W^{-1}\left(\int_{t_0}^{t} \gamma(s)\, ds\right)$$

is the unique solution of (6.6), and it depends continuously on u_0. □

Lemma 6.3 *Let $\gamma(t)$ and $\rho(t)$ be continuous functions on $[0, \infty)$, with $\gamma(t)$ nonnegative and $\rho(t)$ positive. Assume that*

$$\int_{m}^{u_0} \frac{d\tau}{\rho(\tau)} > \int_{t_0}^{T} \gamma(s)\, ds, \tag{6.7}$$

where $t_0 < T$ and $0 < m < u_0$. Then there is a unique solution of

$$u'(t) = -\gamma(t)\rho(u(t)), \quad t \in [t_0, T), \quad u(t_0) = u_0, \tag{6.8}$$

which is $\geq m$ in $[t_0, T)$ and depends continuously on u_0.

Proof. One can separate variables to obtain

$$W(u) = \int_{u}^{u_0} \frac{d\tau}{\rho(\tau)} = \int_{t_0}^{t} \gamma(s)\, ds.$$

The function $W(u)$ is differentiable and decreasing in $\mathbb{R}$, positive in $[m, u_0)$, depends continuously on u_0 and satisfies

$$W(u) \to L = \int_m^{u_0} \frac{d\tau}{\rho(\tau)} > \int_{t_0}^T \gamma(s)\,ds, \quad \text{as } u \to m.$$

Thus, for each $t \in [t_0, T)$ there is a unique $u \in [m, u_0]$ such that

$$u = W^{-1}\left(\int_{t_0}^t \gamma(s)\,ds\right)$$

is the unique solution of (6.8), and it depends continuously on u_0. □

6.2 Extending Solutions

Theorem 6.4 *Assume, in addition to the hypotheses of Theorem 6.1, that*

$$\|g(t,x)\| \le \gamma(t)\rho(\|x-v\|), \quad x \in B_0,\ t \in I_0, \tag{6.9}$$

where $\gamma(t)$ and $\rho(t)$ satisfy the hypotheses of Lemma 6.2 with $T = t_0 + T_1$ and v is a fixed element of X. Let $u(t)$ be the positive solution of

$$u'(t) = \gamma(t)\rho(u(t)),\quad t \in [t_0, T),\quad u(t_0) = u_0 \ge \|x_0 - v\|, \tag{6.10}$$

provided by that lemma. Then the unique solution of (6.4) satisfies

$$\|x(t) - v\| \le u(t), \quad t \in [t_0, T). \tag{6.11}$$

Proof. Assume that there is a $t_1 \in [t_0, T)$ such that

$$u(t_1) < \|x(t_1) - v\|.$$

For $\varepsilon > 0$, let $u_\varepsilon(t)$ be the solution of

$$u'(t) = [\gamma(t) + \varepsilon]\rho(u(t)),\quad t \in [t_0, T),\quad u(t_0) = u_0. \tag{6.12}$$

By Lemma 6.2, a solution exists for $\varepsilon > 0$ sufficiently small. Let

$$w(t) = \|x(t) - v\| - u_\varepsilon(t).$$

Then, we may take ε sufficiently small so that

$$w(t_0) \le 0, \quad w(t_1) > 0.$$

Let t_2 be the largest number in $[t_0, t_1)$ such that $w(t_2) = 0$ and

$$w(t) > 0, \quad t \in (t_2, t_1].$$

For $h > 0$ sufficiently small, we have

$$\frac{w(t_2 + h) - w(t_2)}{h} > 0.$$

Consequently,

$$\begin{aligned} 0 &< \frac{\|x(t_2 + h) - v\| - \|x(t_2) - v\|}{h} - \frac{u_\varepsilon(t_2 + h) - u_\varepsilon(t_2)}{h} \\ &\le \frac{\|x(t_2 + h) - x(t_2)\|}{h} - \frac{u_\varepsilon(t_2 + h) - u_\varepsilon(t_2)}{h} \\ &\to \|x'(t_2)\| - u'_\varepsilon(t_2) \\ &= \|g(t_2, x(t_2))\| - [\gamma(t_2) + \varepsilon]\rho(u_\varepsilon(t_2)) \\ &\le \gamma(t_2)\rho(\|x(t_2) - v\|) - [\gamma(t_2) + \varepsilon]\rho(u_\varepsilon(t_2)) \\ &= -\varepsilon\rho(u_\varepsilon(t_2)) \\ &< 0, \end{aligned} \tag{6.13}$$

as $h \to 0$, since $w(t_2) = 0$. This contradiction proves the theorem. □

6.3 Maximum Intervals

Theorem 6.5 *Let $g(t, x)$ be a continuous map from $\mathbb{R} \times X$ to X, where X is a Banach space. Assume that for each point $(\hat{t}, \hat{x}) \in \mathbb{R} \times X$, there are constants $K, b > 0$ such that*

$$\|g(t, x) - g(t, y)\| \le K\|x - y\|, \quad |t - \hat{t}| < b, \ \|x - \hat{x}\| < b, \ \|y - \hat{x}\| < b. \tag{6.14}$$

Assume also that

$$\|g(t, x)\| \le \gamma(t)\rho(\|x - v\|), \quad x \in X, \ t \in [t_0, T), \tag{6.15}$$

where $T \le \infty$, and $\gamma(t)$, $\rho(t)$ satisfy the hypotheses of Lemma 6.2 with ρ nondecreasing or bounded and v a fixed element of X. Then for each $x_0 \in X$ and $t_0 \in \mathbb{R}$ there is a unique solution $x(t)$ of the equation

$$\frac{dx(t)}{dt} = g(t, x(t)), \quad t \in [t_0, T), \quad x(t_0) = x_0. \tag{6.16}$$

Moreover, $x(t)$ depends continuously on x_0 and satisfies

$$\|x(t) - v\| \leq u(t), \quad t \in [t_0, T), \tag{6.17}$$

where $u(t)$ is the solution of (6.6) in that interval satisfying $u(t_0) = u_0 \geq \|x_0 - v\|$.

Proof. By Theorems 6.1 and 6.4 there is an interval $[t_0, t_0 + m]$, $m > 0$, in which a unique solution of

$$\frac{dx(t)}{dt} = g(t, x(t)), \quad t \in [t_0, t_0 + m], \quad x(t_0) = x_0, \tag{6.18}$$

exists and satisfies

$$\|x(t) - v\| \leq u(t), \quad t \in [t_0, t_0 + m], \tag{6.19}$$

and a unique solution $u(t)$ of

$$u'(t) = \gamma(t)\rho(u(t)), \quad t \in [t_0, T), \quad u(t_0) = u_0 = \|x_0 - v\| \tag{6.20}$$

exist. Let $\tilde{t} \leq T$ be the supremum of all numbers $t_0 + m$ for which this holds. If $t_1 < t_2 < \tilde{t}$, then the solution in $[t_0, t_2]$ coincides with that in $[t_0, t_1]$, since such solutions are unique. Thus a unique solution of (6.18) satisfying (6.19) exists for each t satisfying $t_0 < t < \tilde{t}$. Moreover, we have

$$x(t_2) - x(t_1) = \int_{t_1}^{t_2} g(t, x(t))\, dt.$$

Consequently, when ρ is nondecreasing, we have

$$\begin{aligned} \|x(t_2) - x(t_1)\| &\leq \int_{t_1}^{t_2} \|g(t, x(t))\|\, dt \\ &\leq \int_{t_1}^{t_2} \gamma(t)\rho(\|x(t) - v\|)\, dt \\ &\leq \int_{t_1}^{t_2} \gamma(t)\rho(u(t))\, dt \\ &= u(t_2) - u(t_1). \end{aligned}$$

If $\rho \leq K$, then

$$\|x(t_2) - x(t_1)\| \leq \int_{t_1}^{t_2} \|g(t, x(t))\|\, dt$$

$$\leq \int_{t_1}^{t_2} \gamma(t)\rho(\|x(t) - v\|)\,dt$$

$$\leq \int_{t_1}^{t_2} \gamma(t)K\,dt.$$

Assume that $\tilde{t} < T$. Let t_k be a sequence such that $t_0 < t_k < \tilde{t}$ and $t_k \to \tilde{t}$. Then

$$\|x(t_k) - x(t_j)\| \leq u(t_k) - u(t_j) \to 0$$

when ρ is nondecreasing. Otherwise,

$$\|x(t_k) - x(t_j)\| \leq \int_{t_k}^{t_j} \gamma(t)K\,dt \to 0.$$

Thus $\{x(t_k)\}$ is a Cauchy sequence in X. Since X is complete, $x(t_k)$ converges to an element $x_1 \in X$. Since $\|x(t_k) - v\| \leq u(t_k)$, we see that $\|x_1 - v\| \leq u(\tilde{t})$. Moreover, we note that

$$x(t) \to x_1 \text{ as } t \to \tilde{t}.$$

To see this, let $\varepsilon > 0$ be given. Then there is a k such that

$$\|x(t_k) - x_1\| < \varepsilon, \quad u(\tilde{t}) - u(t_k) < \varepsilon.$$

Then for $t_k \leq t < \tilde{t}$,

$$\begin{aligned}\|x(t) - x_1\| &\leq \|x(t) - x(t_k)\| + \|x(t_k) - x_1\| \\ &\leq u(t) - u(t_k) + \|x(t_k) - x_1\| < 2\varepsilon\end{aligned}$$

when ρ is nondecreasing. Otherwise,

$$\begin{aligned}\|x(t) - x_1\| &\leq \|x(t) - x(t_k)\| + \|x(t_k) - x_1\| \\ &\leq \int_{t_k}^{t} \gamma(s)K\,ds + \|x(t_k) - x_1\| < 2\varepsilon.\end{aligned}$$

We define $x(\tilde{t}) = x_1$. Then, we have a solution of (6.18) satisfying (6.19) in $[0, \tilde{t}]$. By Theorem 6.1, there is a unique solution of

$$\frac{dy(t)}{dt} = g(t, y(t)), \quad y(\tilde{t}) = x_1 \tag{6.21}$$

satisfying $\|y(t)-v\| \leq u(t)$ in some interval $|t-\tilde{t}| < \delta$. By uniqueness, the solution of (6.21) coincides with the solution of (6.18) in the interval $(\tilde{t}-\delta, \tilde{t}]$. Define

$$
\begin{aligned}
z(t) &= x(t), \quad t_0 \leq t < \tilde{t}, \\
z(\tilde{t}) &= x_1, \\
z(t) &= y(t), \quad \tilde{t} < t \leq \tilde{t}+\delta.
\end{aligned}
$$

This gives a solution of (6.18) satisfying (6.19) in the interval $[t_0, \tilde{t}+\delta)$, contradicting the definition of $\tilde{t}$. Hence, $\tilde{t} = T$. □

6.4 Upper and Lower Estimates

Corollary 6.6 *In place of (6.15) assume*

$$\|g(t,x)\| \leq \gamma(t)\rho(d(x,Q)), \quad x \in X, \ t \in [t_0, T), \tag{6.22}$$

where Q is a subset of X. Let $u(t)$ be the solution of (6.6) in that interval satisfying $u(t_0) = u_0 \geq d(x_0, Q)$. Then the solution $x(t)$ of

$$\frac{dx(t)}{dt} = g(t, x(t)), \quad t \in [t_0, T), \quad x(t_0) = d(x_0, Q), \tag{6.23}$$

satisfies

$$d(x(t), Q) \leq u(t), \quad t \in [t_0, T).$$

Proof. Let $\varepsilon > 0$ be given, and let $v_\varepsilon \in Q$ be such that

$$\|x_0 - v_\varepsilon\| \leq d(x_0, Q) + \varepsilon.$$

Let $u_\varepsilon(t)$ be the solution of

$$u'_\varepsilon(t) = \gamma(t)\rho(u_\varepsilon(t)), \quad t \in [t_0, T), \quad u_\varepsilon(t_0) = u_0 = \|x_0 - v_\varepsilon\|. \tag{6.24}$$

Since

$$\|g(t,x)\| \leq \gamma(t)\rho(d(x,Q)) \leq \gamma(t)\rho(\|x - v_\varepsilon\|), \quad x \in X, \ t \in [t_0, T), \tag{6.25}$$

there is a solution $x_\varepsilon(t)$ of

$$\frac{dx(t)}{dt} = g(t, x(t)), \quad t \in [t_0, T), \quad x(t_0) = x_0 - v_\varepsilon. \tag{6.26}$$

which satisfies

$$\|x_\varepsilon(t) - v_\varepsilon\| \le u_\varepsilon(t), \quad t \in [t_0, T). \tag{6.27}$$

By the continuous dependence of solutions on the initial conditions, $u_\varepsilon(t) \to u(t)$ uniformly in $[t_0, T_0]$ as $\varepsilon \to 0$ for any $T_0 < T$. For the same reason $x_\varepsilon(t)$ converges to the solution of (6.16). But this implies

$$d(x(t), Q) \le u(t), \quad t \in [t_0, T).$$

□

Corollary 6.7 *Under the same hypotheses, for each $T_0 < T$ there is a constant C depending only on T_0, Q and $\|x_0\|$ such that*

$$\|x(t)\| \le C, \quad 0 \le t \le T_0$$

for the solution of (6.16).

Proof. Let v be any point of Q such that

$$\|v\| \le \inf_{w \in Q} \|w\| + 1.$$

Then $g(t, x)$ satisfies (6.15). If we take $u_0 = \|x_0 - v\|$, then

$$\|x(t) - v\| \le u(T_0), \quad t \in [t_0, T_0],$$

or

$$\|x(t)\| \le u(T_0) + \|v\| \le u(T_0) + \inf_{w \in Q} \|w\| + 1$$

by Theorem 6.5. Since u_0 depends on Q and $\|x_0\|$, the same is true of $u(T_0)$. □

6.5 Comparison of Solutions

Theorem 6.8 *Let ρ, γ satisfy the hypotheses of Lemma 6.3, with ρ locally Lipschitz continuous. Let $u(t)$ be the solution of (6.8) satisfying $u(t_0) = u_0$, and let $h(t)$ be a continuous function satisfying*

$$h(t) \ge h(s) - \int_s^t \gamma(r)\rho(h(r))\,dr, \quad t_0 \le s < t < T, \quad h(t_0) \ge u_0. \tag{6.28}$$

Then

$$u(t) \leq h(t), \quad t \in [t_0, T). \tag{6.29}$$

Proof. Assume that there is a point t_1 in the interval such that

$$h(t_1) < u(t_1).$$

Let

$$y(t) = u(t) - h(t), \quad t \in [t_0, T).$$

Then, $y(t_0) \leq 0$ and $y(t_1) > 0$. Let τ be the largest point $< t_1$ such that $y(\tau) = 0$. Then

$$y(t) > 0, \quad t \in (\tau, t_1]. \tag{6.30}$$

Moreover, by (6.8) and (6.28) we have

$$y(t) \leq -\int_\tau^t \gamma(s)[\rho(u(s)) - \rho(h(s))]\, ds \leq L \int_\tau^t y(s)\, ds, \tag{6.31}$$

where L is the Lipschitz constant for ρ at $u(\tau)$ times the maximum of γ in the interval. Let

$$w(t) = \int_\tau^t y(s)\, ds.$$

Then

$$[e^{-Lt} w(t)]' = e^{-Lt}[y(t) - Lw(t)] \leq 0, \quad t \in [\tau, t_1].$$

Consequently,

$$e^{-Lt} w(t) \leq e^{-L\tau} w(\tau) = 0, \quad t \in [\tau, t_1].$$

Hence,

$$y(t) \leq Lw(t) \leq 0, \quad t \in [\tau, t_1],$$

contradicting (6.30). This completes the proof. □

Corollary 6.9 *Assume ρ, γ satisfy the hypotheses of Lemma 6.3, with ρ locally Lipschitz continuous and*

$$\|g(t, x)\| \leq \gamma(t)\rho(d(x, Q)), \quad x \in X,\ t \in [t_0, T), \tag{6.32}$$

where Q is a subset of X. Let $x_0 \in X$, and let $u(t)$ be the solution of (6.8) in that interval satisfying $u(t_0) = u_0 = d(x_0, Q)$. Then the solution $x(t)$ of

$$\frac{dx(t)}{dt} = g(t, x(t)), \quad t \in [t_0, T), \quad x(t_0) = x_0, \tag{6.33}$$

satisfies

$$d(x(t), Q) \geq u(t), \quad t \in [t_0, T).$$

Proof. By Corollary 6.6 there is a unique solution $x(t)$ of (6.33). Let $h(t) = d(x(t), Q)$. We have

$$\|x(t) - x(s)\| = \| \int_s^t g(r, x(r))\, dr \| \leq \int_s^t \gamma(r)\rho(h(r))\, dr.$$

Consequently,

$$\|x(s) - v - [x(t) - v]\| \leq \int_s^t \gamma(r)\rho(h(r))\, dr$$

for any $v \in Q$. Thus,

$$h(s) \leq \|x(s) - v\| \leq \|[x(t) - v]\| + \int_s^t \gamma(r)\rho(h(r))\, dr, \quad v \in Q.$$

This gives,

$$h(s) \leq h(t) + \int_s^t \gamma(r)\rho(h(r))\, dr,$$

which is (6.28). The result follows from Theorem 6.8. □

6.6 Summary

Summarizing some of the theorems, we have the following.

Theorem 6.10 *Let $g(t, x)$ be a continuous map from $\mathbb{R} \times X$ to X, where X is a Banach space. Assume that for each point $(\hat{t}, \hat{x}) \in \mathbb{R} \times X$, there are constants $K, b > 0$ such that*

$$\|g(t, x) - g(t, y)\| \leq K\|x - y\|, \quad |t - \hat{t}| < b, \ \|x - \hat{x}\| < b, \ \|y - \hat{x}\| < b, \tag{6.34}$$

and

$$\|g(t,x)\| \le \rho(d(x,Q)), \quad x \in X, \ t \in [t_0,\infty), \tag{6.35}$$

where Q is a subset of X, with ρ nondecreasing or bounded. Assume

$$\int_0^{u_0} \frac{d\tau}{\rho(\tau)} = \int_{u_0}^{\infty} \frac{d\tau}{\rho(\tau)} = \infty \tag{6.36}$$

where $u_0 > 0$. Then for each $x_0 \in X$ and $t_0 \in \mathbb{R}$ there is a unique solution $x(t)$ of the equation

$$\frac{dx(t)}{dt} = g(t,x(t)), \quad t \in [t_0,\infty), \quad x(t_0) = x_0. \tag{6.37}$$

Moreover, $x(t)$ depends continuously on x_0 and satisfies

$$u_1(t) \le d(x(t),Q) \le u_2(t), \quad t \in [t_0,\infty), \tag{6.38}$$

where $u_1(t)$ is the solution of (6.8) in $[t_0,\infty)$ satisfying $u(t_0) = u_0 = d(x_0,Q)$, and $u_2(t)$ is the solution of (6.6) in $[t_0,\infty)$ satisfying $u(t_0) = u_0 = d(x_0,Q)$.

6.7 Notes and Remarks

We added new material to [192].

Chapter 7
Schrödinger Equations

7.1 Introduction

We now consider some applications of the materials presented in Chaps. 1–6. We wish to show how powerful these methods are in obtaining results better than those given by other methods. In Chaps. 7–9 we deal with some problems involving Schrödinger equations.

We consider semilinear elliptic partial differential equations of the form

$$\mathcal{A}u = f(x, u), u \in D \tag{7.1}$$

in unbounded domains. Included is the case of the Schrödinger operator $\mathcal{A} = -\Delta + V(x)$ on $D = H^1(\mathbb{R}^n)$, where $V(x)$ is a given potential. One wishes to find nontrivial solutions and, in particular, the so-called "minimizing solutions." These are solutions that minimize the corresponding energy functional. If they are not trivial, they are called "ground state solutions."

The existence of solutions depends on both the linear operator $\mathcal{A}$ and the nonlinear term $f(x, u)$. We shall study the problem for the case when $\mathcal{A}$ is self-adjoint, having a nonempty resolvent set, and $f(x, u)$ is superlinear. The results are stated in the next section and proved in Sect. 7.5.

7.2 Superlinear Problems

Let $\mathcal{A}$ be a self-adjoint operator on $L^2(\mathbb{R}^n)$. We assume that $\sigma(\mathcal{A})$ is not the whole of $\mathbb{R}$. For convenience, we take $0 \in \rho(\mathcal{A})$. Thus, there is an interval $(a, b) \subset \rho(\mathcal{A})$, where $a < 0 < b$. We let $D = D(|\mathcal{A}|^{(1/2)})$. With the scalar product $(u, v)_D = (|\mathcal{A}|^{(1/2)}u, |\mathcal{A}|^{(1/2)}v)$, it becomes a Hilbert space. We let

M. Schechter, *Critical Point Theory*,
https://doi.org/10.1007/978-3-030-45603-0_7

$$N = E(-\infty, a], \quad M = E[b, \infty)$$

be the negative and positive invariant subspaces of $\mathcal{A}$. Hence,

$$(\mathcal{A}v, v) \le a\|v\|^2, \quad v \in D \cap N,$$

and

$$(\mathcal{A}w, w) \ge b\|w\|^2, \quad w \in D \cap M.$$

We assume that $C_0^\infty(\mathbb{R}^n) \subset D \subset H^{m,2}(\mathbb{R}^n)$ for some $m > 0$. In particular,

$$\|u\|_{m,2} \le C\|u\|_D, \quad u \in D. \tag{7.2}$$

Let q be any number satisfying

$$\begin{aligned} 2 < q &< 2^* := 2n/(n-2m), \quad && 2m < n \\ 2 < q &< \infty, && n \le 2m. \end{aligned}$$

We assume that D is compact in $L^q_{loc}(\mathbb{R}^n)$ and

$$\|u\|_q \le C\|u\|_D, \quad u \in D, \tag{7.3}$$

where $\|\cdot\|_q$ is the norm of $L^q(\mathbb{R}^n)$. We also assume that

$$u_y(x) = u(x+y), \;\; y \in \mathbb{Z}^n \Longrightarrow \|u_y\|_D = \|u\|_D. \tag{7.4}$$

Let $f(x,t)$ be a Carathéodory function on $\mathbb{R}^n \times \mathbb{R}$ satisfying

$$f(x+y,t) = f(x,t), \quad x \in \mathbb{R}^n, \; y \in \mathbb{Z}^n, \; t \in \mathbb{R}, \tag{7.5}$$

$$|f(x,t)| \le C(|t|^{q-1} + 1), \quad x \in \mathbb{R}^n, \; t \in \mathbb{R}, \tag{7.6}$$

and

$$|f(x,t)| \le \sigma|t|, \quad |t| < \delta, \; x \in \mathbb{R}^n, \; t \in \mathbb{R}, \tag{7.7}$$

for some $\sigma < min[-a, b], \; \delta > 0$. Assume that

$$F(x,t) := \int_0^t f(x,s)\,ds$$

satisfies

$$F(x,t) \ge 0, \quad x \in \mathbb{R}^n, \; t \in \mathbb{R}, \tag{7.8}$$

and that

$$F(x,t)/t^2 \to \infty \; as \; t^2 \to \infty, \tag{7.9}$$

$$2F(x,t+s) - 2F(x,t) - (2rs - (r-1)^2 t) f(x,t) \tag{7.10}$$
$$\geq -\, W(x), \quad x \in \mathbb{R}^n, \; s,t \in \mathbb{R}, \; r \in [0,1],$$

where $W(x) \in L^1(\mathbb{R}^n)$. We shall prove the following:

Theorem 7.1 *Under the above hypotheses there is a nontrivial solution of*

$$\mathcal{A}u = f(x,u), u \in D. \tag{7.11}$$

Theorem 7.2 *Let $\mathcal{M}$ be the collection of solutions of (7.11). Then there is a nontrivial solution that minimizes the energy functional*

$$G(u) = \|w\|_D^2 - \|v\|_D^2 - 2\int_{\mathbb{R}^n} F(x,u), \quad u \in D \tag{7.12}$$

over $\mathcal{M}\backslash\{0\}$, where w, v are the projections of u onto M, N, respectively.

Such solutions are called *ground state solutions*.

The chapter is organized as follows. In Sect. 7.3 we describe applications to Schrödinger equations on $\mathbb{R}^n$. We present some lemmas in Sect. 7.4. Proofs are given in Sect. 7.5.

7.3 The Schrödinger Equation

Here we wish to compare our results with those of other researchers. In [241], Szulkin and Weth proved the following theorem:

Theorem 7.3 *Let $\mathcal{A} = -\Delta + V(x)$ on $H^1(\mathbb{R}^n)$. Assume*

1. *V is continuous, 1-periodic in $x_1, \cdots, x_k$ and $0 \in \rho(\mathcal{A})$,*
2. *$f(x,t)$ is continuous, 1-periodic in $x_1, \cdots, x_k$ and $|f(x,t)| \leq C(1+|t|^{p-1})$ for some $p \in (2, 2^*)$, $2^* := 2n/(n-2)$, $n > 2$, $2^* := \infty$, $n \leq 2$.*
3. *$f(x,t) = o(t)$ as $t \to 0$ uniformly in x.*
4. *$F(x,t)/t^2 \to \infty$ as $t^2 \to \infty$ uniformly in x.*
5. *$f(x,t)/|t|$ is strictly increasing in t on $(-\infty, 0)$ and $(0, \infty)$.*

Then (7.11) has a nontrivial ground state solution.

(Cf. also Yang [287] who obtained the same results using a different method.) Note that hypotheses 1–5 imply

1. $F(x,t) > 0, \quad t \neq 0,$
2. $2F(x,t) \leq \varepsilon t^2, \quad |t| < \delta, \ x \in \mathbb{R}^n, \ t \in \mathbb{R},$
3. $H(x,t) := tf(x,t) - 2F(x,t) > 0, \quad t \neq 0,$
4. $tf(x,t) > 0, \ t \neq 0,$
5. $sf(x,t) = tf(x,s), \ st \neq 0 \Longrightarrow s = t, \quad s,t \in \mathbb{R}, \ x \in \mathbb{R}^n,$
6. $F(x,t+s) - F(x,t) - sf(x,t) > 0, \quad x \in \mathbb{R}^n, \ s,t \in \mathbb{R}, \ s \neq 0.$

We note that Theorem 7.3 follows from Theorems 7.1 and 7.2. In fact, we have the following.

Theorem 7.4 *Let $\mathcal{A} = -\Delta + V(x)$ on $H^1(\mathbb{R}^n)$. Assume*

1. *V is continuous, 1-periodic in $x_1, \cdots, x_k$ and $0 \in \rho(\mathcal{A})$,*
2. *$f(x,t)$ is continuous, 1-periodic in $x_1, \cdots, x_k$ and*
$$|f(x,t)| \leq C(|t|^{p-1} + 1)$$
for some $p \in (2, 2^*), \ 2^* := 2n/(n-2), \ n > 2, \ 2^* := \infty, \ n \leq 2.$
3. $$|f(x,t)| \leq \sigma |t|, \quad |t| < \delta, \ x \in \mathbb{R}^n, \ t \in \mathbb{R},$$
for some $\sigma < min[-a, b], \ \delta > 0.$
4. $$F(x,t) \geq 0, \quad x \in \mathbb{R}^n, \ t \in \mathbb{R}.$$
5. $$F(x,t)/t^2 \to \infty \text{ as } t^2 \to \infty$$
uniformly in x.
6. $$\begin{aligned} 2F(x,t+s) - 2F(x,t) - (2rs - (r-1)^2 t) f(x,t) \\ \geq - W(x), \quad x \in \mathbb{R}^n, \ s,t \in \mathbb{R}, \ r \in [0,1], \end{aligned}$$
where $W(x) \in L^1(\mathbb{R}^n)$.

Then (7.11) has a nontrivial ground state solution.

7.4 Some Lemmas

Before proving our main theorem, we shall prove a few lemmas. We pick Λ such that $1 < \Lambda < 2$. For $1 \leq \lambda \leq \Lambda$, define

$$G_\lambda(u) = \lambda\|w\|_D^2 - \|v\|_D^2 - 2\int_{\mathbb{R}^n} F(x,u) = \lambda I(u) - J(u), \quad u \in D, \tag{7.13}$$

where we write $u = v + w,\ v \in N,\ w \in M$.

Lemma 7.5 *Let $\rho_k = \|u_k\|_D \to \infty$. Assume that $\tilde{u}_k = u_k/\rho_k \to \tilde{u}$ a.e. If $\tilde{u} \not\equiv 0$, then*

$$\int_{\mathbb{R}^n} F(x,u_k)/\rho_k^2 \to \infty. \tag{7.14}$$

Proof. Let Ω_0 be the subset of $\mathbb{R}^n$ where $\tilde{u} \neq 0$. If the measure of Ω_0 is positive, then

$$\int_{\mathbb{R}^n} F(x,u_k)/\rho_k^2 \geq \int_{\Omega_0} \frac{F(x,u_k)}{u_k^2}\tilde{u}_k^2 \to \infty,$$

since the integrand is bounded below and $u_k^2 \to \infty$ on Ω_0. □

Lemma 7.6

$$I(u) + J(u) \to \infty \quad \text{as } \|u\|_D \to \infty.$$

Proof. We have

$$I(u) + J(u) = \|w\|_D^2 + \|v\|_D^2 + 2\int_{\mathbb{R}^n} F(x,u) = \|u\|_D^2(1 + 2\int_{\mathbb{R}^n} F(x,u)/\|u\|_D^2).$$

Apply Lemma 7.5. □

Lemma 7.7

$$v_k = Pu_k \to v \text{ weakly in } D, \quad w_k = (I-P)u_k \to w \text{ strongly in } D \tag{7.15}$$

implies

$$G_\lambda{}'(v_k + w_k) \to G_\lambda{}'(v + w) \text{ weakly in } D, \tag{7.16}$$

where P is the projection of D onto N.

Proof. Since the u_k are bounded in D, there is a renamed subsequence converging to a limit u weakly in D, strongly in $L^q_{loc}(\mathbb{R}^n)$ and a.e. in $\mathbb{R}^n$. Let $\varepsilon > 0$ and $h \in D$ be given. Let K be a bounded subset of $\mathbb{R}^n$ such that

$$\int_{\mathbb{R}^n \setminus K} |h(x)|^q dx < \varepsilon^q.$$

By (7.6)

$$\int_{\mathbb{R}^n \setminus K} |f(x, u_k)||h(x)| \le C \int_{\mathbb{R}^n \setminus K} |u_k(x)|^{q-1} |h(x)| dx \le \varepsilon C \|u_k\|_q^{q-1} \le \varepsilon C'.$$

On the other hand,

$$\int_K f(x, u_k)h(x)dx \to \int_K f(x, u)h(x)dx \ as\ k \to \infty.$$

Thus,

$$\int_{\mathbb{R}^n} f(x, u_k)h(x)dx \to \int_{\mathbb{R}^n} f(x, u)h(x)dx \ as\ k \to \infty.$$

But then,

$$\begin{aligned}(G'_\lambda(u_k), h)/2 &= \lambda(w_k, h)_D - (v_k, h)_D - \int_{\mathbb{R}^n} f(x, u_k(x))h(x) \\ &\to \lambda(w, h)_D - (v, h)_D - \int_{\mathbb{R}^n} f(x, u(x))h(x) \\ &= (G'_\lambda(u), h)/2.\end{aligned}$$

This gives (7.16). □

Lemma 7.8 *For each $\rho > 0$ sufficiently small there is an $\varepsilon > 0$ such that*

$$G_\lambda(w) \ge \varepsilon, \quad w \in M, \ \|w\|_D = \rho, \ 1 \le \lambda \le \Lambda. \tag{7.17}$$

Proof. By (7.6) and (7.7),

$$2 \int_{|w|<\delta} |F(x, w)| \le \sigma \int_{|w|<\delta} w^2$$

and

$$\begin{aligned}\int_{|w|>\delta} |F(x, w)| &\le C \int_{|w|>\delta} (|w|^q + |w|) \\ &\le C \int_{|w|>\delta} (|w|^q + |w|^q/\delta^{q-1})\end{aligned}$$

$$\leq C \int_{|w|>\delta} |w|^q$$
$$\leq C \|w\|_q^q.$$

Consequently,

$$G_\lambda(w) \geq \|w\|_D^2 - \sigma \int_{|w|<\delta} w^2 - C \int_{|w|>\delta} (|w|^q + |w|) \geq (b - \sigma - C' \|w\|_q^{q-2}) \|w\|_D^2.$$

We take $\|w\|_D^2$ sufficiently small. □

Lemma 7.9 *Let*

$$Q_R = \{v + sw_0 : \|v\|_D^2 + s^2 \leq R^2,\ v \in N,\ s \geq 0\}, \tag{7.18}$$

where $w_0 \in M$ and $\|w_0\|_D = 1$. Then there is an $R > 0$ such that

$$G_\lambda(u) \leq 0, \quad u \in \partial Q_R,\ 1 \leq \lambda \leq \Lambda. \tag{7.19}$$

Proof. If not, $\exists R_k \to \infty,\ 1 \leq \lambda_k \leq \Lambda,\ u_k = v_k + s_k w_0 \in \partial Q_{R_k}$, such that $G_{\lambda_k}(u_k) > 0$. If $s_k = 0$, then

$$G_{\lambda_k}(v_k) = -\|v_k\|_D^2 - 2 \int_{\mathbb{R}^n} F(x, v_k) \leq -\|v_k\|_D^2 \leq 0.$$

Hence, $s_k \neq 0$ and

$$\|v_k\|_D^2 + s_k^2 = R_k^2.$$

Let $\tilde{u}_k = u_k / R_k = \tilde{v}_k + \tilde{s}_k w_0$. Then

$$\|\tilde{v}_k\|_D^2 + \tilde{s}_k^2 = 1.$$

There are renamed subsequences such that $\tilde{s}_k \to \tilde{s},\ \lambda_k \to \lambda$, and there is a renamed subsequence such that $\tilde{u}_k = u_k/R_k = \tilde{v}_k + \tilde{s}_k w_0 \to \tilde{u}$ *a.e.* Since,

$$0 < G_{\lambda_k}(u_k)/R_k^2 = \lambda_k \tilde{s}_k^2 - \|\tilde{v}_k\|_D^2 - 2 \int_{\mathbb{R}^n} F(x, u_k)/R_k^2,$$

we have by hypothesis

$$\lambda \tilde{s}^2 - (1 - \tilde{s}^2) \geq 0,$$

or

$$\tilde{s}^2 \geq \frac{1}{1 + \lambda} \geq \frac{1}{1 + \Lambda} > 0.$$

Thus, $\tilde{u} \not\equiv 0$. Lemma 7.5 implies

$$\int_{\mathbb{R}^n} F(x, u_k)/R_k^2 \to \infty. \tag{7.20}$$

Since,

$$0 < G_\lambda(u_k)/R_k^2 = \lambda \tilde{s}_k^2 - \|\tilde{v}_k\|_D^2 - 2\int_{\mathbb{R}^n} F(x, u_k)/R_k^2,$$

this produces a contradiction, and the lemma follows. □

Remark 7.10 Let $0 < \rho < R$ be such that Lemmas 8.8 and 7.9 hold. Then

$$\sup_A G_\lambda < \inf_B G_\lambda, \forall \lambda \in [1, \Lambda],$$

where $A = \partial Q_R$ and $B = \{w \in M : \|w\|_D = \rho.\}$

Lemma 7.11 *Let $\mathcal{K}$ be the collection of sets*

$$K = \{\sigma(1)A : \sigma \in \tilde{\Sigma}\},$$

where

$$\tilde{\Sigma} = \{\sigma \in \Sigma_w : G(\sigma(t)u) \leq G(u), \quad t \in I\}.$$

Then we have

(1) For almost all $\lambda \in [1, \Lambda]$ there exists a constant $k_0(\lambda) := k_0$ (depending only on λ) such that for each $\delta > 0$ there exists a $K \in \mathcal{K}$ such that

$$\sup_K G_\lambda \leq a(\lambda) + \delta$$

and

$$\|u\| \leq k_0 \quad \text{whenever} \quad u \in K \ \text{ and } \ G_\lambda(u) \geq a(\lambda) - \delta. \tag{7.21}$$

(2) For almost all $\lambda \in \Lambda$ there exists a bounded sequence $u_k(\lambda) \in D$ such that

$$\|G_\lambda'(u_k)\| \to 0, \quad G_\lambda(u_k) \to a(\lambda), \qquad \text{as } k \to \infty, \tag{7.22}$$

where

$$a(\lambda) := \inf_{K \in \mathcal{K}} \sup_K G_\lambda \in [\inf_B G_\lambda, \sup_{Q_R} G_\lambda].$$

Proof. Replace N by $N \oplus \{w_0\}$ and M by $[N \oplus \{w_0\}]^\perp$. Define

$$F(v+sw_0+w) = v+\|sw_0+w\|w_0, \quad v \in N\oplus\{w_0\}, \quad s \in \mathbb{R}, \quad w \in [N\oplus\{w_0\}]^\perp.$$

Then F satisfies the hypotheses of Theorem 5.7. Moreover,

$$B = F^{-1}(\rho w_0) = \{sw_0 + w \in M : \|sw_0 + w\|^2 = s^2 + \|w\|_D^2 = \rho^2\}.$$

The result now follows from Theorem 5.16. There it was shown that

$$B \cap K \neq \phi, \quad K \in \mathcal{K}. \tag{7.23}$$

Consequently,

$$b_0(\lambda) = \inf_B G_\lambda \leq \inf_{B\cap K} G_\lambda \leq \sup_{B\cap K} G_\lambda \leq \sup_K G_\lambda, \quad K \in \mathcal{K}.$$

Thus, $b_0(\lambda) \leq a(\lambda)$. On the other hand, $Q_R \in \mathcal{K}$. Hence,

$$0 < \varepsilon \leq \inf_B G_\lambda \leq a(\lambda) \leq \sup_{Q_R} G_\lambda.$$

□

Lemma 7.12 *Let $r > 0$ and $q \in [2, 2^*)$, where $2^* = 2n/(n-2)$. If $\{u_k\}$ is a bounded sequence in $E := H^1(\mathbb{R}^n)$, and*

$$\sup_{y\in\mathbb{R}^n} \int_{B(y,r)} |u_k|^q dx \to 0, \quad k \to \infty,$$

where $B(y,r) := \{u \in E : \|u - y\| \leq r\}$, then $u_k \to 0$ in $L^p(\mathbb{R}^n)$ for $q < p < 2^*$.

Proof. We consider $n \geq 3$ and make use of the fact that

$$\int_{B(y,r)} |u(x)|^q dx \leq C\Big(\int_{B(y,r)} (u^2 + |\nabla u|^2)dx\Big)^{q/2}, \quad 2 \leq q \leq 2^*, \ u \in H^1(\mathbb{R}^n).$$

Choose

$$p_1 = q\frac{2^* - p}{2^* - q}, \ p_2 = 2^*\frac{p - q}{2^* - q}, \ t = \frac{2^* - q}{2^* - p} > 1, \ t' = \frac{t}{t-1} > 1.$$

Then $p_1 t = q, \ p_2 t' = 2^*, \ 1/t + 1/t' = 1, \ p_1 + p_2 = p$. By Hölder's inequality, we have

$$
\begin{aligned}
&\int_{B(y,r)} |u_k|^p dx \\
&\leq \Big(\int_{B(y,r)} |u_k|^{p_1 t} dx\Big)^{1/t} \Big(\int_{B(y,r)} |u_k|^{p_2 t'} dx\Big)^{1/t'} \\
&\leq c\Big(\int_{B(y,r)} |u_k|^q dx\Big)^{1/t} \Big(\int_{B(y,r)} |u_k|^{2^*} dx\Big)^{1/t'} \\
&\leq c\Big(\int_{B(y,r)} |u_k|^q dx\Big)^{1/t} \Big(\int_{B(y,r)} (u_k^2 + |\nabla u_k|^2) dx\Big)^{p_2/2}.
\end{aligned}
$$

Covering $\mathbb{R}^n$ by balls with radius r in such a way that each point of $\mathbb{R}^n$ is contained in at most $n+1$ balls, we have

$$
\int_{\mathbb{R}^n} |u_k|^p dx \leq (n+1)c \sup_{y \in \mathbb{R}^n} \Big(\int_{B(y,r)} |u_k|^q dx\Big)^{1/t},
$$

which implies the conclusion of the lemma. □

Lemma 7.13 *For almost every $\lambda \in [1, \Lambda]$ there is a $u_\lambda \in D \backslash \{0\}$ such that $-C \leq G_\lambda(u_\lambda) \leq a(\lambda)$, $G'_\lambda(u_\lambda) = 0$, where*

$$
C = \int_{\mathbb{R}^n} W(x).
$$

Proof. For each $\lambda \in [1, \Lambda]$ for which the conclusion of Lemma 7.11 holds, there is a sequence $\{u_k\} \subset D$ satisfying (7.22). Since the u_k are bounded in D, there is a renamed subsequence converging to a limit u_λ weakly in D, strongly in $L^q_{loc}(\mathbb{R}^n)$ and a.e. in $\mathbb{R}^n$. For any $\varphi \in C_0^\infty(\mathbb{R}^n)$, we have

$$
(G'_\lambda(u_k), \varphi)/2 = \lambda(w_k, \varphi)_D - (v_k, \varphi)_D - \int_{\mathbb{R}^n} f(x, u_k(x))\varphi(x) \to 0.
$$

Hence,

$$
(G'_\lambda(u_\lambda), \varphi)/2 = \lambda(w_\lambda, \varphi)_D - (v_\lambda, \varphi)_D - \int_{\mathbb{R}^n} f(x, u_\lambda(x))\varphi(x) = 0,
$$

showing that $G'_\lambda(u_\lambda) = 0$. Moreover, since u_k is bounded in D,

$$
\int_{\mathbb{R}^n} H(x, u_k(x)) = G_\lambda(u_k) - (G'_\lambda(u_k), u_k)/2 \to a(\lambda),
$$

where

$$
H(x,t) := tf(x,t) - 2F(x,t).
$$

Note that $H(x, u_k(x)) \geq -W(x)$ *a.e.* by (7.10). Since

$$H(x, u_k(x)) \to H(x, u_\lambda(x)) \; a.e.,$$

we have

$$a(\lambda) = \lim \int_{\mathbb{R}^n} H(x, u_k(x)) \geq \int_{\mathbb{R}^n} H(x, u_\lambda(x)) = G_\lambda(u_\lambda).$$

To show that $u_\lambda \neq 0$, note that

$$G_\lambda(u_k) = \lambda \|w_k\|_D^2 - \|v_k\|_D^2 - 2\int_{\mathbb{R}^n} F(x, u_k). \tag{7.24}$$

Consequently,

$$\lambda \|w_k\|_D^2 \geq a(\lambda) - \varepsilon, \quad k > N,$$

for some $N, \varepsilon > 0$. Then we have

$$\begin{aligned}(G'_\lambda(u_k), \hat{u}_k)/2 &= \lambda(w_k, \hat{u}_k)_D - (v_k, \hat{u}_k)_D \\ &\quad - \int_{\mathbb{R}^n} f(x, u_k(x))\hat{u}_k(x) \\ &\to 0,\end{aligned}$$

where $\hat{u}_k = w_k - v_k$. Thus,

$$\begin{aligned}\lambda \|w_k\|_D^2 + \|v_k\|_D^2 + o(1) &= \int_{\mathbb{R}^n} f(x, u_k)\hat{u}_k = \int_{|u_k|<\delta} + \int_{|u_k|>\delta} \\ &\leq \sigma \|u_k\| \cdot \|\hat{u}_k\| + C\|u_k\|_q^{q-1}\|\hat{u}_k\|_q.\end{aligned}$$

Hence,

$$\varepsilon \|u_k\|_D^2 \leq \|u_k\|_D^2 - \sigma \|u_k\|^2 \leq C'\|u_k\|_q^{q-1}\|u_k\|_D$$

for some $\varepsilon > 0$. Since $u_k \neq 0$, this shows that $\|u_k\|_q \geq c > 0$. If there is an $r > 0$ such that

$$\sup_y \int_{B(y,r)} u_k^2 \to 0 \quad as \; k \to \infty, \tag{7.25}$$

then $u_k(x) \to 0$ in $L^p(\mathbb{R}^n)$ for $2 < p < 2^*$ (Lemma 7.12). Thus (7.25) does not hold. In that case there are $\alpha > 0, \; y \in \mathbb{R}^n$ such that

$$\int_{B(y)} u_k^2 \geq \alpha, \tag{7.26}$$

where $B(y) = B(y, 1)$. Since $u_k \to u_\lambda$ in $L^2_{loc}(\mathbb{R}^n)$, we see that $u_\lambda \neq 0$. □

Lemma 7.14 *For every $\lambda \in [1, \Lambda]$ there are sequences $\lambda_k \to \lambda$ and $\{u_k\} \subset D\backslash\{0\}$ such that $-C \le G_{\lambda_k}(u_k) \le a(\lambda_k),\ G'_{\lambda_k}(u_k) = 0$.*

Proof. Apply Lemma 7.13. □

Lemma 7.15

$$\int_{\mathbb{R}^n} [F(x,u) - F(x,rw) + (r^2 w - \frac{1+r^2}{2} u) f(x,u)] \le C,$$
$$u \in D,\ w \in M,\ r \in [0,1],\ \|w\|_D \le \|u\|_D,$$

where the constant C does not depend on u, w, r.

Proof. This follows from (7.10) if we take $t = u$, and $s = rw - u$. □

Lemma 7.16 *If u satisfies $G'_\lambda(u) = 0$ for some $\lambda \in [1, \Lambda]$, then there is a constant C independent of u, λ such that*

$$G_\lambda(rw) - G_\lambda(u) \le C \tag{7.27}$$

for all $r \in [0, 1]$, where w is the projection of u onto M.

Proof. For such u,

$$(G'_\lambda(u), g)/2 = \lambda(w, g_1)_D - (v, g_2)_D - \int g f(x,u) = 0$$

for every $g \in D$, where g_1, g_2 are the projections of g onto M, N, respectively. Let $u = v + w$, where $v \in N,\ w \in M$. Thus,

$$\begin{aligned} G_\lambda(rw) - G_\lambda(u) = {} & \lambda(r^2 - 1)\|w\|_D^2 \\ & + \lambda(w, g_1)_D + [\|v\|_D^2 - (v, g_2)_D] \\ & + \int [2F(x,u) - 2F(x,rw) - g f(x,u)]. \end{aligned}$$

Take

$$g = (r^2 + 1)v - (r^2 - 1)w = (r^2 + 1)u - 2r^2 w.$$

Then we have

$$\begin{aligned} G_\lambda(rw) - G_\lambda(u) = {} & -r^2\|v\|_D^2 \\ & + \int_{\mathbb{R}^n} [2F(x,u) - 2F(x,rw) - ((r^2+1)u - 2r^2 w) f(x,u)] \\ \le {} & C \end{aligned}$$

by Lemma 7.15. □

Lemma 7.17 *There exists a constant $C_1 > \varepsilon > 0$ such that $\varepsilon \leq a(\lambda) \leq C_1$ for $1 \leq \lambda \leq \Lambda$.*

Proof. Lemma 7.11. □

Lemma 7.18 *Any sequence $\{u_k\}$ satisfying*

$$1 \leq \lambda_k \leq \Lambda, \quad |G_{\lambda_k}(u_k)| \leq C, \quad G'_{\lambda_k}(u_k) = 0$$

has a subsequence bounded in D.

Proof. Write $u_k = w_k + v_k$, where $w_k \in M,\ v_k \in N$. Assume that $\rho_k = \|u_k\|_D \to \infty$, and let $\tilde{u}_k = u_k/\rho_k$. Then $\|\tilde{u}_k\|_D = 1$. We may assume that $\lambda_k \to \lambda$. Thus, there is a renamed subsequence such that $\tilde{u}_k \to \tilde{u}$ weakly in D, strongly in $L^2_{loc}(\mathbb{R}^n)$ and a.e. in $\mathbb{R}^n$. If

$$\sup_y \int_{B(y)} \tilde{w}_k^2 \to 0 \quad as\ k \to \infty, \tag{7.28}$$

then $\tilde{w}_k(x) \to 0$ in $L^p(\mathbb{R}^n)$ for $2 < p < 2^*$ (Lemma 7.12). Note that

$$G_{\lambda_k}(u_k) = \lambda_k \|w_k\|_D^2 - \|v_k\|_D^2 - 2\int_{\mathbb{R}^n} F(x, u_k). \tag{7.29}$$

Then we have

$$\begin{aligned}(G'_{\lambda_k}(u_k), \hat{u}_k)/2 &= \lambda_k (w_k, \hat{u}_k)_D - (v_k, \hat{u}_k)_D \\ &\quad - \int_{\mathbb{R}^n} f(x, u_k(x))\hat{u}_k(x) \\ &\to 0,\end{aligned}$$

where $\hat{u}_k = w_k - v_k$. Thus,

$$\begin{aligned}\lambda_k \|w_k\|_D^2 + \|v_k\|_D^2 + o(1) = \int_{\mathbb{R}^n} f(x, u_k)\hat{u}_k &= \int_{|u_k|<\delta} + \int_{|u_k|>\delta} \\ &\leq \sigma \|u_k\| \cdot \|\hat{u}_k\| + C\|u_k\|_q^{q-1} \|\hat{u}_k\|_q.\end{aligned}$$

Hence,

$$\varepsilon \|u_k\|_D^2 \leq \|u_k\|_D^2 - \sigma \|u_k\|^2 \leq C' \|u_k\|_q^{q-1} \|u_k\|_D$$

for some $\varepsilon > 0$. Thus, $\|u_k\|_D^2 \leq \|u_k\|_q^q \to 0$.

By (7.6) and (7.7),

$$2\int_{|u|<\delta}|F(x,u)| \leq \sigma \int_{|u|<\delta} u^2, \quad u \in D$$

and

$$\begin{aligned}\int_{|u|>\delta}|F(x,u)| &\leq C\int_{|u|>\delta}(|u|^q+|u|)\\ &\leq C\int_{|u|>\delta}(|u|^q+|u|^q/\delta^{q-1})\\ &\leq C\int_{|u|>\delta}|u|^q\\ &\leq C\|u\|_q^q, \quad u \in D.\end{aligned}$$

Thus,

$$\int_{\mathbb{R}^n}|F(x,u)| \leq \sigma\|u\|^2 + C\|u\|_q^q, \quad u \in D.$$

Let $c > 0$ be a fixed constant. Then

$$\int_{\mathbb{R}^n}|F(x,c\tilde{w}_k)| \leq c^2 C\|\tilde{w}_k\|_D^2 + Cc^q\|\tilde{w}_k\|_q^q \to 0. \tag{7.30}$$

Let $r_k = c/\rho_k \to 0$. By Lemma 7.16

$$G_{\lambda_k}(r_k w_k) - G_{\lambda_k}(u_k) \leq C. \tag{7.31}$$

Hence,

$$G_{\lambda_k}(c\tilde{w}_k) \leq C'. \tag{7.32}$$

Moreover,

$$-C/\rho_k^2 \leq \lambda_k\|\tilde{w}_k\|_D^2 - (1-\|\tilde{w}_k\|_D^2) - 2\int_{\mathbb{R}^n}F(x,u_k)/\rho_k^2.$$

Hence,

$$(\lambda_k+1)\|\tilde{w}_k\|_D^2 \geq 1 - C/\rho_k^2.$$

In view of (7.32)

$$
\begin{aligned}
C' \geq G_{\lambda_k}(c\tilde{w}_k) &= \lambda_k c^2 \|\tilde{w}_k\|_D^2 - 2\int_{\mathbb{R}^n} F(x, c\tilde{w}_k) \\
&\geq \frac{\lambda_k c^2}{\lambda_k + 1}(1 - \frac{C}{\rho_k^2}) - 2\int_{\mathbb{R}^n} F(x, c\tilde{w}_k) \\
&\to \frac{\lambda c^2}{\lambda + 1}.
\end{aligned}
$$

This means that we can take c so large that $G_{\lambda_k}(c\tilde{w}_k) > C'$, contrary to (7.32).

On the other hand, if (7.28) does not hold, then there is a sequence $y_k \in \mathbb{Z}^n$ and a renamed subsequence such that

$$
\int_{B(y_k)} \tilde{w}_k^2 \geq \alpha > 0 \tag{7.33}
$$

for each k. Let $u_k'(x) = u_k(x - y_k)$. Then $\|u_k'\| = \|u_k\|$, $\|u_k'\|_D = \|u_k\|_D$, and

$$
\int_{\mathbb{R}^n} F(x, u_k') = \int_{\mathbb{R}^n} F(x, u_k).
$$

Also $G_{\lambda_k}(u_k') = G_{\lambda_k}(u_k)$ and $G_{\lambda_k}'(u_k') = 0$. Thus, there is a renamed subsequence of $\tilde{u}_k'$ converging to a limit $\tilde{u}'$ weakly in D, strongly in $L_{loc}^q(\mathbb{R}^n)$ and a.e. in $\mathbb{R}^n$. By (7.33) there is a $y \in \mathbb{Z}^n$ such that

$$
\int_{B(y)} |\tilde{w}_k'|^2 \geq \alpha > 0 \tag{7.34}
$$

for each k. Since $\tilde{u}_k' \to \tilde{u}'$ in $L_{loc}^q(\mathbb{R}^n)$, this shows that $\tilde{u}' \not\equiv 0$. Thus,

$$
\begin{aligned}
-C/\rho_k^2 &\leq \lambda_k \|\tilde{w}_k'\|_D^2 - \|\tilde{v}_k'\|_D^2 \\
&\quad - 2\int_{\mathbb{R}^n} F(x, u_k')/\rho_k^2 \\
&\to -\infty
\end{aligned}
$$

by Lemma 7.5. Again, (7.32) is contradicted. Hence, in fact, the ρ_k are bounded. □

7.5 Proofs of the Theorems

Proof of Theorem 7.1 By Lemma 7.14, there are sequences $1 < \lambda_k \to 1$, $\{u_k\} \subset D$ such that $G_{\lambda_k}(u_k) \leq a(\lambda_k)$, $G_{\lambda_k}'(u_k) = 0$. By Lemma 7.18, the sequence $\{u_k\}$ is bounded in D. Thus, there is a renamed subsequence converging to a limit u weakly in D, strongly in $L^q_{loc}(\mathbb{R}^n)$ and a.e. in $\mathbb{R}^n$. For any $\varphi \in C_0^\infty(\mathbb{R}^n)$ we have

$$(G'_{\lambda_k}(u_k), \varphi)/2 = \lambda_k (w_k, \varphi)_D - (v_k, \varphi)_D - \int_{\mathbb{R}^n} f(x, u_k(x))\varphi(x) = 0.$$

Hence, in the limit,

$$(G'(u), \varphi)/2 = (w, \varphi)_D - (v, \varphi)_D - \int_{\mathbb{R}^n} f(x, u(x))\varphi(x) = 0,$$

showing that $G'(u) = 0$.

If (7.7) holds, then we have

$$\begin{aligned}(G'_{\lambda_k}(u_k), \hat{u}_k)/2 &= \lambda_k (w_k, \hat{u}_k)_D - (v_k, \hat{u}_k)_D \\ &\quad - \int_{\mathbb{R}^n} f(x, u_k(x))\hat{u}_k(x) \\ &= 0,\end{aligned}$$

where $\hat{u}_k = w_k - v_k$. Thus,

$$\begin{aligned}\lambda_k \|w_k\|_D^2 + \|v_k\|_D^2 = \int_{\mathbb{R}^n} f(x, u_k)\hat{u}_k &= \int_{|u_k|<\delta} + \int_{|u_k|>\delta} \\ &\leq \sigma \|u_k\| \cdot \|\hat{u}_k\| + C\|u_k\|_q^{q-1} \|\hat{u}_k\|_q.\end{aligned}$$

Hence,

$$\varepsilon \|u_k\|_D^2 \leq \|u_k\|_D^2 - \sigma \|u_k\|^2 \leq C' \|u_k\|_q^{q-1} \|\hat{u}_k\|_D$$

for some $\varepsilon > 0$. Since $u_k \neq 0$, this shows that $\|u_k\|_q \geq c > 0$. Thus

$$\sup_y \int_{B(y)} \tilde{u}_k^2 \to 0 \quad as\ k \to \infty \tag{7.35}$$

cannot hold. From this we imply that $u \not\equiv 0$. □

Proof of Theorem 7.2 By Theorem 7.1, $\mathcal{M}\backslash\{0\} \neq \phi$. Let

$$\beta = \inf_{\mathcal{M}\backslash\{0\}} G$$

and let $\{u_k\}$ be a sequence in $\mathcal{M}\backslash\{0\}$ such that

$$G(u_k) \to \beta.$$

By Lemma 7.18 there is a renamed subsequence bounded in D. Then we have

$$(G'_{\lambda_k}(u_k), \hat{u}_k)/2 = \lambda_k(w_k, \hat{u}_k)_D - (v_k, \hat{u}_k)_D - \int_{\mathbb{R}^n} f(x, u_k(x))\hat{u}_k(x) = 0,$$

where $\hat{u}_k = w_k - v_k$. Thus,

$$\lambda_k \|w_k\|_D^2 + \|v_k\|_D^2 = \int_{\mathbb{R}^n} f(x, u_k)\hat{u}_k = \int_{|u_k|<\delta} + \int_{|u_k|>\delta}$$
$$\leq \sigma \|u_k\| \cdot \|\hat{u}_k\| + C\|u_k\|_q^{q-1} \|\hat{u}_k\|_q.$$

Hence,

$$\varepsilon \|u_k\|_D^2 \leq \|u_k\|_D^2 - \sigma \|u_k\|^2 \leq C' \|u_k\|_q^{q-1} \|\hat{u}_k\|_D$$

for some $\varepsilon > 0$. Since $u_k \neq 0$, this shows that $\|u_k\|_q \geq c > 0$. Thus (7.35) cannot hold. From this we imply that $u \not\equiv 0$. Now

$$G(u_k) = G(u_k) - (G'(u_k), u_k)/2 = \int_{\mathbb{R}^n} H(x, u_k(x)).$$

But $H(x, u_k(x)) \geq -W(x)$ a.e. and $H(x, u_k(x)) \to H(x, u(x))$ a.e. Hence,

$$G(u) = \int_{\mathbb{R}^n} H(x, u(x)) \leq \liminf \int_{\mathbb{R}^n} H(x, u_k(x)) = \liminf G(u_k) = \beta.$$

This shows that $u \in \mathcal{M}$. □

7.6 Notes and Remarks

This chapter follows [212] with adjustments at the end. Many authors have studied the problem under various stipulations (cf., e.g., [2, 5, 14, 19, 47, 79, 126, 127, 167, 228, 241, 264, 276, 287, 292] and references quoted in them.) Prior to [212] the strongest results appear to be those of Szulkin-Weth [241]. They prove the existence of ground state solutions under hypotheses weaker than those previously assumed. They followed the approach of Pankov [167]. Yang [287] was able to obtain the same results using a different method (based on the approach of [229]) which is much simpler. Among other things, they assumed

1. $f(x, t) = o(t)$ *as* $t \to 0$ uniformly in x.
2. $F(x, t) := \int_0^t f(x, s)\, ds > 0, \ x \in \mathbb{R}^n, \ t \in \mathbb{R}\backslash\{0\}$.
3. $tf(x, t) - 2F(x, t) > 0, \ x \in \mathbb{R}^n, \ t \in \mathbb{R}\backslash\{0\}$.
4. $f(x, t)/|t|$ is strictly increasing in t on $(-\infty, 0)$ and $(0, \infty)$.

Chapter 8
Zero in the Spectrum

8.1 Introduction

In the previous chapter we noted that in our study of semilinear elliptic partial differential equations of the form

$$\mathcal{A}u = f(x, u), u \in D \tag{8.1}$$

in unbounded domains, we required that the resolvent set of $\mathcal{A}$ not be empty. For convenience, we assumed that it contain 0. This allowed us to choose an interval $(a, b) \subset \rho(\mathcal{A})$, where $a < 0 < b$. We let $D = D(|\mathcal{A}|^{(1/2)})$. With the scalar product $(u, v)_D = (|\mathcal{A}|^{(1/2)}u, |\mathcal{A}|^{(1/2)}v)$, it became a Hilbert space. We let

$$N = E(-\infty, a], \quad M = E[b, \infty)$$

be the negative and positive invariant subspaces of $\mathcal{A}$. Then

$$(\mathcal{A}v, v) \leq a\|v\|^2, \quad v \in D \cap N,$$

and

$$(\mathcal{A}w, w) \geq b\|w\|^2, \quad w \in D \cap M.$$

The hypotheses of our theorems depended on a and b using the fact that 0 was embedded in $\rho(\mathcal{A})$. The purpose of the present chapter is to study the situation when 0 is a boundary point of $\rho(\mathcal{A})$ and the arguments do not work.

M. Schechter, *Critical Point Theory*,
https://doi.org/10.1007/978-3-030-45603-0_8

8.2 Zero a Boundary Point of $\rho(\mathcal{A})$

Let $\mathcal{A}$ be a self-adjoint operator on $L^2(\mathbb{R}^n)$. We assume that the interval $(0, b) \subset \rho(\mathcal{A})$, the resolvent set of $\mathcal{A}$. We allow the endpoints to be in $\sigma(\mathcal{A})$. Let c be a point in $(0, b)$. We assume that $C_0^\infty(\mathbb{R}^n) \subset \tilde{D} := D(|\mathcal{A} - c|^{(1/2)}) \subset H^{m,2}(\mathbb{R}^n)$ for some $m > 0$, which need not be an integer. In particular,

$$\|u\|_{m,2} \leq C\|u\|_{\tilde{D}}, \quad u \in D. \tag{8.2}$$

We let

$$N = E(-\infty, 0], \quad M = E[b, \infty)$$

be the negative and positive invariant subspaces of $\mathcal{A}$. Hence,

$$(\mathcal{A}v, v) \leq 0, \quad v \in \tilde{D} \cap N,$$

and

$$(\mathcal{A}w, w) \geq b\|w\|^2, \quad w \in \tilde{D} \cap M.$$

Let q be any number satisfying

$$\begin{aligned} &2 < q < 2^* := 2n/(n - 2m), && 2m < n \\ &2 < q < \infty, && n \leq 2m. \end{aligned}$$

We introduce the norm squared

$$\|u\|_D^2 := (\mathcal{A}w, w) - (\mathcal{A}v, v) + \|v\|_q^2, \; u = w + v, \quad w \in \tilde{D} \cap M, \; v \in \tilde{D} \cap N.$$

We let D be the completion of $H^{m,2}(\mathbb{R}^n)$ with respect to the norm $\|u\|_D$. We assume that D is compact in $L^q_{loc}(\mathbb{R}^n)$ and

$$\|u\|_q \leq C\|u\|_D, \quad u \in D, \tag{8.3}$$

where $\|\cdot\|_q$ is the norm of $L^q(\mathbb{R}^n)$.

We also assume that

$$u_y(x) = u(x + y), \quad y \in \mathbb{Z}^n \Longrightarrow \|u_y\|_D = \|u\|_D. \tag{8.4}$$

Let $f(x, t)$ be a Carathéodory function on $\mathbb{R}^n \times \mathbb{R}$ satisfying

$$f(x + y, t) = f(x, t), \quad x \in \mathbb{R}^n, \; y \in \mathbb{Z}^n, \; t \in \mathbb{R}, \tag{8.5}$$

$$|f(x,t)| \leq C(|t|^{q-1} + \psi(x)), \quad x \in \mathbb{R}^n, \ t \in \mathbb{R}, \tag{8.6}$$

where $\psi(x) \in C_0^\infty(\mathbb{R}^n)$, and

$$|f(x,t)| \leq \sigma |t|^{q-1}, \quad |t| < \delta, \ x \in \mathbb{R}^n, \ t \in \mathbb{R}, \tag{8.7}$$

for some $\sigma > 0, \ \delta > 0$. Assume that

$$F(x,t) := \int_0^t f(x,s)\,ds$$

satisfies

$$F(x,t) \geq 0, \quad x \in \mathbb{R}^n, \ t \in \mathbb{R}, \tag{8.8}$$

and that

$$F(x,t)/t^2 \to \infty \ as \ t^2 \to \infty, \tag{8.9}$$

In addition, assume

$$\begin{aligned} &2F(x,t+s) - 2F(x,t) - (2rs - (r-1)^2 t) f(x,t) \\ &\quad \geq -W(x) \in L^1(\mathbb{R}^n), \quad x \in \mathbb{R}^n, \ s,t \in \mathbb{R}, \ r \in [0,1]. \end{aligned} \tag{8.10}$$

We shall prove:

Theorem 8.1 *Under the above hypotheses there is a nontrivial solution of*

$$\mathcal{A}u = f(x,u), \quad u \in D. \tag{8.11}$$

Moreover, there is a minimum energy solution, i.e., a solution that minimizes the energy functional

$$G(u) = (\mathcal{A}u, u) - 2\int_{\mathbb{R}^n} F(x,u), \quad u \in D, \tag{8.12}$$

over the set $\mathcal{M}$ of solutions of (8.11).

We consider $u \in D$ a solution of (8.11) if $G'(u) = 0$. This follows from the fact that

$$(G'(u), \varphi)/2 = (w, \mathcal{A}\varphi) + (v, \mathcal{A}\varphi) - \int_{\mathbb{R}^n} f(x, u(x))\varphi(x) = 0$$

for any $\varphi \in C_0^\infty(\mathbb{R}^n)$.

8.3 The Schrödinger Equation

As an example of an operator $\mathcal{A}$ satisfying the hypotheses of Theorem 8.1, we take the Schrödinger operator $\mathcal{A} = -\Delta + V(x)$ on $H^1(\mathbb{R}^n)$. If we assume that V is continuous, 1-periodic in $x_1, \cdots, x_k$, then the spectrum of $\mathcal{A}$ is not the whole of $\mathbb{R}$ and consists of closed, disjoint intervals (cf. [181]).

Theorem 8.2 *Let $\mathcal{A} = -\Delta + V(x)$ on $H^1(\mathbb{R}^n)$. Assume*

1. *V is continuous, 1-periodic in $x_1, \cdots, x_k$.*
2. $(0, b) \subset \rho(\mathcal{A})$.
3. *$f(x,t)$ is continuous, 1-periodic in $x_1, \cdots, x_k$ and*

$$|f(x,t)| \leq C(|t|^{q-1} + \psi(x)), \quad x \in \mathbb{R}^n,\ t \in \mathbb{R},$$

   for some $q \in (2, 2^*)$, where $\psi(x) \in C_0^\infty(\mathbb{R}^n)$.
4. $|f(x,t)| \leq \sigma |t|^{q-1}, \quad |t| < \delta,\ x \in \mathbb{R}^n,\ t \in \mathbb{R}$, *for some* $\sigma > 0,\ \delta > 0$.
5. $2F(x,t) \geq 0, \quad x \in \mathbb{R}^n,\ t \in \mathbb{R}$.
6. $F(x,t)/t^2 \to \infty$ *as* $t^2 \to \infty$,
7. $2F(x,t+s) - 2F(x,t) - (2rs - (r-1)^2 t) f(x,t) \geq -W(x) \in L^1(\mathbb{R}^n)$, $x \in \mathbb{R}^n,\ s,t \in \mathbb{R},\ r \in [0,1]$.

Then (8.11) has a nontrivial solution and a minimum energy solution.

Remark 8.3 The following three hypotheses

1. $f(x,t) = o(t)$ *as* $t \to 0$ uniformly in x.
2. $F(x,t)/t^2 \to \infty$ *as* $t^2 \to \infty$ uniformly in x.
3. *$f(x,t)/|t|$ is strictly increasing in t on $(-\infty, 0)$ and $(0, \infty)$.*

imply the following statements (cf. [241])

1. $F(x,t) > 0, \quad t \neq 0$,
2. $2F(x,t) \leq \varepsilon t^2, \quad |t| < \delta,\ x \in \mathbb{R}^n,\ t \in \mathbb{R}$,
3. $tf(x,t) > 0, \quad t \neq 0$,
4. $H(x,t) := tf(x,t) - 2F(x,t) > 0, \quad t \neq 0$,
5. $sf(x,t) = tf(x,s),\ st \neq 0 \implies s = t, \quad s,t \in \mathbb{R},\ x \in \mathbb{R}^n$,
6. $F(x,t+s) - F(x,t) - sf(x,t) > 0, \quad x \in \mathbb{R}^n,\ s,t \in \mathbb{R},\ s \neq 0$,
7. $2F(x,t+s) - 2F(x,t) - (2rs - (r-1)^2 t) f(x,t) > 0, \quad x \in \mathbb{R}^n,\ s,t \in \mathbb{R}\backslash\{0\},\ r \in [0,1]$.

In particular, they imply our hypotheses.

8.4 Some Lemmas

Before proving our main theorem, we shall prove a few lemmas. We pick Λ such that $1 < \Lambda < 2$. For $1 \leq \lambda \leq \Lambda$, define

$$G_\lambda(u) = \lambda(\mathcal{A}w, w) + (\mathcal{A}v, v) - 2\int_{\mathbb{R}^n} F(x, u) = \lambda I(u) - J(u), \quad u \in D, \tag{8.13}$$

where

$$I(u) = (\mathcal{A}w, w)$$

and

$$J(u) = -(\mathcal{A}v, v) + 2\int_{\mathbb{R}^n} F(x, u),$$

where we write $u = v + w,\ v \in N,\ w \in M$. Then $G_\lambda(u) \in C^1$ on D and

$$(G'_\lambda(u), g)/2 = \lambda(\mathcal{A}w, g) + (\mathcal{A}v, g) - \int_{\mathbb{R}^n} f(x, u)g\, dx, \quad g \in D. \tag{8.14}$$

Note that G_λ is $|\cdot|_w$-upper semicontinuous; G'_λ is weakly sequentially continuous on D. To see this, suppose $|u_j - u|_w \to 0$. Then $|v_j - v|_w \to 0$ and $\|w_j - w\|_D \to 0$. Since $\|v_j\|_D$ is bounded, there is a subsequence converging weakly to v in D. Thus,

$$\|v\|_D \leq \liminf \|v_j\|_D, \quad \|w\|_D = \lim \|w_j\|_D,$$

$$\int_{\mathbb{R}^n} F(x, u)\, dx = \lim \int_{\mathbb{R}^n} F(x, u_j)\, dx.$$

Consequently, $G_\lambda(u) \geq \limsup G_\lambda(u_j)$. Moreover, if $u_j \rightharpoonup u$ in D, then

$$\begin{aligned}(G'_\lambda(u_j), \varphi)/2 &= \lambda(\mathcal{A}w_j, \varphi) + (\mathcal{A}v_j, \varphi) - \int_{\mathbb{R}^n} f(x, u_j(x))\varphi(x)\\ &\to \lambda(\mathcal{A}w, \varphi) + (\mathcal{A}v, \varphi) - \int_{\mathbb{R}^n} f(x, u(x))\varphi(x)\\ &= (G'_\lambda(u), \varphi)/2.\end{aligned}$$

Lemma 8.4 *Let $R_k = \|u_k\|_D \to \infty$. Assume that $\tilde{u}_k = u_k/R_k \to \tilde{u}$ a.e. If $\tilde{u} \not\equiv 0$, then*

$$\int_{\mathbb{R}^n} F(x, u_k)/R_k^2 \to \infty. \tag{8.15}$$

Proof. Let Ω_0 be the subset of $\mathbb{R}^n$ where $\tilde{u} \neq 0$. The measure of Ω_0 is positive. Consequently,

$$2\int_{\mathbb{R}^n} F(x, u_k)/R_k^2 = 2\int_{\Omega_0} \frac{F(x, u_k)}{u_k^2}\tilde{u}_k^2 \to \infty$$

since the integrand is bounded below and $u_k^2 \to \infty$ a.e. on Ω_0. □

Lemma 8.5

$$I(u) + J(u) \to \infty \quad \text{as } \|u\|_D \to \infty.$$

Proof. We have

$$\begin{aligned} I(u) + J(u) &= (\mathcal{A}w, w) - (\mathcal{A}v, v) + 2\int_{\mathbb{R}^n} F(x, u) \\ &= \|u\|_D^2(1 - \|v\|_q^2/\|u\|_D^2) \\ &\quad + 2\int_{\mathbb{R}^n} F(x, u)/\|u\|_D^2. \end{aligned}$$

Assume that $R_k = \|u_k\|_D \to \infty$, and let $\tilde{u}_k = u_k/R_k$. Write $u_k = w_k + v_k$, where $w_k \in M,\ v_k \in N$. Then $\|\tilde{u}_k\|_D = 1$. Thus, there is a renamed subsequence such that $\tilde{u}_k \to \tilde{u}$ weakly in D, strongly in $L^2_{loc}(\mathbb{R}^n)$, and a.e. in $\mathbb{R}^n$. If

$$\sup_y \int_{B(y)} \tilde{u}_k^2 \to 0 \quad as\ k \to \infty, \tag{8.16}$$

then $\tilde{u}_k(x) \to 0$ in $L^q(\mathbb{R}^n)$. In this case

$$I(u_k) + J(u_k) \geq \|u_k\|_D^2(1 - \|v_k\|_q^2/\|u_k\|_D^2) \to \infty.$$

Otherwise, there is a sequence $y_k \in \mathbb{Z}^n$ and a renamed subsequence such that

$$\int_{B(y_k)} \tilde{u}_k^2 \geq \alpha > 0 \tag{8.17}$$

for each k. Then there is a $y \in \mathbb{R}^n$ such that

$$\int_{B(y)} \tilde{u}_k^2 \geq \alpha - \varepsilon. \tag{8.18}$$

Since $\tilde{u}_k \to \tilde{u}$ in $L^2_{loc}(\mathbb{R}^n)$, we see that $\tilde{u} \neq 0$. But then by Lemma 8.4,

$$\int_{\mathbb{R}^n} F(x, u_k)/R_k^2 \geq 2 \int_{\Omega_0} \frac{F(x, u_k)}{u_k^2} \tilde{u}_k^2 \to \infty.$$

□

Lemma 8.6 *There are positive constants η, r such that*

$$G_\lambda(w) \geq \eta, \quad w \in M, \ \|w\|_D = r, \ 1 \leq \lambda \leq \Lambda. \tag{8.19}$$

Proof. By (7.6) and (7.7),

$$2 \int_{|w|<\delta} F(x, w) \leq C \int_{|w|<\delta} w^q$$

and

$$\begin{aligned}\int_{|w|>\delta} |F(x, w)| &\leq C \int_{|w|>\delta} (|w|^q + |w|) \\ &\leq C \int_{|w|>\delta} (|w|^q + |w|^q/\delta^{q-1}) \\ &\leq C \int_{|w|>\delta} |w|^q \\ &\leq C\|w\|_q^q.\end{aligned}$$

Consequently,

$$G_\lambda(w) \geq \|w\|_D^2 - C \int_{|w|<\delta} w^q - C \int_{|w|>\delta} (|w|^q + |w|) \geq (1 - C'\|w\|_q^{q-2})\|w\|_D^2.$$

We take $\|w\|_D^2$ sufficiently small. □

Lemma 8.7 *Let*

$$Q_R = \{v + sw_0 : \Lambda s^2(\mathcal{A}w_0, w_0) - (\mathcal{A}v, v) + \|v\|_q^2 \leq R^2, \ v \in N, \ s \geq 0\}, \tag{8.20}$$

where $w_0 \in M$, $\|w_0\| = 1$. Then there is an $R > 0$ such that

$$G_\lambda(u) \leq 0, \quad u \in \partial Q_R \quad 1 \leq \lambda \leq \Lambda. \tag{8.21}$$

Proof. If not, there exist $R_k \to \infty$, $u_k = v_k + s_k w_0 \in \partial Q_{R_k}$, such that $G_\lambda(u_k) > 0$. If $s_k = 0$, then

$$G_\lambda(v_k) = (\mathcal{A}v_k, v_k) - 2 \int_{\mathbb{R}^n} F(x, v_k) \leq 0.$$

Hence, we must have $s_k \neq 0$ and

$$\Lambda s_k^2(\mathcal{A}w_0, w_0) - (\mathcal{A}v_k, v_k) + \|v_k\|_q^2 = R_k^2.$$

Then

$$G_\lambda(u_k) = (\lambda + \Lambda)s_k^2(\mathcal{A}w_0, w_0) + \|v_k\|_q^2 - R_k^2 - 2\int_{\mathbb{R}^n} F(x, u_k) \geq 0.$$

Let $\tilde{u}_k = u_k/R_k = \tilde{v}_k + \tilde{s}_k w_0$. Then $\|\tilde{u}_k\|_D = 1$. Consequently, there are renamed subsequences such that $\tilde{s}_k \to \tilde{s}$, and $\tilde{u}_k = u_k/R_k = \tilde{v}_k + \tilde{s}_k w_0 \to \tilde{u}$ weakly in D, strongly in $L^2_{loc}(\mathbb{R}^n)$, and a.e. in $\mathbb{R}^n$. If

$$\sup_y \int_{B(y)} \tilde{u}_k^2 \to 0 \quad as\ k \to \infty, \tag{8.22}$$

we have by Lemma 7.12 that $\tilde{u}_k(x) \to 0$ in $L^q(\mathbb{R}^n)$. In this case,

$$\limsup G_\lambda(u_k)/R_k^2 \leq (\lambda + \Lambda)\tilde{s}^2(\mathcal{A}w_0, w_0) - 1.$$

Since $G_\lambda(u_k) > 0$ for each k, this shows us that $\tilde{s} \neq 0$. Thus $\tilde{u} \neq 0$, and

$$\int_{\mathbb{R}^n} F(x, u_k)/R_k^2 \to \infty$$

in view of Lemma 8.4. But then

$$G_\lambda(u_k)/R_k^2 = (\lambda + \Lambda)\tilde{s}_k^2(\mathcal{A}w_0, w_0) + \|\tilde{v}_k\|_q^2 - 1 - 2\int_{\mathbb{R}^n} F(x, u_k)/R_k^2 \to -\infty.$$

On the other hand, if there is a sequence $y_k \in \mathbb{Z}^n$ and a renamed subsequence such that

$$\int_{B(y_k)} \tilde{u}_k^2 \geq \alpha > 0 \tag{8.23}$$

for each k, then there is a $y \in \mathbb{R}^n$ such that

$$\int_{B(y)} \tilde{u}_k^2 \geq \alpha - \varepsilon. \tag{8.24}$$

Since $\tilde{u}_k \to \tilde{u}$ in $L^2_{loc}(\mathbb{R}^n)$, we see that $\tilde{u} \neq 0$. But then by Lemma 8.4,

$$\int_{\mathbb{R}^n} F(x, u_k)/R_k^2 \geq 2\int_{\Omega_0} \frac{F(x, u_k)}{u_k^2}\tilde{u}_k^2 \to \infty.$$

Hence,

$$G_\lambda(u_k)/R_k^2 = (\lambda + \Lambda)\tilde{s}_k^2(\mathcal{A}w_0, w_0) + \|\tilde{v}_k\|_q^2 - 1 - 2\int_{\mathbb{R}^n} F(x, u_k)/R_k^2 \to -\infty.$$

This shows that no such sequence exists, and the lemma is proved. □

Lemma 8.8 *For almost all* $\lambda \in [1, \Lambda]$, *there exists a sequence* (u_k) *such that*

$$\sup_k \|u_k\|_D < \infty, \quad G'_\lambda(u_k) \to 0, \quad G_\lambda(u_k) \to a(\lambda); \tag{8.25}$$

where

$$a(\lambda) := \inf_{\Gamma \in \Phi} \sup_{u \in \mathcal{A}} G_\lambda(\Gamma(t)u) \in [\inf_B G_\lambda, \sup_{Q_R} G_\lambda],$$

Q_R *is defined by (8.20),*

$$A := \partial Q_R, \quad B := M \cap \partial \mathcal{B}_r,$$

and $0 < r < R$.

Proof. Q_R is the closure of a bounded, convex, open set in $N \oplus \{w_0\}$. It follows from Lemmas 8.6 and 8.7 that for each $\lambda \in [1, \Lambda]$,

$$\sup_A G_\lambda < \inf_B G_\lambda.$$

We apply Corollary 5.21 and Remark 5.8. □

Lemma 8.9 *For almost every* $\lambda \in [1, \Lambda]$ *there is a* $u_\lambda \in D \backslash \{0\}$ *such that* $-C \leq G_\lambda(u_\lambda) \leq a(\lambda)$, $G'_\lambda(u_\lambda) = 0$.

Proof. For each $\lambda \in [1, \Lambda]$ for which the conclusion of Lemma 8.8 holds, there is a sequence $\{u_k\} \subset D$ satisfying (8.25). Since the u_k are bounded in D, there is a renamed subsequence converging to a limit u_λ weakly in D, strongly in $L^q_{loc}(\mathbb{R}^n)$, and a.e. in $\mathbb{R}^n$. For any $g \in D$, we have

$$(G'_\lambda(u_k), g)/2 = \lambda(\mathcal{A}w_k, g) + (\mathcal{A}v_k, g) - \int_{\mathbb{R}^n} f(x, u_k) g \, dx \to 0. \tag{8.26}$$

Hence,

$$(G'_\lambda(u_\lambda), g)/2 = \lambda(\mathcal{A}w_\lambda, g) + (\mathcal{A}v_\lambda, g) - \int_{\mathbb{R}^n} f(x, u_\lambda) g \, dx = 0,$$

showing that $G'_\lambda(u_\lambda) = 0$. Moreover, since u_k is bounded in D,

$$\int_{\mathbb{R}^n} H(x, u_k(x)) = G_\lambda(u_k) - (G'_\lambda(u_k), u_k)/2 \to a(\lambda),$$

where

$$H(x, t) := tf(x, t) - 2F(x, t).$$

Note that $H(x, u_k(x)) \geq -W(x)$ $a.e.$ by (8.10). Since

$$H(x, u_k(x)) \to H(x, u_\lambda(x)) \; a.e.,$$

we have

$$a(\lambda) = \lim \int_{\mathbb{R}^n} H(x, u_k(x)) \geq \int_{\mathbb{R}^n} H(x, u_\lambda(x)) = G_\lambda(u_\lambda).$$

Moreover,

$$G_\lambda(u_\lambda) = \int_{\mathbb{R}^n} H(x, u_\lambda(x)) \geq -\int_{\mathbb{R}^N} W(x).$$

To show that $u_\lambda \neq 0$, note that $a(\lambda) > 0$ by Lemma 8.6. Moreover,

$$G_\lambda(u_k) = \lambda(\mathcal{A}w_k, w_k) + (\mathcal{A}v_k, v_k) - 2\int_{\mathbb{R}^n} F(x, u_k) \to a(\lambda). \tag{8.27}$$

Then we have

$$\begin{aligned}(G'_\lambda(u_k), \hat{u}_k)/2 &= \lambda(\mathcal{A}w_k, \hat{u}_k) + (\mathcal{A}v_k, \hat{u}_k) \\ &\quad - \int_{\mathbb{R}^n} f(x, u_k(x))\hat{u}_k(x) \\ &\to 0,\end{aligned}$$

where $\hat{u}_k = w_k - v_k$. If there is an $r > 0$ such that

$$\sup_y \int_{B(y,r)} u_k^2 \to 0 \quad as\ k \to \infty, \tag{8.28}$$

then $u_k(x) \to 0$ in $L^q(\mathbb{R}^n)$ (Lemma 7.12). Then,

$$\begin{aligned}\lambda\|w_k\|_D^2 + \|v_k\|_D^2 + o(1) &= \int_{\mathbb{R}^n} f(x, u_k)\hat{u}_k = \int_{|u_k|<\delta} + \int_{|u_k|>\delta} \\ &\leq C\|u_k\|_q^{q-1}\|\hat{u}_k\|_q.\end{aligned}$$

Hence,

$$\|u_k\|_D^2 \leq C'\|u_k\|_q^{q-1}\|u_k\|_D.$$

Since $u_k \neq 0$, this shows that $\|u_k\|_q \geq c > 0$. This shows that (8.28) does not hold. Consequently, there is a sequence $y_k \in \mathbb{Z}^n$ and a renamed subsequence such that

$$\int_{B(y_k)} u_k^2 \geq \alpha > 0 \tag{8.29}$$

for each k. Then there is a $y \in \mathbb{R}^n$ such that

$$\int_{B(y)} u_k^2 \geq \alpha - \varepsilon. \tag{8.30}$$

Since $u_k \to u_\lambda$ in $L^2_{loc}(\mathbb{R}^n)$, we see that $u_\lambda \neq 0$. □

Lemma 8.10 *For every* $\lambda \in [1, \Lambda]$ *there are sequences* $\lambda_k \to \lambda$ *and* $\{u_k\} \subset D\backslash\{0\}$ *such that* $-C \leq G_{\lambda_k}(u_k) \leq a(\lambda_k),\ G'_{\lambda_k}(u_k) = 0$.

Proof. Apply Lemma 8.9. □

Lemma 8.11

$$\int_{\mathbb{R}^n}[F(x,u) - F(x,rw) + (r^2 w - \frac{1+r^2}{2}u)f(x,u)] \leq C,$$

$$u \in D,\ w \in M,\ r \in [0,1],\ \|w\|_D \leq \|u\|_D,$$

where the constant C does not depend on u, w, r.

Proof. This follows from (8.10) if we take $t = u$, and $s = rw - u$. □

Lemma 8.12 *If* u *satisfies* $G'_\lambda(u) = 0$ *for some* $\lambda \in [1, K]$, *then there is a constant* C *independent of* u, λ, r *such that*

$$G_\lambda(rw) - r^2(\mathcal{A}v, v) - G_\lambda(u) \leq C \tag{8.31}$$

for all $r \in [0, 1]$, *where w,v are the projections of u onto M, N, respectively.*

Proof. For such u, let $u = v + w$, where $v \in N,\ w \in M$. Then

$$(G'_\lambda(u), g)/2 = \lambda(\mathcal{A}w, g_1) + (\mathcal{A}v, g_2) - \int gf(x,u) = 0$$

for every $g \in D$, where g_1, g_2 are the projections of g onto M, N, respectively. Take

$$g = (r^2 + 1)v - (r^2 - 1)w = (r^2 + 1)u - 2r^2 w.$$

Then we have

$$\begin{aligned}
G_\lambda(rw) - r^2(\mathcal{A}v, v) - G_\lambda(u) &= \lambda(r^2 - 1)(\mathcal{A}w, w) - (\mathcal{A}v, v) \\
&+ \lambda(\mathcal{A}w, g_1) + (\mathcal{A}v, g_2) - r^2(\mathcal{A}v, v) \\
&+ \int_{\mathbb{R}^n} [2F(x, u) - 2F(x, rw) - gf(x, u)]\, dx \\
&= \int_{\mathbb{R}^n} [2F(x, u) - 2F(x, rw) - ((r^2 + 1)u - 2r^2 w) f(x, u)]\, dx \\
&\le C
\end{aligned}$$

by Lemma 8.11. □

Lemma 8.13 *There exists constants* $C_1 > \eta > 0$ *such that* $\eta \le a(\lambda) \le C_1$ *for* $1 \le \lambda \le \Lambda$.

Proof. Lemmas 8.6 and 8.8. □

Lemma 8.14 *Any sequence* $\{u_k\} \subset D$ *satisfying*

$$1 \le \lambda_k \le \Lambda, \quad |G_{\lambda_k}(u_k)| \le C, \quad G'_{\lambda_k}(u_k) = 0$$

has a subsequence bounded in D.

Proof. Write $u_k = w_k + v_k$, where $w_k \in M,\ v_k \in N$. Assume that $\rho_k = \|u_k\|_D \to \infty$, and let $\tilde{u}_k = u_k/\rho_k$. Then $\|\tilde{u}_k\|_D = 1$. We may assume that $\lambda_k \to \lambda$. Thus, there is a renamed subsequence such that $\tilde{u}_k \to \tilde{u}$ weakly in D, strongly in $L^2_{loc}(\mathbb{R}^n)$, and a.e. in $\mathbb{R}^n$. If

$$\sup_y \int_{B(y)} \tilde{w}_k^2 \to 0 \quad \text{as } k \to \infty, \tag{8.32}$$

then $\tilde{w}_k(x) \to 0$ in $L^p(\mathbb{R}^n)$ for $2 < p < 2^*$ (Lemma 7.12). Note that

$$G_{\lambda_k}(u_k) = \lambda_k(\mathcal{A}w_k, w_k) + (\mathcal{A}v_k, v_k) - 2\int_{\mathbb{R}^n} F(x, u_k). \tag{8.33}$$

Then we have

$$\begin{aligned}
(G'_{\lambda_k}(u_k), \hat{u}_k)/2 &= \lambda_k(\mathcal{A}w_k, \hat{u}_k) + (\mathcal{A}v_k, \hat{u}_k) \\
&- \int_{\mathbb{R}^n} f(x, u_k(x))\hat{u}_k(x) \\
&= 0,
\end{aligned}$$

where $\hat{u}_k = w_k - v_k$. Thus,

$$\lambda\|w_k\|_D^2 + \|v_k\|_D^2 + o(1) = \int_{\mathbb{R}^n} f(x, u_k)\hat{u}_k = \int_{|u_k|<\delta} + \int_{|u_k|>\delta}$$
$$\leq C\|u_k\|_q^{q-1}\|\hat{u}_k\|_q.$$

This implies

$$\|u_k\|_D^2 \leq C'\|u_k\|_q^{q-1}\|u_k\|_D.$$

Since $u_k \neq 0$, it follows that $\|u_k\|_q \geq c > 0$. Thus, $\|u_k\|_D^2 \leq C\|u_k\|_q^q + o(1) \to 0$.

By (8.6) and (8.7),

$$2\int_{|u|<\delta} |F(x,u)| \leq C\int_{|u|<\delta} u^q, \quad u \in D$$

and

$$\begin{aligned}\int_{|u|>\delta} |F(x,u)| &\leq C\int_{|u|>\delta} (|u|^q + |u|)\\ &\leq C\int_{|u|>\delta} (|u|^q + |u|^q/\delta^{q-1})\\ &\leq C\int_{|u|>\delta} |u|^q\\ &\leq C\|u\|_q^q, \quad u \in D.\end{aligned}$$

These imply,

$$\int_{\mathbb{R}^n} |F(x,u)| \leq C\|u\|_q^q, \quad u \in D.$$

Let $s > 0$ be a fixed constant. Then

$$\int_{\mathbb{R}^n} |F(x, s\tilde{w}_k)| \leq Cs^q\|\tilde{w}_k\|_q^q \to 0. \tag{8.34}$$

By Lemma 8.12

$$G_{\lambda_k}(r_k w_k) - r_k^2(\mathcal{A}v_k, v_k) - G_{\lambda_k}(u_k) \leq C. \tag{8.35}$$

Hence,

$$G_{\lambda_k}(s\tilde{w}_k) - s^2(\mathcal{A}\tilde{v}_k, \tilde{v}_k) \leq C'. \tag{8.36}$$

But

$$\begin{aligned} G_{\lambda_k}(s\tilde{w}_k) - s^2(A\tilde{v}_k, \tilde{v}_k) &= \lambda_k s^2(A\tilde{w}_k, \tilde{w}_k) - s^2(A\tilde{v}_k, \tilde{v}_k) - 2\int_{\mathbb{R}^n} F(x, s\tilde{w}_k) \\ &\geq s^2(\|\tilde{u}_k\|_D^2 - \|\tilde{v}_k\|_q^2) - 2\int_{\mathbb{R}^n} F(x, s\tilde{w}_k) \\ &\to s^2 \end{aligned}$$

by (8.34). This implies

$$\limsup_{k,s}[G_{\lambda_k}(s\tilde{w}_k) - s^2(A\tilde{v}_k, \tilde{v}_k)] = \infty,$$

contrary to (8.36). On the other hand, if (8.32) does not hold, then there is a sequence $y_k \in \mathbb{Z}^n$ and a renamed subsequence such that

$$\int_{B(y_k)} \tilde{w}_k^2 \geq \alpha > 0 \tag{8.37}$$

for each k. Let $u_k'(x) = u_k(x - y_k)$. Then $\|u_k'\| = \|u_k\|$, $\|u_k'\|_D = \|u_k\|_D$, and

$$\int_{\mathbb{R}^n} F(x, u_k') = \int_{\mathbb{R}^n} F(x, u_k).$$

Also $G_{\lambda_k}(u_k') = G_{\lambda_k}(u_k)$ and $G_{\lambda_k}'(u_k') = 0$. Thus, there is a renamed subsequence of $\tilde{u}_k'$ converging to a limit $\tilde{u}'$ weakly in D, strongly in $L_{loc}^q(\mathbb{R}^n)$, and a.e. in $\mathbb{R}^n$. By (8.37) there is a $y \in \mathbb{Z}^n$ such that

$$\int_{B(y)} |\tilde{w}_k'|^2 \geq \alpha > 0 \tag{8.38}$$

for each k. Since $\tilde{u}_k' \to \tilde{u}'$ in $L_{loc}^q(\mathbb{R}^n)$, this shows that $\tilde{u}' \not\equiv 0$. Thus,

$$\begin{aligned} -C/\rho_k^2 &\leq \lambda_k(A\tilde{w}_k', \tilde{w}_k') + (A\tilde{v}_k', \tilde{v}_k') \\ &\quad - 2\int_{\mathbb{R}^n} F(x, u_k')/\rho_k^2 \\ &\to -\infty \end{aligned}$$

by Lemma 8.4. Again, (8.36) is contradicted. Hence, in fact, the ρ_k are bounded. This proves the lemma. □

8.5 Proof of Theorem 8.1

Proof of Theorem 8.1 By Lemma 8.10, there are sequences $1 < \lambda_k \to 1$, $\{u_k\} \subset D\backslash\{0\}$ such that $-C \le G_{\lambda_k}(u_k) \le a(\lambda_k)$, $G'_{\lambda_k}(u_k) = 0$. By Lemma 8.14, the sequence $\{u_k\}$ has a renamed subsequence bounded in D. Thus, there is a renamed subsequence converging to a limit u weakly in D, strongly in $L^q_{loc}(\mathbb{R}^n)$, and a.e. in $\mathbb{R}^n$. For any $\varphi \in C_0^\infty(\mathbb{R}^n)$ we have

$$(G'_{\lambda_k}(u_k), \varphi)/2 = \lambda_k(w_k, \mathcal{A}\varphi) + (v_k, \mathcal{A}\varphi) - \int_{\mathbb{R}^n} f(x, u_k(x))\varphi(x) = 0.$$

Hence, in the limit,

$$(G'(u), \varphi)/2 = (w, \mathcal{A}\varphi) + (v, \mathcal{A}\varphi) - \int_{\mathbb{R}^n} f(x, u(x))\varphi(x) = 0,$$

showing that $G'(u) = 0$. Thus, $\mathcal{M} \neq \phi$.

To show that $u \neq 0$, note that $a(\lambda) > 0$ by Lemma 8.6. Moreover,

$$G_\lambda(u_k) = \lambda(\mathcal{A}w_k, w_k) + (\mathcal{A}v_k, v_k) - 2\int_{\mathbb{R}^n} F(x, u_k) \to a(\lambda). \tag{8.39}$$

Then we have

$$\begin{aligned}(G'_\lambda(u_k), \hat{u}_k)/2 &= \lambda(\mathcal{A}w_k, \hat{u}_k) + (\mathcal{A}v_k, \hat{u}_k)\\ &\quad - \int_{\mathbb{R}^n} f(x, u_k(x))\hat{u}_k(x)\\ &\to 0,\end{aligned}$$

where $\hat{u}_k = w_k - v_k$. If there is an $r > 0$ such that

$$\sup_y \int_{B(y,r)} u_k^2 \to 0 \quad as\ k \to \infty, \tag{8.40}$$

then $u_k(x) \to 0$ in $L^p(\mathbb{R}^n)$ for $2 < p < 2^*$ (Lemma 7.12). Thus,

$$\begin{aligned}\lambda\|w_k\|_D^2 + \|v_k\|_D^2 + o(1) = \int_{\mathbb{R}^n} f(x, u_k)\hat{u}_k &= \int_{|u_k|<\delta} + \int_{|u_k|>\delta}\\ &\le C\|u_k\|_q^{q-1}\|\hat{u}_k\|_q.\end{aligned}$$

This implies

$$\|u_k\|_D^2 \le C'\|u_k\|_q^{q-1}\|u_k\|_D.$$

Since $u_k \neq 0$, it follows that $\|u_k\|_q \geq c > 0$. This shows that (8.40) does not hold. Consequently, there is a sequence $y_k \in \mathbb{Z}^n$ and a renamed subsequence such that

$$\int_{B(y_k)} u_k^2 \geq \alpha > 0 \tag{8.41}$$

for each k. Then there is a $y \in \mathbb{R}^n$ such that

$$\int_{B(y)} u_k^2 \geq \alpha - \varepsilon. \tag{8.42}$$

Since $u_k \to u$ in $L^2_{loc}(\mathbb{R}^n)$, we see that $u \neq 0$.

Let

$$\beta = \inf_{\mathcal{M}} G_\lambda$$

and let $\{u_k\}$ be a sequence in $\mathcal{M}$ such that

$$G_\lambda(u_k) \to \beta.$$

By Lemma 8.14, the sequence is bounded. Hence, there is a renamed subsequence bounded in $D\backslash\{0\}$ and converging to a limit u weakly in D, strongly in $L^q_{loc}(\mathbb{R}^n)$, and a.e. in $\mathbb{R}^n$. For any $\varphi \in C_0^\infty(\mathbb{R}^n)$ we have

$$(G'_\lambda(u_k), \varphi)/2 = (w_k, \mathcal{A}\varphi) + (v_k, \mathcal{A}\varphi) - \int_{\mathbb{R}^n} f(x, u_k(x))\varphi(x) = 0.$$

Hence, in the limit,

$$(G'_\lambda(u), \varphi)/2 = (w, \mathcal{A}\varphi) + (v, \mathcal{A}\varphi) - \int_{\mathbb{R}^n} f(x, u(x))\varphi(x) = 0,$$

showing that $G'_\lambda(u) = 0$, i.e., that $u \in \mathcal{M}$. Now

$$G_\lambda(u_k) = G_\lambda(u_k) - (G'_\lambda(u_k), u_k)/2 = \int_{\mathbb{R}^n} H(x, u_k(x)).$$

But $H(x, u_k(x)) \geq -W(x)$ a.e. and $H(x, u_k(x)) \to H(x, u(x))$ a.e. Hence,

$$G_\lambda(u) = \int_{\mathbb{R}^n} H(x, u(x)) \leq \liminf \int_{\mathbb{R}^n} H(x, u_k(x)) = \liminf G_\lambda(u_k) = \beta.$$

Thus, $G_\lambda(u) = \beta$. □

8.6 Notes and Remarks

This chapter follows [220].

Most researchers studied the case when 0 is in the resolvent set of $\mathcal{A}$, and not in the spectrum. We include the case when 0 is in the spectrum and $f(x,t)$ satisfies $f(x,0) = 0$ and $df(x,0)/dt = 0$. This case is much more difficult than the case when 0 is in the resolvent.

Very few papers studied the case when 0 is in the spectrum of $\mathcal{A}$. The author knows only of [14, 27, 276, 292]. These papers studied only the Schrödinger equation. The first to study this case were H. Berestycki and P. L. Lions [27], who studied the equation

$$\mathcal{A}u = -\Delta u = f(u), \quad u \in H^1(\mathbb{R}^n).$$

Here $V \equiv 0$ and $f(x,t) = f(t)$. The spectrum of $\mathcal{A}$ is $\mathbb{R}^+$. Among their hypotheses are

$$\limsup_{s\to 0^+} \frac{f(s)}{s^l} \leq 0,$$

and

$$\limsup_{s\to +\infty} \frac{f(s)}{s^l} = 0,$$

where $l = (n+2)/(n-2)$. T. Bartsch and Y. Ding [14] studied the case where $(0, b)$ is in the resolvent set and 0 is in the spectrum. Their main assumption was

$$a_1|t|^\mu \leq \gamma F(x,t) \leq tf(x,t), \quad x \in \mathbb{R}^n,\ t \in \mathbb{R},$$

where $a_1 > 0,\ 2 < \gamma \leq \mu < 2^* = 2n/(n-2)$ and

$$F(x,t) := \int_0^t f(x,s)\,ds.$$

M. Willem and W. Zou [276] also studied this problem. Their assumptions included

$$c_1|t|^\mu \leq tf(x,t) \leq c_2|t|^\mu \quad x \in \mathbb{R}^n,\ t \in \mathbb{R},$$

where $2 < \mu < 2^*$,

$$H(x,t) := tf(x,t) - 2F(x,t) > 0, \quad x \in \mathbb{R}^n,\ t \in \mathbb{R}\backslash\{0\},$$

$$\liminf_{t\to 0} \frac{tf(x,t)}{F(x,t)} \geq \gamma_0 > 2, \quad x \in \mathbb{R}^n,$$

$$\liminf_{|t|\to\infty} \frac{H(x,t)}{|t|^{\alpha}} \geq c > 0, \quad x \in \mathbb{R}^n,$$

where $\alpha > 2^*\mu(\mu - 2)/(2^*\mu - 2^* - \mu)$.

M. Yang et al. [292] also considered this situation. Their assumptions include

$$f(x,t) = o(|t|) \text{ as } t \to 0 \text{ uniformly in } x \in \mathbb{R}^n,$$

$$F(x,t) \geq c_0|t|^{\mu}, \quad x \in \mathbb{R}^n,\ t \in \mathbb{R},$$

for $c_0 > 0,\ \mu > 2$,

$$f(x,t)/|t| \text{ is strictly increasing in } t \text{ on } (-\infty, 0) \text{ and } (0,\infty). \tag{8.43}$$

They did not obtain ground state solutions. Hypothesis (8.43) first appeared in [241]. As far as we know, no other research was done on the problem when 0 is in the spectrum.

Chapter 9
Global Solutions

9.1 Introduction

In the previous two chapters, we studied the semilinear Schrödinger equation

$$-\Delta u + V(x)u = f(x, u), \quad u \in H^1(\mathbb{R}^n), \tag{9.1}$$

where $V(x)$ is a given potential. We needed the linear operator $-\Delta u + V(x)u$ to have a nonempty resolvent. To achieve this, we assumed that $V(x)$ was periodic in x. This forced us to assume the same for $f(x, u)$, and we had to deal with several restrictions in our methods. In this chapter we study the equation without making any periodicity assumptions on the potential or on the nonlinear term. But we must be assured that the linear operator has nonempty resolvent. To accomplish this, we make assumptions on $V(x)$ which guarantee that the essential spectrum of $-\Delta u + V(x)u$ is the same as that of $-\Delta u$. In other words, our assumptions are such that the potential does not change the essential spectrum of the linear operator. This results in $[0, \infty)$ being the absolutely continuous part of the spectrum. There may be no negative eigenvalues, a finite number of negative eigenvalues, or an infinite number of negative eigenvalues. If there are an infinite number of negative eigenvalues, they will converge to 0. In each case we obtain nontrivial solutions. We also obtain least energy solutions.

9.2 The Hypotheses

We consider the semilinear Schrödinger equation

$$-\Delta u + V(x)u = f(x, u), \quad u \in H^1(\mathbb{R}^n), \tag{9.2}$$

M. Schechter, *Critical Point Theory*,
https://doi.org/10.1007/978-3-030-45603-0_9

where $V(x)$ is a given potential. One wishes to find solutions and, in particular, the so-called "least energy solutions." These are solutions that minimize the corresponding energy functional. The existence of solutions depends on both the linear operator $\mathcal{A}$ and the nonlinear term $f(x, u)$.

The purpose of the present chapter is to solve the Eq. (9.2) under assumptions on $V(x)$ such that the essential spectrum of $\mathcal{A} = -\Delta + V(x)$ is the same as that of $-\Delta$, i.e., $[0, \infty)$. The situation is different if there are no negative eigenvalues, one negative eigenvalue, or two or more negative eigenvalues. If there are no negative eigenvalues, one can solve under the same hypotheses for $f(x, t)$ that can be used for the equation

$$- \Delta u = f(x, u), \quad u \in H^1(\mathbb{R}^n). \tag{9.3}$$

Otherwise, the hypotheses on $f(x, t)$ need to only take the negative eigenvalues into consideration. We can even deal with the case where the negative eigenvalues converge to 0. We do not need an interval of the form $(-\varepsilon, 0)$ to be free of the spectrum. In each of these cases different methods must be employed, requiring different assumptions on the nonlinear term.

Concerning the function $V(x)$ we make the following assumptions:

(V_1)

$$\sup_y \int_{|x-y|<\delta} |V(x)|\, \omega_2(x-y)\, dx \to 0 \ as \ \delta \to 0,$$

and

(V_2)

$$\int_{|x-y|<1} |V(x)|\, \omega_2(x-y)\, dx \to 0 \ as \ |y| \to \infty,$$

where

$$\omega_s(x) = \begin{cases} |x|^{s-n}, & 0 < s < n \\ 1 - \ln |x|^2, & s = n \\ 1, & s > n. \end{cases}$$

These assumptions imply that there is a forms extension $\mathcal{A}$ of the operator

$$-\Delta u(x) + V(x)u(x)$$

on the space $H = H^{1,2}(\mathbb{R}^n)$ having essential spectrum equal to $[0, \infty)$ and a (possibly empty) discrete, countable negative spectrum consisting of isolated eigenvalues of finite multiplicity with a finite lower bound $-L$

$$-\infty < -L \leq \lambda_0 < \lambda_1 < \lambda_2 < \ldots < \lambda_l < \ldots < 0. \tag{9.4}$$

For each $l > 0$, define the subspaces $M = M_l$ and $N = N_l$ of H as

$$N = \bigoplus_{k<l} E(\lambda_k), \quad M = N^{\perp}, \quad H = M \oplus N .$$

For the operator $\mathcal{A}$ there are three possibilities: (a) it has no negative eigenvalues, (b) it has only one negative eigenvalue, and (c) it has two or more negative eigenvalues. What is interesting is that each of these possibilities must be dealt with differently. We shall study all of them separately.

The following notation will be used throughout the chapter:

$$\|u\|_q := (\int_{\mathbb{R}^n} |u(x)|^q dx)^{1/q}, \quad \|u\| = \|u\|_2,$$

$$(u, v) = \int_{\mathbb{R}^n} u(x)v(x)\, dx, \quad a(u, v) = (\mathcal{A}u, v), \quad a(u) = a(u, u).$$

Let q be any number satisfying

$$2 < q \leq 2^* = 2n/(n-2), \quad n > 2$$

$$2 < q < \infty, \quad n \leq 2$$

and let $f(x, t)$ be a Carathéodory function on $\mathbb{R}^n \times \mathbb{R}$. As usual, this means that $f(x, t)$ is continuous in t for a.e. $x \in \mathbb{R}^n$ and measurable in x for every $t \in \mathbb{R}$. We make the following assumptions:

(A) The function $f(x, t)$ satisfies

$$|f(x, t)| \leq S(x)^q (|t|^{q-1} + W(x))$$

and

$$f(x, t)/S(x)^q = o(|t|^{q-1}) \text{ as } |t| \to \infty,$$

where $S(x) > 0$ is a function in $L^q(\mathbb{R}^n)$ satisfying

$$\|Su\|_q \leq C\|u\|_H, \quad u \in H,$$

and W is a function in $L^{\infty}(\mathbb{R}^n)$.

Our other hypotheses depend only on the primitive of $f(x, t)$:

$$F(x, t) = \int_0^t f(x, s)\, ds.$$

Let

$$G(u) = (\mathcal{A}u, u) - 2\int_{\mathbb{R}^n} F(x, u)\,dx, \quad u \in H. \tag{9.5}$$

It follows that G is a continuously differentiable functional on the whole of H (cf., e.g., [193]). It is easily checked that $u \in H$ is a (weak) solution of (9.2) iff it is a critical point of $G(u)$. Our methods will make use of this fact. They involve finding linking sets A, B which separate G, i.e., are such that

$$\sup_A G \leq \inf_B G.$$

If the spectrum of $\mathcal{A}$ has no gap, i.e., consists only of the interval $[0, \infty)$, the choice of the sets A, B is very limited. If it has one gap, the subspace in the gap is very useful. If it has two gaps, we obtain two useful subspaces. In our situation, if there are no negative eigenvalues, there are no gaps. If there is one negative eigenvalue, there is one gap. If there are two or more negative eigenvalues, there are at least two gaps. Therefore we consider three different situations.

9.3 The Space N_1

We let N_1 be the set of measurable functions $h(x)$ on $\mathbb{R}^n$ satisfying

$$\sup_u \frac{\|hu\|}{\|u\|_H} < \infty.$$

It is clear that bounded functions are in N_1. It can be shown that $L^p(\mathbb{R}^n) \subset N_1$ for $p \geq n$. For a general description of this space, cf. [183]. These functions will be used in our theorems because of the following properties:

Lemma 9.1 *If $g^{-1} \in N_1$, the following statements are true:*

1. *There is a constant C such that*

$$\|u\|_H \leq C\|u\|'_H, \quad u \in H, \tag{9.6}$$

where

$$(\|u\|'_H)^2 = \|\nabla u\|^2 + \|gu\|^2.$$

2. *If A has no negative eigenvalues, then there is an $\varepsilon > 0$ such that*

$$a(u) + \|gu\|^2 \geq \varepsilon\|u\|_H^2, \quad u \in H. \tag{9.7}$$

3. *If $\mathcal{A}$ has only one negative eigenvalue λ_0, then there is an $\varepsilon > 0$ such that*

$$a(u) + \|gu\|^2 \geq \varepsilon \|u\|_H^2, \quad u \in M. \tag{9.8}$$

4. *If $\mathcal{A}$ has negative eigenvalues λ_{l-1}, λ_l, then there is an $\varepsilon > 0$ such that*

$$a(u) - \lambda_l \|u\|^2 + \|gu\|^2 \geq \varepsilon \|u\|_H^2, \quad u \in M. \tag{9.9}$$

If $g \in N_1$, then there is a constant C such that

$$\|u\|_H' \leq C\|u\|_H, \quad u \in H. \tag{9.10}$$

If $g, g^{-1} \in N_1$, then the two norms are equivalent.

Lemma 9.1 will be proved in Sect. 9.9.

9.4 No Negative Eigenvalues

In this case,

$$a(u) = (\mathcal{A}u, u) \geq 0, \quad u \in H.$$

There are no gaps in the spectrum of $\mathcal{A}$.

We have

Theorem 9.2 *Assume*

1. *There is a function $g(x)$ such that $g, g^{-1} \in N_1$ and*

$$2F(x,u) \leq -g(x)^2|u|^2 + W(x), \quad u \in \mathbb{R},\ x \in \mathbb{R}^n,$$

where $W(x) \in L^1(\mathbb{R}^n)$.

Then the Eq. (9.2) has a solution.

Remark 9.3 It is clear from the equation that the solution obtained will be nontrivial if

$$f(x,0) \neq 0.$$

To guarantee that a solution will be nontrivial even when $f(x,0) = 0$, we have the following.

Theorem 9.4 *Assume*

1. *There are constants* $0 < \alpha < 2, \delta > 0$, *and a function* $g(x)$ *such that* $g, g^{-1} \in N_1$ *and*

$$2F(x,u) \leq -g(x)^2|u|^2, \quad u \in \mathbb{R},\ x \in \mathbb{R}^n,\ |u| < \delta,$$
$$\leq -g(x)^2|u|^2 + S(x)^q|u|^\alpha, \quad u \in \mathbb{R},\ x \in \mathbb{R}^n,\ |u| > \delta.$$

2. *There is a locally bounded function* $h(t)$ *such that*

$$2F(x,u) \geq -h(u), \quad u \in \mathbb{R},\ x \in \mathbb{R}^n,$$

and

$$c_0 = \sup_{\mathbb{R}} h(u)/u^2 < \infty.$$

Then the Eq. (9.2) has a nontrivial solution.

9.5 Only One Negative Eigenvalue

Let $\lambda_0 < 0$ be the eigenvalue. In this case

$$a(u) \geq \lambda_0\|u\|^2, \quad u \in H.$$

We can make use of the fact that there is a gap in the spectrum of $\mathcal{A}$. We have

Theorem 9.5 *Assume*

1.

$$2F(x,u) \geq \lambda_0|u|^2, \quad u \in \mathbb{R},\ x \in \mathbb{R}^n.$$

2. *There are constants* $0 < \alpha < 2, \delta > 0$, *and a function* $g(x)$ *such that* $g, g^{-1} \in N_1$ *and*

$$2F(x,u) \leq -g(x)^2|u|^2, \quad u \in \mathbb{R},\ x \in \mathbb{R}^n,\ |u| < \delta,$$
$$\leq -g(x)^2|u|^2 + S(x)^q|u|^\alpha, \quad u \in \mathbb{R},\ x \in \mathbb{R}^n,\ |u| > \delta.$$

Then the Eq. (9.2) has a solution.

To obtain a nontrivial solution we have the following.

Theorem 9.6 *Assume*

1.

$$2F(x, u) \geq \lambda_0 |u|^2, \quad u \in \mathbb{R},\ x \in \mathbb{R}^n.$$

2. *There are constants* $0 < \alpha < 2, \delta > 0$, *and a function* $g(x)$ *such that* $g, g^{-1} \in N_1$ *and*

$$\begin{aligned} 2F(x, u) &\leq -g(x)^2 |u|^2, \quad u \in \mathbb{R},\ x \in \mathbb{R}^n,\ |u| < \delta, \\ &\leq -g(x)^2 |u|^2 + S(x)^q |u|^\alpha, \quad u \in \mathbb{R},\ x \in \mathbb{R}^n,\ |u| > \delta. \end{aligned}$$

3. *There is a locally bounded function* $h(t)$ *such that*

$$2F(x, u) \geq -h(u), \quad u \in \mathbb{R},\ x \in \mathbb{R}^n,$$

and

$$c_0 = \sup_{\mathbb{R}} h(u)/u^2 < \infty.$$

Then the Eq. (9.2) has a nontrivial solution.

9.6 Two or More Negative Eigenvalues

Here again we can make use of the fact that there is more than one gap in the spectrum. This allows us to use more complicated linking methods.

Let $\lambda_{l-1},\ \lambda_l$ be two consecutive negative eigenvalues of $\mathcal{A}$. We have the following.

Theorem 9.7 *Assume*

1.

$$2F(x, u) \geq \lambda_{l-1} |u|^2, \quad u \in \mathbb{R},\ x \in \mathbb{R}^n,$$

2. *There are constants* $0 < \alpha < 2, \delta > 0$, *and a function* $g(x)$ *such that* $g, g^{-1} \in N_1$ *and*

$$\begin{aligned} 2F(x, u) &\leq -g(x)^2 |u|^2 + \lambda_l |u|^2, \quad u \in \mathbb{R},\ x \in \mathbb{R}^n,\ |u| < \delta, \\ &\leq -g(x)^2 |u|^2 + \lambda_l |u|^2 + S(x)^q |u|^\alpha, \quad u \in \mathbb{R},\ x \in \mathbb{R}^n,\ |u| > \delta. \end{aligned}$$

Then the Eq. (9.2) has a solution.

To obtain a nontrivial solution we have the following.

Theorem 9.8 *Assume*

1.

$$2F(x,u) \geq \lambda_{l-1}|u|^2, \quad u \in \mathbb{R},\ x \in \mathbb{R}^n,$$

2. *There are constants* $0 < \alpha < 2, \delta > 0$, *and a function* $g(x)$ *such that* $g, g^{-1} \in N_1$ *and*

$$\begin{aligned} 2F(x,u) &\leq -g(x)^2|u|^2 + \lambda_l|u|^2, \quad u \in \mathbb{R},\ x \in \mathbb{R}^n,\ |u| < \delta, \\ &\leq -g(x)^2|u|^2 + \lambda_l|u|^2 + S(x)^q|u|^\alpha, \quad u \in \mathbb{R},\ x \in \mathbb{R}^n,\ |u| > \delta. \end{aligned}$$

3. *There is a locally bounded function* $h(t)$ *such that*

$$2F(x,u) \geq -h(u), \quad u \in \mathbb{R},\ x \in \mathbb{R}^n,$$

and

$$c_0 = \sup_{\mathbb{R}} h(u)/u^2 < \infty.$$

Then the Eq. (9.2) has a nontrivial solution.

Remark 9.9 Note that the hypothesis of Theorem 9.2 is stronger than hypotheses 1 and 2 of Theorem 9.5. Hypothesis 1 of Theorem 9.5 requires

$$2F(x,u) \geq \lambda_l|u|^2, \quad u \in \mathbb{R},\ x \in \mathbb{R}^n,$$

which is stronger than hypothesis 1 of Theorem 9.7. Consequently, the hypotheses of Theorem 9.4 are stronger than those of Theorem 9.6 which are stronger than those of Theorem 9.8.

9.7 Least Energy Solutions

Let $\mathcal{M}$ be the set of all solutions of (9.2). A solution $\tilde{u}$ is called a "least energy solution" if it minimizes the functional

$$G(u) = a(u) - 2\int_{\mathbb{R}^n} F(x,u)\,dx \tag{9.11}$$

over the set $\mathcal{M}$.

We have the following.

Theorem 9.10 *If we add the following hypothesis to Theorems 9.2–9.8, then Eq. (9.2) has a least energy solution: The function given by*

$$H(x,u) = uf(x,u) - 2F(x,u) \tag{9.12}$$

satisfies

$$H(x,u) \geq -W(x) \in L^1(\mathbb{R}^n), \quad u \in \mathbb{R},\ x \in \mathbb{R}^n. \tag{9.13}$$

We shall prove Theorems 9.2–9.10 in Sect. 9.10. In the next section we describe the construction of the operator $\mathcal{A}$. We obtain the largest self-adjoint extension of $\mathcal{A}_0$ which preserves the essential spectrum.

9.8 The Operator $\mathcal{A}$

The following was proved in [183] (Theorem 10.9, ch. 6, p. 153).

Theorem 9.11 *Let P(D) be an elliptic constant coefficient operator of order 2 on* $\mathbb{R}^n$, *and let V(x) be a function satisfying* (V_1) *and* (V_2). *If* $\rho(P_0)$ *is not empty, then P(D)+V has a forms extension operator* $\mathcal{A}$ *such that* $\sigma_e(\mathcal{A}) = \sigma_e(P_0)$.

Here P_0 is the closure of the operator $P(D)$ restricted to $C_0^\infty(\mathbb{R}^n)$, and σ_e is the essential spectrum. Any point not in the essential spectrum is either a point in the resolvent or an isolated eigenvalue of finite multiplicity. If $P(D) = -\Delta$, then $\sigma(P_0) = [0, \infty)$. Consequently, in our case, the negative spectrum of $\mathcal{A}$ can have at most a countable number of isolated eigenvalues of finite multiplicity having a finite lower bound. The theorem is proved by showing that the bilinear form

$$a(u,v) = (P(D)u, v) + (Vu, v), \quad u, v \in H$$

is bounded and closed on H. Moreover, if the coefficients of $P(D)$ and the function $V(x)$ are real, then the bilinear form is symmetric and $\mathcal{A}$ is self-adjoint.

9.9 Proof of Lemma 9.1

Proof. We have

$$(u,u) = (gu, g^{-1}u) \leq \|gu\| \cdot \|g^{-1}u\| \leq C\|gu\| \cdot \|u\|_H \leq \varepsilon \|u\|_H^2 + K_\varepsilon \|gu\|^2.$$

Hence,

$$\|u\|_H^2 \leq \varepsilon \|u\|_H^2 + \|\nabla u\|^2 + K_\varepsilon \|gu\|^2.$$

This implies (9.6). To prove (9.7), assume that there is a sequence $u^{(k)} \in H$ such that $\|u^{(k)}\|'_H = 1$ and $a(u^{(k)}) + \|gu^{(k)}\|^2 \to 0$, where

$$(\|u\|'_H)^2 = \|\nabla u\|^2 + \|gu\|^2.$$

Since

$$\rho_k = \|u^{(k)}\|_H \le C,$$

there is a renamed subsequence such that $u^{(k)}$ converges to a limit $u \in H$ weakly in H, strongly in $L^2_{loc}(\mathbb{R}^n)$, and a.e. in $\mathbb{R}^n$. Since $a(u^{(k)}) \ge 0$, we have $a(u^{(k)}) \to 0$ and $\|gu^{(k)}\| \to 0$. By our hypothesis on $V(x)$, there is a renamed subsequence of $u^{(k)}$ such that $b(u^{(k)}) \to b(u)$, where $b(u) = (Vu, u)$. Thus, $1+b(u) = 0$, showing that $u \neq 0$. Since $gu^{(k)} \to gu$ a.e., we have $u = 0$, providing a contradiction. The same proof can be used to prove (9.8) if we keep in mind that (9.8) holds only on M. To prove (9.9), assume that there is a sequence $u^{(k)} \in M$ such that $\|u^{(k)}\|'_H = 1$ and $a(u^{(k)}) - \lambda_l \|u^{(k)}\|^2 + \|gu^{(k)}\|^2 \to 0$. Since

$$\rho_k = \|u^{(k)}\|'_H \le C,$$

there is a renamed subsequence such that $u^{(k)}$ converges to a limit $u \in H$ weakly in H, strongly in $L^2_{loc}(\mathbb{R}^n)$, and a.e. in $\mathbb{R}^n$. Also, there is a renamed subsequence such that $\|\nabla u^{(k)}\| \to \nu$ and $\|gu^{(k)}\| \to \tau$, where $\nu^2 + \tau^2 = 1$. Since $a(u^{(k)}) - \lambda_l \|u^{(k)}\|^2 \ge 0$ in M, we have $a(u^{(k)}) - \lambda_l \|u^{(k)}\|^2 \to 0$ and $\|gu^{(k)}\| \to 0$. By our hypothesis on $V(x)$, there is a renamed subsequence of $u^{(k)}$ such that $b(u^{(k)}) \to b(u)$, where $b(u) = (Vu, u)$. Since $a(u^{(k)}) - \lambda_l \|u^{(k)}\|^2 \to 0$, and $a(u^{(k)}) \to \nu^2 + b(u)$, we see that $\nu^2 + b(u) \le 0$. Since $gu^{(k)} \to gu$ a.e. and $\|gu^{(k)}\| \to 0$, we have $\tau = 0$ and $u = 0$. But then, $\nu = 1$ and $1 + b(u) \le 0$, showing that $u \neq 0$, providing a contradiction. □

9.10 Proofs of the Theorems

We now give the proof of Theorem 9.2.

Proof. We define

$$G(u) = a(u) - 2\int_{\mathbb{R}^n} F(x, u(x))\, dx, \quad u \in H. \tag{9.14}$$

By Lemma 9.1 there is an $\varepsilon > 0$ such that

$$a(u) + \|gu\|^2 \ge \varepsilon \|u\|^2_H, \quad u \in H. \tag{9.15}$$

Thus

$$G(u) \to \infty, \quad \|u\|_H \to \infty, \tag{9.16}$$

i.e., $G(u)$ is coercive. Let

$$c = \inf_H G.$$

By Corollary 3.5 (taking $N = \{0\}$ and $M = H$), there is a sequence $\{u^{(k)}\} \subset H$ such that

$$G(u^{(k)}) = a(u^{(k)}) - 2\int_{\mathbb{R}^n} F(x, u^{(k)}(x))\, dx \to c, \tag{9.17}$$

$$(G'(u^{(k)}), z)/2 = a(u^{(k)}, z) - \int_{\mathbb{R}^n} f(x, u^{(k)}) \cdot z(x)\, dx \to 0, \quad z \in H \tag{9.18}$$

and

$$(G'(u^{(k)}), u^{(k)})/2 = a(u^{(k)}) - \int_{\mathbb{R}^n} f(x, u^{(k)}) \cdot u^{(k)}\, dx \to 0. \tag{9.19}$$

Since $G(u)$ is coercive,

$$\rho_k = \|u^{(k)}\|'_H \le C,$$

where

$$(\|u\|'_H)^2 = \|\nabla u\|^2 + \|gu\|^2.$$

Thus there is a renamed subsequence such that $u^{(k)}$ converges to a limit $u \in H$ weakly in H, strongly in $L^2_{loc}(\mathbb{R}^n)$, and a.e. in $\mathbb{R}^n$. From (9.18) we see that

$$(G'(u), z)/2 = a(u, z) - \int_{\mathbb{R}^n} f(x, u(x)) \cdot z(x)\, dx = 0, \quad z \in C_0^\infty(\mathbb{R}^n),$$

from which we conclude easily that u is a solution of (9.2). □

Proof of Theorem 9.4 By Lemma 9.1, there is an $\varepsilon > 0$ such that

$$a(u) + \|gu\|^2 \ge \varepsilon \|u\|_H^2, \quad u \in H. \tag{9.20}$$

Consequently,

$$G(u) \ge \varepsilon\|u\|_H^2 - \int_{|u|>\delta} S(x)^q |u(x)|^\alpha \ge \varepsilon \|u\|_H^2 - C\|u\|_H^q$$

by Hypothesis (A). As a consequence, there are positive constants η, ρ such that

$$G(u) \geq \eta, \quad \|u\|_H = \rho. \tag{9.21}$$

Since $0 \in \sigma_e(\mathcal{A})$, there is a $\varphi \in H$ such that $\|\varphi\| = 1$ and $a(\varphi) = (\mathcal{A}\varphi, \varphi) < \eta/2$. Consequently,

$$G(s\varphi) = s^2 a(\varphi) - 2\int F(x, s\varphi) \leq s^2\eta/2 + \int h(s\varphi).$$

Thus,

$$\limsup_{s\to\infty} G(s\varphi)/s^2 \leq \eta/2 + C,$$

since

$$\limsup_{s\to\infty} \int \frac{h(s\varphi)}{s^2\varphi^2}\varphi^2 \leq C\|\varphi\|^2.$$

This implies that there is a sequence $u^{(k)}$ in H such that

$$G(u^{(k)}) \to c \geq \eta/2, \quad G'(u^{(k)})/(\|u^{(k)}\|_H + 1)^2 \to 0 \tag{9.22}$$

(Example 1a of Chap. 3).

If

$$\rho_k = \|u^{(k)}\|'_H \leq C,$$

there is a renamed subsequence such that $u^{(k)}$ converges to a limit $u \in H$ weakly in H, strongly in $L^2_{loc}(\mathbb{R}^n)$, and a.e. in $\mathbb{R}^n$. In particular we have

$$(G'(u^{(k)}), z)/2 = a(u^{(k)}, z) - \int f(x, u^{(k)}) \cdot z(x)\, dx \to 0, \quad z \in H. \tag{9.23}$$

From this we see that

$$(G'(u), z)/2 = a(u, z) - \int f(x, u(x)) \cdot z(x)\, dx = 0, \quad z \in C_0^\infty(\mathbb{R}^n),$$

from which we conclude easily that u is a solution of (9.2). Moreover, since $\|u^{(k)}\|'_H$ is bounded, (9.22) implies $G(u) = c \geq \eta/2$ (Theorem 3.4.1 of [184]). Since, $G(0) = 0$, we see that $u \neq 0$.

If

$$\rho_k = \|u^{(k)}\|'_H \to \infty,$$

let $\tilde{u}^{(k)} = u^{(k)}/\rho_k$. Then, $\|\tilde{u}^{(k)}\|_H' = 1$. There is a renamed subsequence such that $\tilde{u}^{(k)}$ converges to a function $\tilde{u}(x) \in H$ weakly in H, strongly in $L^2_{loc}(\mathbb{R}^n)$, a.e. in $\mathbb{R}^n$, and such that $\|\nabla\tilde{u}^{(k)}\| \to \nu$ and $\|g\tilde{u}^{(k)}\| \to \tau$, where $\nu^2 + \tau^2 = 1$. Let $b(u, v) = (Vu, v)$, $b(u) = b(u, u)$. Since $\tilde{u}^{(k)}$ converges to $\tilde{u}$ weakly in H, we see that $b(\tilde{u}^{(k)}) \to b(\tilde{u})$. Hence, $\|\nabla\tilde{u}^{(k)}\|^2 + b(\tilde{u}^{(k)}) + \|g\tilde{u}^{(k)}\|^2 \to \nu^2 + b(\tilde{u}) + \tau^2$. Note that

$$\int S^q|u|^\alpha \leq \left(\int |Su|^{\alpha\cdot(q/\alpha)}\right)^{(\alpha/q)}\left(\int S^{(q-\alpha)\cdot q/(q-\alpha)}\right)^{(q-\alpha)/q} \leq \|Su\|_q^\alpha \cdot \|S\|_q^{q-\alpha}.$$

Thus,

$$0 \leftarrow G(u^{(k)})/\rho_k^2 \geq \|\nabla\tilde{u}^{(k)}\|^2 + b(\tilde{u}^{(k)}) + \|g\tilde{u}^{(k)}\|^2 - O(\rho_k^{\alpha-2}) \to \nu^2 + b(\tilde{u}) + \tau^2. \tag{9.24}$$

Thus, $1 + b(\tilde{u}) = \nu^2 + b(\tilde{u}) + \tau^2 \leq 0$. This implies that $b(\tilde{u}) \neq 0$, and consequently $\tilde{u}(x) \not\equiv 0$. But $g\tilde{u}^{(k)} \to g\tilde{u}$ a.e. This means that $\tilde{u} = 0$, a contradiction. Hence, the ρ_k are bounded, and the proof is complete. □

Proof of Theorem 9.5 We follow the proof of Theorem 9.4. By Lemma 9.1 there is an $\varepsilon > 0$ such that (9.8) holds. This implies that there are positive constants η, ρ such that

$$G(u) \geq \eta, \quad \|u\|_H = \rho, \ u \in M \tag{9.25}$$

by the argument given there. Note that

$$G(v) = a(v) - 2\int F(x, v) = \lambda_0\|v\|^2 - 2\int F(x, v) \leq 0, \qquad v \in N$$

by Hypothesis 1. Define $A = M \cap \mathcal{B}_\rho$, $\ B = N$, where

$$\mathcal{B}_\rho = \{u \in H : \|u\|_H < \rho\}.$$

Then A links B (Example 2 of Chap. 3). If $G_1 = -G$, then

$$\sup_A G_1 \leq \inf_B G_1.$$

By Theorem 1.2, there is a sequence $\{u^{(k)}\} \subset H$ such that

$$G(u^{(k)}) = a(u^{(k)}) - 2\int_{\mathbb{R}^n} F(x, u^{(k)}(x))\,dx \to c, \tag{9.26}$$

$$(G'(u^{(k)}), z)/2 = a(u^{(k)}, z) - \int_{\mathbb{R}^n} f(x, u^{(k)}) \cdot z(x)\,dx \to 0, \quad z \in H \tag{9.27}$$

and

$$(G'(u^{(k)}), u^{(k)})/2 = a(u^{(k)}) - \int_{\mathbb{R}^n} f(x, u^{(k)}) \cdot u^{(k)}\, dx \to 0. \tag{9.28}$$

If

$$\rho_k = \|u^{(k)}\|'_H \leq C,$$

where

$$(\|u\|'_H)^2 = \|\nabla u\|^2 + \|gu\|^2,$$

then there is a renamed subsequence such that $u^{(k)}$ converges to a limit $u \in H$ weakly in H, strongly in $L^2_{loc}(\mathbb{R}^n)$, and a.e. in $\mathbb{R}^n$. From (9.27) we see that

$$(G'(u), z)/2 = a(u, z) - \int_{\mathbb{R}^n} f(x, u(x)) \cdot z(x)\, dx = 0, \quad z \in C_0^\infty(\mathbb{R}^n),$$

from which we conclude easily that u is a solution of (9.2).

If

$$\rho_k = \|u^{(k)}\|'_H \to \infty,$$

let $\tilde{u}^{(k)} = u^{(k)}/\rho_k$. Then, $\|\tilde{u}^{(k)}\|'_H = 1$. There is a renamed subsequence such that $\tilde{u}^{(k)}$ converges to a function $\tilde{u}(x) \in H$ weakly in H, strongly in $L^2_{loc}(\mathbb{R}^n)$, a.e. in $\mathbb{R}^n$, and such that $\|\nabla \tilde{u}^{(k)}\| \to \nu$ and $\|g\tilde{u}^{(k)}\| \to \tau$, where $\nu^2 + \tau^2 = 1$. Let $b(u, v) = (Vu, v)$, $b(u) = b(u, u)$. Since $\tilde{u}^{(k)}$ converges to $\tilde{u}$ weakly in H, we see that $b(\tilde{u}^{(k)}) \to b(\tilde{u})$. Hence, $0 \leq \|\nabla \tilde{u}^{(k)}\|^2 + b(\tilde{u}^{(k)}) + \|g\tilde{u}^{(k)}\|^2 \to \nu^2 + b(\tilde{u}) + \tau^2$. Since $a(\tilde{u}^{(k)}) \geq 0$, we have $\nu^2 + b(\tilde{u}) \geq 0$. Note that

$$\int S^q |u|^\alpha \leq \left(\int |Su|^{\alpha \cdot (q/\alpha)} \right)^{(\alpha/q)} \left(\int S^{(q-\alpha) \cdot q/(q-\alpha)} \right)^{(q-\alpha)/q} \leq \|Su\|_q^\alpha \cdot \|S\|_q^{q-\alpha}.$$

Moreover,

$$G(u^{(k)})/\rho_k^2 = \|\nabla \tilde{u}^{(k)}\|^2 + b(\tilde{u}^{(k)}) + \|g\tilde{u}^{(k)}\|^2 + O(\rho_k^{\alpha-2}) \to \nu^2 + b(\tilde{u}) + \tau^2 = 0.$$

This implies $\nu^2 + b(\tilde{u}) = 0$ and $\tau^2 = 0$. But $g\tilde{u}^{(k)} \to g\tilde{u}$ a.e. This means that $\tilde{u} = 0$ and $\nu = 1$, so that

$$1 + b(\tilde{u}) = 0.$$

But this implies that $b(\tilde{u}) \neq 0$, and consequently $\tilde{u}(x) \not\equiv 0$, a contradiction. Hence, the ρ_k are bounded, and the proof is complete. □

Proof of Theorem 9.6 We follow the proof of Theorem 9.4. By Lemma 9.1 there is an $\varepsilon > 0$ such that (9.8) holds. This implies that there are positive constants η, ρ such that

$$G(u) \geq \eta, \quad \|u\|_H = \rho, \ u \in M \tag{9.29}$$

as in the proof of Theorem 9.4. Note that

$$G(v) = a(v) - 2\int F(x, v) = \lambda_0 \|v\|^2 - 2\int F(x, v) \leq 0, \quad v \in N$$

by Hypothesis 1. Since $\mathcal{A}$ is not invertible on M, there is a $\varphi \in M$ such that $\|\varphi\| = 1$ and $a(\varphi) = (\mathcal{A}\varphi, \varphi) < \eta$. For $R > \rho$, let

$$A_R = [N \cap \mathcal{B}_R] \cup \{v + s\varphi : s \geq 0, \ \|v + s\varphi\| = R\}$$

$$B = M \cap \partial\mathcal{B}_\rho,$$

where

$$\mathcal{B}_\rho = \{u \in H : \|u\|_H < \rho\}.$$

By Example 3 of Chap. 1, $A_R, \ B$ link each other. Now

$$\begin{aligned} G(v + s\varphi) &\leq \ a(v) + s^2\eta + \int \frac{h(v + s\varphi)}{(v + s\varphi)^2}(v + s\varphi)^2 \\ &\leq \ \lambda_0\|v\|^2 + s^2\eta + c_0\|v + s\varphi\|^2 \\ &\leq \ (c_0 + \lambda_0)\|v\|^2 + (c_0 + \eta)s^2 \\ &\leq \ (2c_0 + \lambda_0 + \eta)R^2. \end{aligned}$$

We can now apply Example 1a of Chap. 3 to conclude that there is a sequence $u^{(k)}$ in H such that

$$G(u^{(k)}) \to c \geq \eta, \quad G'(u^{(k)})/(\|u^{(k)}\|_H + 1)^2 \to 0. \tag{9.30}$$

We can now follow the proofs of Theorem 9.5 to reach the desired conclusion. □

Proof of Theorem 9.7 Define the subspaces M and N of H as before:

$$N \ = \ \bigoplus_{k<l} E(\lambda_k)\,, \quad M \ = \ N^\perp\,, \quad H \ = \ M \oplus N\,.$$

Let

$$G(u) = a(u) - 2\int_{\mathbb{R}^n} F(x,u)\,dx. \tag{9.31}$$

We note that Hypothesis 1 implies

$$G(v) \le 0, \quad v \in N. \tag{9.32}$$

In fact, we have

$$G(u) = a(u) - 2\int_{\mathbb{R}^n} F(x,u)\,dx \le \int_{\mathbb{R}^n} [\lambda_{l-1}u^2 - 2F(x,u)]\,dx \le 0, \quad u \in N.$$

In view of inequality (9.9), we see that there are positive constants η, ρ such that

$$G(u) \ge \eta, \quad \|u\|_H = \rho, \ u \in M. \tag{9.33}$$

Take

$$A = \partial\mathcal{B}_\rho \cap M,$$
$$B = N,$$

where

$$\mathcal{B}_\rho = \{u \in H : \|u\|_H < \rho\}.$$

By Example 2 of Chap. 3, A links B. Moreover,

$$\sup_A[-G] \le 0 \le \inf_B[-G]. \tag{9.34}$$

Hence, we may apply Theorem 1.2 to conclude that that there is a sequence $\{u^{(k)}\} \subset H$ such that

$$G(u^{(k)}) = a(u^{(k)}) - 2\int_{\mathbb{R}^n} F(x,u^{(k)}(x))\,dx \to c \le 0, \tag{9.35}$$

$$(G'(u^{(k)}), z)/2 = a(u^{(k)}, z) - \int_{\mathbb{R}^n} f(x,u^{(k)})\cdot z(x)\,dx \to 0, \quad z \in H \tag{9.36}$$

and

$$(G'(u^{(k)}), u^{(k)})/2 = a(u^{(k)}) - \int_{\mathbb{R}^n} f(x,u^{(k)})\cdot u^{(k)}\,dx \to 0. \tag{9.37}$$

If

$$\rho_k = \|u^{(k)}\|'_H \leq C,$$

there is a renamed subsequence such that $u^{(k)}$ converges to a limit $u \in H$ weakly in H strongly in $L^2_{loc}(\mathbb{R}^n)$, and a.e. in $\mathbb{R}^n$. From (9.36) we see that

$$(G'(u), z)/2 = a(u, z) - \int_{\mathbb{R}^n} f(x, u(x)) \cdot z(x)\, dx = 0, \quad z \in C_0^\infty(\mathbb{R}^n),$$

from which we conclude easily that u is a solution of (9.2). We can now follow the proof of Theorem 9.4 until we reach (9.24). In our case this becomes

$$G(u^{(k)})/\rho_k^2 \geq \|\nabla \tilde{u}^{(k)}\|^2 + b(\tilde{u}^{(k)}) + \|g\tilde{u}^{(k)}\|^2 - \lambda_l \|\tilde{u}^{(k)}\|^2 - O(\rho_k^{\alpha-2}). \tag{9.38}$$

Since $\lambda_l \leq 0$,

$$G(u^{(k)})/\rho_k^2 \to 0,$$

and

$$\|\nabla \tilde{u}^{(k)}\|^2 + b(\tilde{u}^{(k)}) + \|g\tilde{u}^{(k)}\|^2 \to \nu^2 + b(\tilde{u}) + \tau^2,$$

we again have $1 + b(\tilde{u}) \leq 0$, which leads to the desired conclusion. □

Proof of Theorem 9.8 We follow the proof of Theorem 9.7. We note that Hypothesis 1 implies

$$G(v) \leq 0, \quad v \in N. \tag{9.39}$$

In fact, we have

$$G(u) = a(u) - 2\int_{\mathbb{R}^n} F(x, u)\, dx \leq \int_{\mathbb{R}^n} [\lambda_{l-1} u^2 - 2F(x, u)]\, dx \leq 0, \quad u \in N.$$

In view of inequality (9.9), we see that there are positive constants η, ρ such that

$$G(u) \geq \eta, \quad \|u\|_H = \rho, \ u \in M. \tag{9.40}$$

Let φ be an eigenfunction of $\mathcal{A}$ corresponding to λ_l and satisfying $\|\varphi\| = 1$. For $R > \rho$, let

$$A_R = [N \cap \mathcal{B}_R] \cup \{v + s\varphi : s \geq 0,\ \|v + s\varphi\| = R\}$$

$$B = M \cap \partial \mathcal{B}_\rho,$$

where

$$\mathcal{B}_\rho = \{u \in H : \|u\|_H < \rho\}.$$

By Example 3 of Chap. 3, A_R, B link each other. Now

$$\begin{aligned} G(v+s\varphi) &\le a(v) + s^2\lambda_l + \int \frac{h(v+s\varphi)}{(v+s\varphi)^2}(v+s\varphi)^2 \\ &\le \lambda_{l-1}\|v\|^2 + s^2\lambda_l + c_0\|v+s\varphi\|^2 \\ &\le (c_0+\lambda_{l-1})\|v\|^2 + (c_0+\lambda_l)s^2 \\ &\le (2c_0+\lambda_{l-1}+\lambda_l)R^2. \end{aligned}$$

We can now apply Example 3a of Chap. 3 to conclude that there is a sequence $u^{(k)}$ in H such that

$$G(u^{(k)}) \to c \ge \eta/2, \quad G'(u^{(k)})/(\|u^{(k)}\|_H + 1)^2 \to 0. \tag{9.41}$$

We can now follow the proof of Theorem 9.4 until we reach (9.24). In our case this becomes

$$G(u^{(k)})/\rho_k^2 \ge \|\nabla\tilde{u}^{(k)}\|^2 + b(\tilde{u}^{(k)}) + \|g\tilde{u}^{(k)}\|^2 - \lambda_l\|\tilde{u}^{(k)}\|^2 - O(\rho_k^{\alpha-2}). \tag{9.42}$$

Since $\lambda_l \le 0$,

$$G(u^{(k)})/\rho_k^2 \to 0,$$

and

$$\|\nabla\tilde{u}^{(k)}\|^2 + b(\tilde{u}^{(k)}) + \|g\tilde{u}^{(k)}\|^2 \to \nu^2 + b(\tilde{u}) + \tau^2,$$

we again have $1 + b(\tilde{u}) \le 0$, which leads to the desired conclusion. □

Proof of Theorem 9.10 Let

$$\alpha = \inf_{\mathcal{M}} G(u).$$

There is a sequence $\{u^{(k)}\} \in \mathcal{M}$ such that

$$G(u^{(k)}) = a(u^{(k)}) - 2\int_{\mathbb{R}^n} F(x, u^{(k)}(x))\,dx \to \alpha, \tag{9.43}$$

$$(G'(u^{(k)}), z)/2 = a(u^{(k)}, z) - \int_{\mathbb{R}^n} f(x, u^{(k)}) \cdot z(x)\,dx = 0, \quad z \in H \tag{9.44}$$

and

$$(G'(u^{(k)}), u^{(k)})/2 = a(u^{(k)}) - \int_{\mathbb{R}^n} f(x, u^{(k)}) \cdot u^{(k)}\, dx = 0. \tag{9.45}$$

Thus,

$$\int_{\mathbb{R}^n} H(x, u^{(k)}(x))\, dx = G(u^{(k)}) \to \alpha.$$

In view of assumption (9.12), we see that $\alpha \geq -\int W > -\infty$. In view of the arguments given in the proofs of Theorems 9.2–9.8, we see that

$$\rho_k = \|u^{(k)}\|_H \leq C.$$

Hence, there is a renamed subsequence such that $u^{(k)}$ converges to a limit $u \in H$, weakly in H, strongly in $L^2_{loc}(\mathbb{R}^n)$, and a.e. in $\mathbb{R}^n$. From (9.44) we see that

$$(G'(u), z)/2 = a(u, z) - \int_{\mathbb{R}^n} f(x, u(x)) \cdot z(x)\, dx = 0, \quad z \in C_0^\infty(\mathbb{R}^n),$$

from which we conclude easily that u is a solution of (9.2). Hence, $u \in \mathcal{M}$. Moreover,

$$G(u) = \int_{\mathbb{R}^n} H(x, u(x))\, dx \leq \liminf \int_{\mathbb{R}^n} H(x, u^{(k)}(x))\, dx = \liminf G(u^{(k)}) = \alpha.$$

Thus, $G(u) = \alpha$. If $\alpha \neq 0$, then $u \neq 0$. This completes the proof. □

9.11 Radial Solutions

We now consider the semilinear radial Schrödinger equation

$$-\Delta u + V(r)u = f(r, u), \quad u \in H^1(\mathbb{R}^n), \tag{9.46}$$

where $V(r)$ is a given potential and $r = |x|$. In this case $V(r)$ and $f(r, t)$ depend only on $r = |x|$ and we are looking for solutions that depend only on $r = |x|$. As in the general case, one wishes to find solutions and, in particular, solutions that minimize the corresponding energy functional. If $u(x) \equiv 0$ is a solution, one wishes to find other solutions as well. As before, the existence of solutions depends on both the linear operator $\mathcal{A}$ and the nonlinear term $f(r, u)$.

In the present section we solve the Eq. (9.46) under assumptions on $V(r)$ such that the essential spectrum of

$$-\Delta u + V(r)u$$

is the same as that of $-\Delta u$, i.e., $[0, \infty)$.

Concerning the function $V(r)$ we make the following assumptions:

$V(r) \in L^1_{loc}(\mathbb{R})$ and satisfies

$$\int_0^1 |V(r)|^p r^{n-1} dr < \infty \ \ for\ some\ p > n/2, \tag{9.47}$$

and

$$\int_R^{R+1} |V(r)|\, dr \to 0\ as\ R \to \infty. \tag{9.48}$$

These assumptions imply that there is a forms extension $\mathcal{A}$ of the operator

$$-\Delta u(r) + V(r)u(r)$$

on the space $H = H^{1,2}(\mathbb{R}^n)$ having essential spectrum equal to $[0, \infty)$ and a (possibly empty) discrete, countable negative spectrum consisting of isolated eigenvalues of finite multiplicity with a finite lower bound $-L$

$$-\infty < -L \le \lambda_0 < \lambda_1 < \lambda_2 < \ldots < \lambda_l < \ldots < 0 \tag{9.49}$$

(cf. Lemma 9.28).

For each $l > 0$, define the subspaces $M = M_l$ and $N = N_l$ of H as

$$N = \bigoplus_{k<l} E(\lambda_k)\,, \quad M = N^\perp\,, \quad H = M \oplus N\,.$$

As in the general case, for the operator $\mathcal{A}$ there are three possibilities: (a) it has no negative eigenvalues, (b) it has only one negative eigenvalue, and (c) it has two or more negative eigenvalues. What is interesting is that each of these possibilities can be dealt with differently. As before, we shall study all of them separately.

Let q be any number satisfying

$$\begin{aligned} &2 < q \le 2^* = 2n/(n-2), \quad && n > 2 \\ &2 < q < \infty, && n \le 2 \end{aligned} \tag{3.1.3}$$

and let $f(r, t)$ be a Carathéodory function on $\mathbb{R} \times \mathbb{R}$. By the Sobolev inequality,

$$\|u\|_q \le C\|u\|_H, \quad u \in H,$$

where

$$\|u\|_q := (\int_{\mathbb{R}^n} |u(|x|)|^q dx)^{1/q} = (c_n \int_{\mathbb{R}} |u(r)|^q r^{n-1} dr)^{1/q}, \ \|u\| = \|u\|_2.$$

We make the following assumptions:

(A) The function $f(r, t)$ satisfies

$$|f(r, t)| \leq S(r)^q (|t|^{q-1} + W(r))$$

and

$$f(r, t)/S(r)^q = o(|t|^{q-1}) \text{ as } |t| \to \infty,$$

where $S(r) > 0$ is a function in $L^q(\mathbb{R}^n)$ satisfying

$$\|Su\|_q \leq C\|u\|_H, \quad u \in H \tag{9.50}$$

and W is a function in $L^\infty(\mathbb{R}^n)$ (a sufficient condition for (9.50) to hold is given in Lemma 9.29).

Our other hypotheses depend only on the primitive of $f(r, t)$:

$$F(r, t) = \int_0^t f(r, s)\, ds.$$

9.12 No Negative Eigenvalues

In this case,

$$a(u) = (\mathcal{A}u, u) \geq 0, \quad u \in H.$$

We let N_1 be the set of measurable functions $h(r)$ satisfying

$$\sup_u \frac{\|hu\|}{\|u\|_H} < \infty.$$

For a general description of this space, cf. [183]. Sufficient conditions for a radial function to be in N_1 are given in Lemma 9.25. We have the following.

Theorem 9.12 *Assume*

1. *There is a function $g(r)$ such that $g, g^{-1} \in N_1$ and*

$$2F(r, u) \leq -g(r)^2|u|^2 + W(r), \quad u \in \mathbb{R}, \ r \in \mathbb{R}^+,$$

where $W(r) \in L^1(\mathbb{R}^+, r^{n-1})$.

Then the Eq. (9.46) has a solution.

Remark 9.13 It is clear from the equation that the solution obtained will be nontrivial if

$$f(r, 0) \neq 0.$$

To guarantee that a solution will be nontrivial even when $f(r, 0) = 0$, we have the following.

Theorem 9.14 *Assume*

1. *There are positive constants* $\alpha < 2, \delta$ *and a function* $g(r)$ *such that* $g, g^{-1} \in N_1$ *and*

$$\begin{aligned} 2F(r,u) &\le -g(r)^2|u|^2, \quad u \in \mathbb{R},\ r \in \mathbb{R}^+,\ |u| < \delta, \\ &\le -g(r)^2|u|^2 + S(r)^q|u|^\alpha, \quad u \in \mathbb{R},\ r \in \mathbb{R}^+,\ |u| > \delta. \end{aligned}$$

2. *There is a locally bounded function* $h(t)$ *such that*

$$2F(r,u) \ge -h(u), \quad u \in \mathbb{R},\ r \in \mathbb{R}^+,$$

and

$$c_0 = \sup_{\mathbb{R}} h(u)/u^2 < \infty.$$

Then the Eq. (9.46) has a nontrivial solution.

9.13 Only One Negative Eigenvalue

Let $\lambda_0 < 0$ be the eigenvalue. In this case

$$a(u) \ge \lambda_0 \|u\|^2, \quad u \in H.$$

We can make use of the fact that there is a gap in the spectrum of $\mathcal{A}$. We have

Theorem 9.15 *Assume*

1.

$$2F(r,u) \ge \lambda_0 |u|^2, \quad u \in \mathbb{R},\ r \in \mathbb{R}^+.$$

2. *There are positive constants* $\alpha < 2, \delta$ *and a function* $g(r)$ *such that* $g, g^{-1} \in N_1$ *and*

$$2F(r,u) \le -g(r)^2|u|^2, \quad u \in \mathbb{R},\ r \in \mathbb{R}^+,\ |u| < \delta,$$
$$\le -g(r)^2|u|^2 + S(r)^q|u|^\alpha, \quad u \in \mathbb{R},\ r \in \mathbb{R}^+,\ |u| > \delta.$$

Then the Eq. (9.46) has a solution.

To obtain a nontrivial solution we have the following.

Theorem 9.16 *Assume*

1.

$$2F(r,u) \ge \lambda_0|u|^2, \quad u \in \mathbb{R},\ r \in \mathbb{R}^+.$$

2. There are positive constants $\alpha < 2, \delta$ *and a function* $g(r)$ *such that* $g, g^{-1} \in N_1$ *and*

$$2F(r,u) \le -g(r)^2|u|^2, \quad u \in \mathbb{R},\ r \in \mathbb{R}^+,\ |u| < \delta,$$
$$\le -g(r)^2|u|^2 + S(r)^q|u|^\alpha, \quad u \in \mathbb{R},\ r \in \mathbb{R}^+,\ |u| > \delta.$$

3. There is a locally bounded function $h(t)$ *such that*

$$2F(r,u) \ge -h(u), \quad u \in \mathbb{R},\ r \in \mathbb{R}^+,$$

and

$$c_0 = \sup_{\mathbb{R}} h(u)/u^2 < \infty.$$

Then the Eq. (9.46) has a nontrivial solution.

9.14 Two or More Negative Eigenvalues

Here again we can make use of the fact that there are gaps in the spectrum.

Let λ_{l-1}, λ_l be two consecutive negative eigenvalues of $\mathcal{A}$. We have the following.

Theorem 9.17 *Assume*

1.

$$2F(r,u) \ge \lambda_{l-1}|u|^2, \quad u \in \mathbb{R},\ r \in \mathbb{R}^+,$$

2. There are positive constants $\alpha < 2, \delta$ *and a function* $g(r)$ *such that* $g, g^{-1} \in N_1$ *and*

$$2F(r,u) \le -g(r)^2|u|^2 + \lambda_l|u|^2, \quad u \in \mathbb{R},\ r \in \mathbb{R}^+,\ |u| < \delta,$$
$$\le -g(r)^2|u|^2 + \lambda_l|u|^2 + S(r)^q|u|^\alpha, \quad u \in \mathbb{R},\ r \in \mathbb{R}^+,\ |u| > \delta.$$

Then the Eq. (9.46) has a solution.

To obtain a nontrivial solution we have the following.

Theorem 9.18 *Assume*

1.

$$2F(r,u) \ge \lambda_{l-1}|u|^2, \quad u \in \mathbb{R},\ r \in \mathbb{R}^+,$$

2. *There are positive constants $\alpha < 2$, δ and a function $g(r)$ such that $g, g^{-1} \in N_1$ and*

$$2F(r,u) \le -g(r)^2|u|^2 + \lambda_l|u|^2, \quad u \in \mathbb{R},\ r \in \mathbb{R}^+,\ |u| < \delta,$$
$$\le -g(r)^2|u|^2 + \lambda_l|u|^2 + S(r)^q|u|^\alpha, \quad u \in \mathbb{R},\ r \in \mathbb{R}^+,\ |u| > \delta.$$

3. *There is a locally bounded function $h(t)$ such that*

$$2F(r,u) \ge -h(u), \quad u \in \mathbb{R},\ r \in \mathbb{R}^+,$$

and

$$c_0 = \sup_{\mathbb{R}} h(u)/u^2 < \infty.$$

Then the Eq. (9.46) has a nontrivial solution.

9.15 Least Energy Solutions

Let $\mathcal{M}$ be the set of all solutions of (9.46). A solution $\tilde{u}$ is called a "least energy solution" if it minimizes the functional

$$G(u) = a(u) - 2\int_{\mathbb{R}^n} F(x,u)\,dx \tag{9.51}$$

over the set $\mathcal{M}$.

We have the following.

Theorem 9.19 *If we add the following hypothesis to Theorems 9.12–9.18, then Eq. (9.46) has a least energy solution: The function given by*

$$H(x, u) = uf(x, u) - 2F(x, u) \tag{9.52}$$

satisfies

$$H(x, u) \geq -W(x) \in L^1(\mathbb{R}^n), \quad u \in \mathbb{R},\ x \in \mathbb{R}^n. \tag{9.53}$$

The proofs of Theorems 9.12–9.19 are almost identical to those of Theorems 9.2–9.10 and will be omitted. We shall only prove those lemmas which differ from the lemmas proved above. In the next section we describe the construction of the operator $\mathcal{A}$. We obtain the largest self-adjoint extension of $\mathcal{A}_0$ which preserves the essential spectrum.

9.16 The Operator $\mathcal{A}$

In constructing the operator $\mathcal{A}$ we shall make use of the following considerations.

We define a bilinear form $a_0(\cdot, \cdot)$ on the set $L^2(\mathbb{R}^+, r^{n-1}) \times L^2(\mathbb{R}^+, r^{n-1})$,

$$a_0(u, v) = (u', v') + (u, v). \tag{9.54}$$

The domain of the bilinear form is the set $D(a_0) = H$. Thus we can define an operator $\mathcal{A}_0$ such that $u \in D(\mathcal{A}_0)$ if and only if $u \in D(a_0)$ and there exists $f \in L^2(\mathbb{R}^+, r^{n-1})$ such that

$$a_0(u, v) = (f, v), \quad v \in D(a_0). \tag{9.55}$$

If u and f satisfy this condition we say $\mathcal{A}_0 u = f$.

Lemma 9.20 *The operator $\mathcal{A}_0$ is a self-adjoint operator from $L^2(\mathbb{R}^+, r^{n-1})$ to $L^2(\mathbb{R}^+, r^{n-1})$. It is one-to-one and onto.*

Proof. Let $f \in L^2(\mathbb{R}, \mathbb{R}^n)$. Then

$$(v, f) \leq \|v\| \cdot \|f\| \leq \|v\|_H \|f\|, \quad v \in H.$$

Thus (v, f) is a bounded linear functional on H. Since H is complete, there is a $u \in H$ such that

$$(u, v)_H = (f, v), \quad v \in H.$$

Consequently, $u \in D(\mathcal{A}_0)$ and $\mathcal{A}_0 u = f$. Moreover, if $\mathcal{A}_0 u = 0$, then

$$(u, v)_H = 0, \quad v \in H.$$

Thus, $u = 0$. Hence, $\mathcal{A}_0$ is one-to-one and onto.

Next, we show that $\mathcal{A}_0$ is self-adjoint. Consider any $u, f \in L^2(\mathbb{R}^+, r^{n-1})$, and suppose for any $v \in D(\mathcal{A}_0)$,

$$(u, \mathcal{A}_0 v) = (f, v) . \tag{9.56}$$

Since $\mathcal{A}_0$ is onto and $f \in L^2(\mathbb{R}^+, r^{n-1})$, there exists $w \in D(\mathcal{A}_0)$ such that $\mathcal{A}_0 w = f$. Then

$$(u - w, \mathcal{A}_0 v) = (f, v) - (\mathcal{A}_0 w, v) = 0 .$$

Since $u - w \in L^2(\mathbb{R}^+, r^{n-1})$, we can find a $v \in D(\mathcal{A}_0)$ such that $\mathcal{A}_0 v = u - w$, and

$$\|u - w\|^2 = 0.$$

This implies $u = w$ in the space $L^2(\mathbb{R}^+, r^{n-1})$, and therefore $u \in D(\mathcal{A}_0)$. Hence, $\mathcal{A}_0 u = \mathcal{A}_0 w = f$. □

The following was proved in [182]:

Theorem 9.21 *Let $b(\cdot, \cdot)$ be a Hermitian bilinear form such that $D(a_0) \subset D(b)$ and for some positive real number K,*

$$|b(u)| \leq K a_0(u) \quad u \in D(a_0), \tag{9.57}$$

where $b(u) = b(u, u)$ and $a_0(u) = a_0(u, u)$. Assume that every sequence $(u_k) \subset D(a)$ which satisfies

$$\|u_k\|^2 + a_0(u_k) \leq C \tag{9.58}$$

has a subsequence (v_j) such that

$$b(v_j - v_k) \to 0 . \tag{9.59}$$

Assume also that if (9.58) and (9.59) hold and $v_j \to 0$ in the $L^2(\mathbb{R}^+, r^{n-1})$ norm, then $b(v_j) \to 0$. Define

$$a(u, v) = a_0(u, v) + b(u, v) . \tag{9.60}$$

and let $\mathcal{A}$, $\mathcal{A}_0$ be the operators associated with a, a_0, respectively. Then

$$\sigma_e(\mathcal{A}) = \sigma_e(\mathcal{A}_0). \tag{9.61}$$

We shall show in Lemma 9.28 that Hypotheses (9.47) and (9.48) imply (9.61).

9.17 Some Lemmas

Lemma 9.22 *If $g^{-1} \in N_1$, then there is a constant C such that*

$$\|u\|_H^2 \le C(\|\nabla u\|^2 + \|gu\|^2), \quad u \in H. \tag{9.62}$$

If $\mathcal{A}$ has no negative eigenvalues, then there is an $\varepsilon > 0$ such that

$$a(u) + \|gu\|^2 \ge \varepsilon \|u\|_H^2, \quad u \in H. \tag{9.63}$$

If it has only one negative eigenvalue λ_0, then there is an $\varepsilon > 0$ such that

$$a(u) + \|gu\|^2 \ge \varepsilon \|u\|_H^2, \quad u \in M. \tag{9.64}$$

If it has negative eigenvalues λ_{l-1}, λ_l, then there is an $\varepsilon > 0$ such that

$$a(u) - \lambda_l \|u\|^2 + \|gu\|^2 \ge \varepsilon \|u\|_H^2, \quad u \in M. \tag{9.65}$$

Proof of Lemma 9.22 See the proof of Lemma 9.1. □

Lemma 9.23 *If $I \subset [0, \infty)$, $|I| = 1$, and $m \ge 0$, then*

$$\sup_I r^m u(r)^2 \le C \int_I [r^{m-1} u(r)^2 + r^m (u(r)^2 + u'(r)^2)]\, dr, \quad u \in C^1(\mathbb{R}^+). \tag{9.66}$$

Proof. If $r_1 < r_2 \in I$, then

$$\begin{aligned}
r_2^{2m} u(r_2)^2 - r_1^{2m} u(r_1)^2 &= \int_{r_1}^{r_2} [(r^m u(r))^2]' dr \\
&= \int_{r_1}^{r_2} 2r^m u(r) [r^m u(r)]' dr \\
&= \int_{r_1}^{r_2} 2r^m u(r) [r^m u'(r) + m r^{m-1} u(r)]\, dr \\
&= \int_{r_1}^{r_2} [2r^{2m} u(r) u'(r) + 2m r^{2m-1} u(r)^2]\, dr \\
&\le \int_{r_1}^{r_2} [r^{2m} (u(r)^2 + u'(r)^2) + 2m r^{2m-1} u(r)^2]\, dr \\
&\le \int_I [r^{2m} (u(r)^2 + u'(r)^2) + 2m r^{2m-1} u(r)^2]\, dr.
\end{aligned}$$

Take

$$r_1^{2m}u(r_1)^2 = \int_I r^{2m}u(r)^2\,dr.$$

Then

$$r_2^{2m}u(r_2)^2 \le \int_I [r^{2m}(2u(r)^2 + u'(r)^2) + 2mr^{2m-1}u(r)^2]\,dr.$$

Replace $2m$ by m. Since r_2 was arbitrary, the result follows. □

Lemma 9.24 *Under the same hypotheses*

$$|r_2^m u(r_2) - r_1^m u(r_1)| \le C|r_2 - r_1|^{1/2}\Big(\int_I r^{2m-2}(u(r)^2 + u'(r)^2)\,dr\Big)^{1/2}.$$

Proof.

$$\begin{aligned}
r_2^m u(r_2) - r_1^m u(r_1) &= \int_{r_1}^{r_2} [r^m u(r)]'\,dr \\
&= \int_{r_1}^{r_2} [mr^{m-1}u(r) + r^m u'(r)]\,dr \\
&\le \Big(\int_{r_1}^{r_2} dr\Big)^{1/2}\Big(\int_{r_1}^{r_2} [mr^{m-1}u(r)]^2 dr\Big)^{1/2} \\
&\quad + \Big(\int_{r_1}^{r_2} r dr\Big)^{1/2}\Big(\int_{r_1}^{r_2} [r^{m-1}u'(r)]^2 dr\Big)^{1/2} \\
&\le C|r_2 - r_1|^{1/2}\Big(\int_I r^{2m-2}(u(r)^2 + u'(r)^2)\,dr\Big)^{1/2}.
\end{aligned}$$

□

Lemma 9.25 *If g satisfies*

$$\int_0^1 g(r)^n\, r^{n-1} dr < \infty \tag{9.67}$$

and

$$\sup_{R\ge 1} \int_R^{R+1} g(r)^2 dr < \infty, \tag{9.68}$$

then $g \in N_1$. In particular, any nonzero constant is in N_1.

Proof. By Lemma 9.23,

$$\sup_I r^m u(r)^2 \leq C \int_I [r^{m-1}u(r)^2 + r^m (u(r)^2 + u'(r)^2)]\,dr, \quad u \in C^1(\mathbb{R}^+). \tag{9.69}$$

Let $I_k = [k, k+1]$. In particular, we have for $k \geq 1$

$$\sup_{I_k} r^{n-1}u(r)^2 \leq C \int_{I_k} r^{n-1}(u(r)^2 + u'(r)^2)\,dr, \quad u \in C^1(\mathbb{R}^+). \tag{9.70}$$

Hence, by (9.68),

$$\begin{aligned}\int_{I_k} g(r)^2 u(r)^2 r^{n-1} dr &\leq C \int_{I_k} r^{n-1}(u(r)^2 + u'(r)^2)\,dr \int_{I_k} g(r)^2 dr\\ &\leq C' \int_{I_k} r^{n-1}(u(r)^2 + u'(r)^2)\,dr.\end{aligned}$$

Thus,

$$\begin{aligned}\int_1^\infty g(r)^2 u(r)^2 r^{n-1} dr &= \sum_{k=1}^\infty \int_{I_k} g(r)^2 u(r)^2 r^{n-1} dr\\ &\leq C'' \sum_{k=1}^\infty \int_{I_k} r^{n-1}(u(r)^2 + u'(r)^2)\,dr\\ &= C'' \int_1^\infty r^{n-1}(u(r)^2 + u'(r)^2)\,dr.\end{aligned}$$

By the Sobolev inequality

$$\int_0^\infty |u(r)|^p r^{n-1} dr \leq C\|u\|_H^p, \quad 2 \leq p \leq 2^*.$$

Consequently,

$$\begin{aligned}\int_{I_0} g(r)^2 u(r)^2 r^{n-1} dr &\leq \Big(\int_0^1 |g(r)|^n r^{n-1} dr\Big)^{2/n} \Big(\int_0^1 |u(r)|^{2^*} r^{n-1} dr\Big)^{2/2^*}\\ &\leq \Big(\int_0^1 |g(r)|^n r^{n-1} dr\Big)^{2/n} \|u\|_H^2.\end{aligned}$$

Combining the two inequalities gives the desired result. □

Lemma 9.26 *If, in addition,*

$$\int_R^{R+1} g(r)^2 dr \to 0 \ as \ R \to \infty, \tag{9.71}$$

and

$$\int_0^1 |g(r)|^p r^{n-1} dr < \infty \ \ for \ some \ p > n, \tag{9.72}$$

then $g \in CN_1$.

Here CN_1 represents the set of functions multiplication by which is a compact mapping of H into $L^2(\mathbb{R}^+, r^{n-1})$.

Proof. Let $\varepsilon > 0$, and take $R > 0$ so large that

$$\int_{I_k} g(r)^2 dr < \varepsilon, \quad k \geq R, \ \ I_k = [k, k+1].$$

Then by Lemma 9.24

$$\begin{aligned}
\int_R^\infty g(r)^2 u(r)^2 r^{n-1} dr &= \sum_{k=R}^\infty \int_{I_k} g(r)^2 u(r)^2 r^{n-1} dr \\
&\leq C \sum_{k=R}^\infty \int_{I_k} (u'(r)^2 + u(r)^2) r^{n-1} dr \int_{I_k} g(r)^2 dr \\
&\leq \varepsilon C \sum_{k=R}^\infty \int_{I_k} (u'(r)^2 + u(r)^2) r^{n-1} dr \\
&= \varepsilon C \int_R^\infty (u'(r)^2 + u(r)^2) r^{n-1} dr.
\end{aligned}$$

Moreover,

$$\begin{aligned}
\sup_{1 \leq r \leq R} r^{n-1} u(r)^2 &\leq C \sup_{1 \leq k \leq R} \int_{I_k} (u'(r)^2 + u(r)^2) r^{n-1} dr \\
&\leq C \int_1^R (u'(r)^2 + u(r)^2) r^{n-1} dr.
\end{aligned}$$

Also

$$|r_2^m u(r_2) - r_1^m u(r_1)| \leq C |r_2 - r_1|^{1/2} \Big(\sup_{1 \leq k \leq R} \int_{I_k} r^{n-1} (u(r)^2 + u'(r)^2) \, dr \Big)^{1/2}.$$

Let $u_j(r)$ be a bounded sequence in H. Then for $r, r_1, r_2 \in [1, R]$, we have

$$|r^m u_j(r)| \le C\|u_j\|_H$$

and

$$|r_2^m u_j(r_2) - r_1^m u_j(r_1)| \le C|r_2 - r_1|^{1/2}\|u_j\|_H,$$

where $m = (n-1)/2$. Thus the sequence $r^m u_j(r)$ is uniformly bounded and equicontinuous. Consequently, there is a renamed subsequence such that $r^m u_j(r)$ converges uniformly in $[1, R]$. This implies

$$\int_1^R g(r)^2[u_i(r) - u_j(r)]^2 r^{n-1} dr$$

$$\le \sup_{1\le r\le R} r^{n-1}|u_i(r) - u_j(r)|^2 \int_1^R g(r)^2 dr \to 0, \quad i, j \to \infty.$$

Since the u_k are bounded in H, there is a renamed subsequence such that u_k converges weakly in H, strongly in $L^s(\mathbb{R}^+, r^{n-1})$, for $2 < s < 2^*$, and a.e. in $\mathbb{R}^+$. Thus if (9.72) holds, we have

$$\begin{aligned}\int_0^1 g(r)^2[u_j(r) - u_k(r)]^2 r^{n-1} dr &\le \Big(\int_0^1 |g(r)|^p r^{n-1} dr\Big)^{2/p} \\ &\times \Big(\int_0^1 |u_j(r) - u_k(r)|^s r^{n-1} dr\Big)^{2/s} \\ &\to 0 \; as \; j, k \to \infty,\end{aligned}$$

with $p = 2(s/2)' > n$.

Since

$$\int_1^R g(r)^2[u_j(r) - u_k(r)]^2 r^{n-1} dr \to 0, \quad i, j \to \infty$$

and

$$\int_R^\infty g(r)^2 u_k(r)^2 r^{n-1} dr < \varepsilon, \quad k \ge N,$$

the result follows. □

Remark 9.27 We can replace (9.67) with

$$\int_0^1 g(r)^2 r^{-1} dr < \infty. \tag{9.73}$$

Proof. If $I_0 = [0, 1]$, then

$$\sup_{I_0} r^n u(r)^2 \leq C \int_{I_0} r^{n-1}(u(r)^2 + u'(r)^2)\, dr, \quad u \in H. \tag{9.74}$$

If u_k are bounded in H, there is a renamed subsequence such that u_k converges weakly in H and a.e. in $\mathbb{R}^+$. By (9.74)

$$g(r)^2[u_j(r) - u_k(r)]^2 r^{n-1} \leq Cg(r)^2 r^{-1} \in L^1(I_0)$$
$$\to 0 \ a.e.\ as\ j, k \to \infty.$$

Hence,

$$\int_0^1 g(r)^2[u_j(r) - u_k(r)]^2 r^{n-1} dr \to 0.$$

□

Lemma 9.28 *Let the operator $\mathcal{A}_0$ be defined as in the previous section. Define*

$$a_0(u, v) = (\mathcal{A}_0 u, v), \quad b(u, v) = (Vu, v), \quad u, v \in H.$$

If $\mathcal{A}$ is defined by (9.60), then (9.61) holds.

Proof. Define $g(r) = |V(r)|^{1/2}$. By (9.47), (9.48) and Lemma 9.26, $g \in CN_1$. Hence, the hypotheses of Theorem 9.21 are satisfied. In particular, if (9.58) holds, then there is a subsequence and a function $w \in L^2(\mathbb{R}^+, r^{n-1})$ such that $\|gv_j - w\| \to 0$. Hence, (9.59) holds. If $\|v_j\| \to 0$, we see from $(gv_j, u) = (v_j, gu), \ u \in H$, that $(w, u) = 0, \ u \in H$. Thus $w = 0$ and $b(v_j, v_j) \to 0$. The conclusion comes from Theorem 9.21. □

Lemma 9.29 *A sufficient condition for $S(r)$ to satisfy*

$$\|Su\|_q \leq C\|u\|_H, \quad u \in H \tag{9.75}$$

is

$$\int_1^\infty S(r)^q dr < \infty \tag{9.76}$$

and

$$\int_0^1 |S(r)|^p r^{n-1} dr < \infty \quad for\ some\ p \geq 2^* q/(2^* - q). \tag{9.77}$$

Proof. By Lemma 9.23,

$$\sup_I r^m u(r)^2 \leq C \int_I [r^{m-1} u(r)^2 + r^m (u(r)^2 + u'(r)^2)]\, dr, \quad u \in C^1(\mathbb{R}^+). \tag{9.78}$$

Let $I_k = [k, k+1]$. In particular, we have for $k \geq 1$

$$\sup_{I_k} r^{n-1} u(r)^2 \leq C \int_{I_k} r^{n-1}(u(r)^2 + u'(r)^2)\, dr, \quad u \in C^1(\mathbb{R}^+). \tag{9.79}$$

Hence,

$$\begin{aligned}\int_{I_k} S(r)^q u(r)^q r^{n-1} dr &\leq \int_{I_k} S(r)^q [u(r)^2 r^{n-1}]^{(q/2)} dr \\ &\leq C\Big(\int_{I_k} r^{n-1}(u(r)^2 + u'(r)^2)\, dr\Big)^{(q/2)} \int_{I_k} S(r)^q dr \\ &\leq C \|u\|_H^q \int_{I_k} S(r)^q dr.\end{aligned}$$

Thus,

$$\begin{aligned}\int_1^\infty S(r)^q u(r)^q r^{n-1} dr &= \sum_{k=1}^\infty \int_{I_k} S(r)^q u(r)^q r^{n-1} dr \\ &\leq C'' \sum_{k=1}^\infty \int_{I_k} S(r)^q dr \cdot \|u\|_H^q \\ &= C'' \int_1^\infty S(r)^q dr \cdot \|u\|_H^q.\end{aligned}$$

By the Sobolev inequality

$$\int_0^\infty |u(r)|^p r^{n-1} dr \leq C \|u\|_H^p, \quad 2 \leq p \leq 2^*.$$

Consequently,

$$\begin{aligned}\int_{I_0} S(r)^q u(r)^q r^{n-1} dr &\leq \Big(\int_0^1 |S(r)|^{qt} r^{n-1} dr\Big)^{1/t} \Big(\int_0^1 |u(r)|^{qt'} r^{n-1} dr\Big)^{1/t'} \\ &\leq \Big(\int_0^1 |S(r)|^p r^{n-1} dr\Big)^{1/t} \|u\|_H^p,\end{aligned}$$

where $t = p/(p-q)$ and $t' = p/q$. (We need $qt' \leq 2^*$ and consequently $tq \geq 2^* q/(2^* - q)$.) Combining the two inequalities gives the desired result. □

9.18 Proofs of the Theorems

We now give the following.

Proof of Theorem 9.12 See the proof of Theorem 9.2. □

Proof of Theorem 9.14 See the proof of Theorem 9.4. □

Proof of Theorem 9.15 See the proof of Theorem 9.5. □

Proof of Theorem 9.16 See the proof of Theorem 9.6. □

Proof of Theorem 9.17 See the proof of Theorem 9.7. □

Proof of Theorem 9.18 See the proof of Theorem 9.8. □

Proof of Theorem 9.19 See the proof of Theorem 9.10. □

9.19 Notes and Remarks

Many authors have studied the problem for the Schrödinger equation (9.2) under various stipulations (cf., e.g., [2, 5, 14, 19, 47, 79, 126, 127, 142, 167, 200, 228, 231, 241, 264, 276, 287, 292] and references quoted in them). In almost all cases it was required to stipulate that the spectrum of the linear operator $\mathcal{A}u = -\Delta u + V(x)u$ have a gap. This caused writers to make various assumptions on the potential $V(x)$ to guarantee that this is the case. However, many of these assumptions caused the nature of the spectrum to be far different from that of $-\Delta u$. Thus, any theorem proved for

$$\mathcal{A}u = -\Delta u + V(x)u$$

did not hold for $\mathcal{A}u = -\Delta u$.

Some authors assumed

$$\inf_{\mathbb{R}^n} V(x) > 0; \quad \lim_{|x|\to\infty} V(x) = \infty.$$

Others assumed that there exists a constant B such that $V(x) \leq B$ for all $x \in \mathbb{R}^n$, $V(x) \to B \; as \; |x| \to \infty$ and $\sigma(-\Delta + V(x)) > 0$ together with other assumptions. Another approach assumes that for every $M > 0$ the set $\Omega = \{x \in \mathbb{R}^n : V(x) < M\}$ has finite Lebesgue measure. Others assumed that $V(x)$ is in some combination of $L^p(\mathbb{R}^n)$ spaces. In each case the growth of $f(x,t)$ is controlled by the growth of $V(x)$. In most cases the resulting spectrum of $\mathcal{A} = -\Delta + V(x)$ is discrete, consisting only of isolated eigenvalues of finite multiplicity tending to $+\infty$. All of these assumptions cause restrictions on the nonlinear term depending on V. In most cases the hypothesis

$$\mu F(x,t) \le tf(x,t), \quad |x| > R,$$

is used, where $\mu > 2$ and

$$F(x,t) = \int_0^t f(x,s)\,ds.$$

A different approach is to assume that the potential is periodic in the coordinates of $\mathbb{R}^n$ and then apply concentration compactness methods. In this case the resulting spectrum of $\mathcal{A} = -\Delta + V(x)$ is absolutely continuous and consists of a finite number of disjoint closed intervals. In order to apply this method, $f(x,t)$ must be periodic in x as well. In the few publications where 0 is permitted to be in $\sigma(-\Delta + V(x))$, an interval of the form $(-\varepsilon, 0)$ is required to be free of the spectrum.

Remark 9.30 Our theorems differ from results of several authors, including [17, 65, 107, 165, 180], who gave various conditions on the function $V(x)$ to ensure that the spectrum of $\mathcal{A}$ is discrete. We do not.

H. Berestycki and P. L. Lions [27] studied the equation

$$Au = -\Delta u = f(u), \quad u \in H^1(\mathbb{R}^n), \quad n > 2.$$

Here $V \equiv 0$ and $f(r,t) = f(t)$ is continuous and odd. Among their hypotheses are

$$-\infty < \liminf_{s \to 0^+} \frac{f(s)}{s} \le \limsup_{s \to 0^+} \frac{f(s)}{s} < 0,$$

and

$$\limsup_{s \to +\infty} \frac{f(s)}{s^l} \le 0,$$

where $l = (n+2)/(n-2)$. They showed that for this case (9.46) has a $C^2(\mathbb{R}^n)$ solution $u(r) = u(|x|)$ having exponential decay at infinity.

Chapter 10
Second Order Hamiltonian Systems

10.1 Introduction

We consider the system

$$-\ddot{x}(t) = B(t)x(t) + \nabla_x V(t, x(t)), \tag{10.1}$$

where

$$x(t) = (x_1(t), \cdots, x_n(t)) \tag{10.2}$$

is a map from $I = [0, T]$ to $\mathbb{R}^n$ such that each component $x_j(t)$ is a periodic function in H^1 with period T, and the function $V(t, x) = V(t, x_1, \cdots, x_n)$ is continuous from $\mathbb{R}^{n+1}$ to $\mathbb{R}$ with

$$\nabla_x V(t, x) = (\partial V/\partial x_1, \cdots, \partial V/\partial x_n) \in C(\mathbb{R}^{n+1}, \mathbb{R}^n). \tag{10.3}$$

For each $x \in \mathbb{R}^n$, the function $V(t, x)$ is periodic in t with period T.

We shall study this problem under several sets of assumptions. First, we make no assumption on $\nabla_x V(t, x)$ other than (10.3). This allows both sublinear and superlinear problems. The elements of the symmetric matrix $B(t)$ are to be real-valued functions $b_{jk}(t) = b_{kj}(t)$. Our assumption on $B(t)$ is as follows:

$(B1)$ Each component of $B(t)$ is an integrable function on I, i.e., for each j and k, $b_{jk}(t) \in L^1(I)$.

This assumption implies that there is an extension $\mathcal{D}$ of the operator

$$\mathcal{D}_0 x = -\ddot{x}(t) - B(t)x(t)$$

M. Schechter, *Critical Point Theory*,
https://doi.org/10.1007/978-3-030-45603-0_10

having a discrete, countable spectrum consisting of isolated eigenvalues of finite multiplicity with a finite lower bound $-L$

$$-\infty < -L \leq \lambda_0 < \lambda_1 < \lambda_2 < \ldots < \lambda_l < \ldots . \tag{10.4}$$

Let λ_l be the first positive eigenvalue of $\mathcal{D}$. We allow $\lambda_{l-1} = 0$. Let H be the set of vector functions $x(t)$ described above. It is a Hilbert space with norm satisfying

$$\|x\|_H^2 = \sum_{j=1}^{n} \|x_j\|_{H^1}^2.$$

We also write

$$\|x\|^2 = \sum_{j=1}^{n} \|x_j\|^2,$$

where $\|\cdot\|$ is the $L^2(I)$ norm. Define the subspaces M and N of H as,

$$N = \bigoplus_{k<l} E(\lambda_k), \quad M = N^{\perp}, \quad H = M \oplus N,$$

where $E(\lambda_k)$ is the eigenspace of λ_k. Let

$$G(x) = d(x) - 2\int_I V(t, x)\, dt, \tag{10.5}$$

where $d(x) = (\mathcal{D}x, x)$ (cf. Sect. 10.4). Let

$$x(t) = w(t) + v(t),\ w(t) \in M,\ v(t) \in N.$$

We write

$$G_\lambda(x) = \lambda d(w) + d(v) - 2\int_I V(t, x)\, dt, \quad 0 < \lambda < \infty. \tag{10.6}$$

We let $\mathcal{D}_\lambda$ be the operator corresponding to $d_\lambda(x) = \lambda d(w) + d(v)$.

Solutions of

$$\mathcal{D}_\lambda x(t) = \nabla_x V(t, x(t)) \tag{10.7}$$

are generalized solutions of (10.1) when $\lambda = 1$. We introduced the parameter λ to make use of the monotonicity trick of Chap. 4. This requires us to work in an interval of a parameter. This allows us to obtain solutions under very weak hypotheses. However, we obtain solutions for almost every value of the parameter. We can then obtain solutions for all values of the parameter by introducing appropriate mild assumptions.

10.2 The Case $\lambda_0 \leq 0$

In this case $l > 0$ and $\dim N > 0$. We have the following.

Theorem 10.1 *Assume*

1.

$$2V(t,x) \geq \lambda_{l-1}|x|^2 - W(t), \quad t \in I, \ x \in \mathbb{R}^n$$

and

$$2V(t,x) + \nu_l|x|^2 \to \infty, \quad |x| \to \infty,$$

where $\nu_l = \min[\lambda_l, |\lambda_{l-1}|]$ *and* $W(t) \in L^1(I)$.

2.

$$2V(t,x) \leq \mu|x|^2 + W(t), \quad t \in I, \ x \in \mathbb{R}^n,$$

where $\mu > 0$ *and* $W(t) \in L^1(I)$.

Then the system

$$\mathcal{D}_\lambda x(t) = \nabla_x V(t, x(t)) \tag{10.8}$$

has a solution for almost all $\lambda \in [\mu/\lambda_l, \infty)$.

Theorem 10.2 *Assume*

1.

$$2V(t,x) \geq \lambda_{l-1}|x|^2, \quad t \in I, \ x \in \mathbb{R}^n,$$

and

$$2V(t,x) + \nu_l|x|^2 \to \infty, \quad |x| \to \infty,$$

where $\nu_l = \min[\lambda_l, |\lambda_{l-1}|]$.

2. *There are positive constants* μ *and* m *such that*

$$2V(t,x) \leq \mu|x|^2, \quad |x| \leq m, \ x \in \mathbb{R}^n.$$

Then the system (10.8) has a solution for almost all $\lambda \in [\mu/\lambda_l, \infty)$.

Let $\mathcal{M}_\lambda$ be the set of all solutions of

$$\mathcal{D}_\lambda x(t) = \nabla_x V(t, x(t)). \tag{10.9}$$

A solution x is called a "least energy solution" if it minimizes the functional

$$G_\lambda(x) = d_\lambda(x) - 2\int_I V(t,x)\,dt \tag{10.10}$$

over the set $\mathcal{M}_\lambda$. A nontrivial least energy solution is called a "ground state solution."

Theorem 10.3 *If, in addition, we assume the following: There are a constant $\gamma > \mu$ and a function $W(t) \in L^1(I)$ such that*

$$2V(t,x) \geq \gamma|x|^2 - W(t), \quad t \in I,\ x \in \mathbb{R}^n,$$

then the system (10.8) has a nontrivial solution for almost all $\lambda \in (\mu/\lambda_l, \gamma/\lambda_l)$.

The advantage of these theorems is that we obtain solutions under very weak hypotheses. In fact, we make no assumption on $\nabla_x V(t,x)$ other than (10.3). The disadvantage is that we do not obtain a solution for any particular value of λ. If we wish to prove existence for every such λ, we will have to make assumptions concerning $\nabla_x V(t,x)$ as well. We now present additional hypotheses which guarantee existence of solutions for all values of λ in a given interval. We have the following.

Theorem 10.4 *Assume*

1.

$$2V(t,x) \geq \lambda_{l-1}|x|^2, \quad t \in I,\ x \in \mathbb{R}^n.$$

2.

$$V(t,x)/|x|^2 \to \infty, \quad \text{as } |x| \to \infty.$$

3. There are positive constants μ and m such that

$$2V(t,x) \leq \mu|x|^2, \quad |x| \leq m,\ x \in \mathbb{R}^n.$$

4. There is a function $W(t) \in L^1(I)$ such that

$$\begin{aligned} 2V(t,x+y) - 2V(t,x) - (2ry - (r-1)^2x)\cdot\nabla_x V(x,t) \\ \geq -W(t), \quad t \in I,\ x,y \in \mathbb{R}^n,\ r \in [0,1]. \end{aligned} \tag{10.11}$$

Then the system (10.8) has a nontrivial solution for all values of λ satisfying $\lambda \in [\mu/\lambda_l, \infty)$.

Theorem 10.5 *The conclusions of Theorem 10.4 hold if we replace Hypothesis 4 with*

4′. There are a constant C and a function $W(t) \in L^1(I)$ such that

$$H(t, \theta x) \le C(H(t, x) + W(t)), \quad 0 \le \theta \le 1, \ t \in I, \ x \in \mathbb{R}^n,$$

where

$$H(t, x) = \nabla_x V(t, x) \cdot x - 2V(t, x). \tag{10.12}$$

Let $\mathcal{M}_\lambda$ be the set of all solutions of

$$\mathcal{D}_\lambda x(t) = \nabla_x V(t, x(t)). \tag{10.13}$$

A solution x is called a "least energy solution" if it minimizes the functional

$$G_\lambda(x) = d_\lambda(x) - 2\int_I V(t, x)\, dt \tag{10.14}$$

over the set $\mathcal{M}_\lambda$. A nontrivial least energy solution is called a "ground state solution."

We have

Theorem 10.6 *Under the hypotheses of Theorem 10.4, system (10.13) has a ground state solution for each $\lambda \in [\mu/\lambda_l, \infty)$.*

Remark 10.7 Since $\lambda_{l-1} \le 0$, these theorems allow linear, sublinear, and superlinear growth at infinity for the problem (10.13). In particular, potentials of the form $b(t)|x|^p$ are included. The original problem (10.1) corresponds to the problem (10.13) when $\lambda = 1$. It will not be included in the interval $\lambda \in [\mu/\lambda_l, \infty)$ unless $\mu \le \lambda_l$. The monotonicity trick enables one to solve problems such as (10.13) for almost all λ in an interval when the hypotheses are too weak to solve it for definite values of λ. By adding hypotheses, one is able to solve them for specific values of λ.

10.3 The Case $\lambda_0 > 0$

In this case $l = 0$ and $N = \{0\}, \ M = H$. We have the following.

Theorem 10.8 *Assume*

1.

$$2V(t, x) + \lambda_0|x|^2 \to \infty, \quad |x| \to \infty.$$

2.

$$2V(t,x) \le \mu|x|^2 + W(t), \quad t \in I,\ x \in \mathbb{R}^n,$$

where $\mu > 0$ and $W(t) \in L^1(I)$.

Then the system (10.8) has a solution for almost all $\lambda \in [\mu/\lambda_0, \infty)$.

Theorem 10.9 *Assume*

1.

$$2V(t,x) + \lambda_0|x|^2 \to \infty, \quad |x| \to \infty.$$

2. There are positive constants μ and m such that

$$2V(t,x) \le \mu|x|^2, \quad |x| \le m,\ x \in \mathbb{R}^n.$$

Then the system (10.8) has a solution for almost all $\lambda \in [\mu/\lambda_0, \infty)$.

Theorem 10.10 *If, in addition, we assume the following:*
There are a constant $\gamma > \mu$ and a function $W(t) \in L^1(I)$ such that

$$2V(t,x) \ge \gamma|x|^2 - W(t), \quad t \in I,\ x \in \mathbb{R}^n,$$

then the system (10.8) has a nontrivial solution for almost all $\lambda \in (\mu/\lambda_0, \gamma/\lambda_0)$.

Theorem 10.11 *Assume*

1.

$$V(t,x)/|x|^2 \to \infty, \quad \text{as } |x| \to \infty.$$

2. There are positive constants μ and m such that

$$2V(t,x) \le \mu|x|^2, \quad |x| \le m,\ x \in \mathbb{R}^n.$$

3. There is a function $W(t) \in L^1(I)$ such that

$$\begin{aligned} 2V(t,x+y) - 2V(t,x) - (2ry - (r-1)^2 x) \cdot \nabla_x V(x,t) & \\ \ge -W(t), \quad t \in I,\ x,y \in \mathbb{R}^n,\ r \in [0,1]. & \end{aligned} \tag{10.15}$$

Then the system (10.8) has a nontrivial solution for all values of λ satisfying $\lambda \in [\mu/\lambda_0, \infty)$.

Theorem 10.12 *The conclusions of Theorem 10.11 hold if we replace Hypothesis 3 with*

3′. There are a constant C and a function $W(t) \in L^1(I)$ such that

$$H(t, \theta x) \leq C(H(t, x) + W(t)), \quad 0 \leq \theta \leq 1, \ t \in I, \ x \in \mathbb{R}^n,$$

where

$$H(t, x) = \nabla_x V(t, x) \cdot x - 2V(t, x). \tag{10.16}$$

Let $\mathcal{M}_\lambda$ be the set of all solutions of

$$\mathcal{D}_\lambda x(t) = \nabla_x V(t, x(t)). \tag{10.17}$$

Theorem 10.13 *Under the hypotheses of Theorem 10.11, system (10.13) has a ground state solution for each $\lambda \in [\mu/\lambda_0, \infty)$.*

We shall prove Theorems 10.1–10.10 in Sect. 10.5 and Theorems 10.4–10.6, 10.11–10.13 in Sect. 10.7. We use linking and sandwich methods of critical point theory and then apply the monotonicity trick.

10.4 The Operator $\mathcal{D}$

In proving our theorems we shall make use of the following considerations.

We define a bilinear form $a(\cdot, \cdot)$ on the set $L^2(I, \mathbb{R}^n) \times L^2(I, \mathbb{R}^n)$,

$$a(u, v) = (\dot{u}, \dot{v}) + (u, v). \tag{10.18}$$

The domain of the bilinear form is the set $D(a) = H$, consisting of those periodic $x(t) = (x_1(t), \cdots, x_n(t)) \in L^2(I, \mathbb{R}^n)$ having weak derivatives in $L^2(I, \mathbb{R}^n)$. H is a dense subset of $L^2(I, \mathbb{R}^n)$. Note that H is a Hilbert space with scalar product $(u, v)_H = a(u, v)$. Thus we can define an operator $\mathcal{A}$ such that $u \in D(\mathcal{A})$ if and only if $u \in D(a)$ and there exists $g \in L^2(I, \mathbb{R}^n)$ such that

$$a(u, v) = (g, v), \quad v \in D(a). \tag{10.19}$$

If u and g satisfy this condition we say $\mathcal{A}u = g$.

Lemma 10.14 *The operator $\mathcal{A}$ is a self-adjoint Fredholm operator from $L^2(I, \mathbb{R}^n)$ to $L^2(I, \mathbb{R}^n)$. It is one-to-one and onto.*

Proof. Let $f \in L^2(I, \mathbb{R}^n)$. Then

$$(v, f) \leq \|v\| \cdot \|f\| \leq \|v\|_H \|f\|, \quad v \in H.$$

Thus (v, f) is a bounded linear functional on H. Since H is complete, there is a $u \in H$ such that

$$(u, v)_H = (f, v), \quad v \in H.$$

Consequently, $u \in D(\mathcal{A})$ and $\mathcal{A}u = f$. Moreover, if $\mathcal{A}u = 0$, then

$$(u, v)_H = 0, \quad v \in H.$$

Thus, $u = 0$. Hence, $\mathcal{A}$ is one-to-one and onto.

For any two functions $x, y \in D(\mathcal{A})$,

$$(\mathcal{A}x, y) = (\dot{x}, \dot{y}) + (x, y) = (x, \mathcal{A}y). \tag{10.20}$$

Thus, $\mathcal{A}$ is symmetric. It is now easy to show that $D(\mathcal{A}) \subset D(a)$ is also a dense subset of $L^2(I, \mathbb{R}^n)$. In fact, if $f \in L^2(I, \mathbb{R}^n)$ satisfies $(f, v) = 0 \; \forall v \in D(\mathcal{A})$, then $w = \mathcal{A}^{-1} f$ satisfies $(w, \mathcal{A}v) = (\mathcal{A}w, v) = 0 \; \forall \; v \in D(\mathcal{A})$. Since $\mathcal{A}$ is onto, $w = 0$. Hence, $f = \mathcal{A}w = 0$.

Next, we show that $\mathcal{A}$ is self-adjoint. Consider any $u, f \in L^2(I, \mathbb{R}^n)$, and suppose for any $v \in D(\mathcal{A})$,

$$(u, \mathcal{A}v) = (f, v). \tag{10.21}$$

Since $\mathcal{A}$ is onto and $f \in L^2(I, \mathbb{R}^n)$, there exists $w \in D(\mathcal{A})$ such that $\mathcal{A}w = f$. Then using (10.20),

$$(u - w, \mathcal{A}v) = (f, v) - (\mathcal{A}w, v) = 0.$$

Since $u - w \in L^2(I, \mathbb{R}^n)$, we can find a $v \in D(\mathcal{A})$ such that $\mathcal{A}v = u - w$, and

$$\|u - w\|^2 = 0.$$

This implies $u = w$ in the space $L^2(I, \mathbb{R}^n)$, and therefore $u \in D(\mathcal{A})$. Hence, $\mathcal{D}u = \mathcal{A}w = f$. □

Lemma 10.15 *The essential spectrum of $\mathcal{A}$ is the null set.*

The essential spectrum of a linear operator is the set of all points of the spectrum which are not isolated eigenvalues of finite multiplicity.

Proof. By (10.14), $\mathcal{A}$ is linear, self-adjoint, and onto $L^2(I, \mathbb{R}^n)$.

Next, we note that

$$\| \mathcal{A}^{-1} u \| \leq \|u\|.$$

To see this, let $f = \mathcal{A}u$. Then $u = \mathcal{A}^{-1} f$, and

$$(u, v)_H = (f, v), \quad v \in H.$$

Thus,

$$\|u\|_H^2 \leq \|f\| \cdot \|u\| \leq \|f\| \cdot \|u\|_H.$$

Hence, $\|u\| \leq \|f\|$.

Now we show that the inverse operator $\mathcal{A}^{-1}$ is compact on $L^2(I, \mathbb{R}^n)$. Let (u_k) be a bounded sequence in $L^2(I, \mathbb{R}^n)$, and let $C > 0$ satisfy for each k, $\|u_k\| \leq C$. By applying the inverse operator, let (x_k) be the sequence such that for each k, $\mathcal{A}x_k = u_k$. From the above statements, for each k, $\|x_k\| \leq C$. From the definition of the operator $\mathcal{A}$, for any $x \in D(\mathcal{A})$,

$$(\mathcal{A}x, x) = (\dot{x}, \dot{x}) + (x, x) = \|x\|_H^2 \geq 0.$$

Hence, $\mathcal{K} = \mathcal{A}^{-1}$ is a positive compact operator, and the eigenvalues μ_k of $\mathcal{K}$ are denumerable and have 0 as their only possible limit point. The eigenfunctions ϕ_k of $\mathcal{K}$ are also eigenfunctions of $\mathcal{K}^{-1} = \mathcal{A}$ and satisfy

$$\mathcal{A}\phi_k = \frac{1}{\mu_k}\phi_k.$$

Since the values μ_k are bounded and have no limit point except 0, there are no limit points of the set $(1/\mu_k)$ and the essential spectrum of $\mathcal{A}$ is the null set. □

We will use two theorems of Schechter [182] on bilinear forms to prove Lemma 10.18.

Theorem 10.16 *Let $a(\cdot, \cdot)$ be a closed Hermitian bilinear form with dense domain in $L^2(I, \mathbb{R}^n)$. If for some real number N,*

$$a(u, u) + N\|u\|^2 \geq 0, \tag{10.22}$$

then the operator $\mathcal{A}$ associated with $a(\cdot, \cdot)$ is self-adjoint and $\sigma(A) \subset [-N, \infty)$.

A bilinear form $a(u, v)$ is closed if $a(u_j - u_k) \to 0$, $u_k \to u$ imply $u \in D(a)$ and $a(u_k - u) \to 0$.

Theorem 10.17 *Suppose $a(\cdot, \cdot)$ is a bilinear form satisfying the hypotheses of Theorem 10.16. Let $b(\cdot, \cdot)$ be a Hermitian bilinear form such that $D(a) \subset D(b)$ and for some positive real number K, for any $u \in D(a)$,*

$$|b(u, u)| \leq Ka(u, u). \tag{10.23}$$

Assume that every sequence $(u_k) \subset D(a)$ which satisfies

$$\|u_k\|^2 + a(u_k, u_k) \leq C \tag{10.24}$$

has a subsequence (v_j) *such that*

$$b(v_j - v_k,\ v_j - v_k) \rightarrow 0\,. \tag{10.25}$$

Assume also that if (10.24) and (10.25) *hold and* $v_j \rightarrow 0$ *in the* $L^2(I, \mathbb{R}^n)$ *norm, then* $b(v_j, v_j) \rightarrow 0$. *Set*

$$c(u, v) = a(u, v) + b(u, v)\,. \tag{10.26}$$

and let $\mathcal{A}, \mathcal{C}$ *be the operators associated with* a, c, *respectively. Then*

$$\sigma_e(\mathcal{A}) = \sigma_e(\mathcal{C})\,.$$

Let

$$b(u, v) = -\sum_{j=1}^{n}\sum_{k=1}^{n}\int_0^T \left(b_{jk}(t) + \delta_{jk}\right) u_k(t) v_j(t)\, dt \tag{10.27}$$

and

$$d(u, v) = a(u, v) + b(u, v). \tag{10.28}$$

We shall prove the following.

Lemma 10.18 *The operator* $\mathcal{D}$ *associated with the bilinear form* $d(\cdot, \cdot)$ *under assumption (B1) is self-adjoint. Its essential spectrum is the null set and there exists a finite real value L such that* $\sigma(\mathcal{D}) \subset [-L, \infty)$. $\mathcal{D}$ *has a discrete, countable spectrum consisting of isolated eigenvalues of finite multiplicity with a finite lower bound -L*

$$-\infty < -L \le \lambda_0 < \lambda_1 < \lambda_2 < \ldots < \lambda_l < \ldots\,. \tag{10.29}$$

To show the bilinear form $b(\cdot, \cdot)$ is Hermitian, we can use the symmetry of the matrix $B(t) + I$ to rearrange the order of the finite summation,

$$\begin{aligned} b(u, v) &= -\sum_{j=1}^{n}\sum_{k=1}^{n}\int_0^T \left(b_{jk}(t) + \delta_{jk}\right) u_k(t) v_j(t)\, dt \\ &= -\sum_{k=1}^{n}\sum_{j=1}^{n}\int_0^T \left(b_{jk}(t) + \delta_{jk}\right) v_j(t) u_k(t)\, dt \\ &= -\sum_{k=1}^{n}\sum_{j=1}^{n}\int_0^T \left(b_{kj}(t) + \delta_{kj}\right) v_j(t) u_k(t)\, dt \end{aligned}$$

$$= b(v, u) .$$

Since

$$\|u\|_\infty = \|u\|_{L^\infty(I,\mathbb{R}^n)} \le C\|u\|_H, \quad u \in H, \tag{10.30}$$

the magnitude of $b(u) = b(u, u)$ is bounded by a multiple of the bilinear form $a(\cdot, \cdot)$ and satisfies (10.23),

$$\begin{aligned} |b(u)| &\le K_B\|u\|_\infty^2 && (10.31)\\ &\le K_B (M\|u\|_H)^2 \\ &\le K_B \cdot M^2\|u\|_H^2 = Ka(u) . \end{aligned}$$

Consider a sequence $(x_k) \subset D(\mathcal{A})$ which is bounded by a constant C in the H norm. Then each term of the sequence satisfies

$$\|x_k\|^2 + a(x_k) = 2(x_k, x_k) + (\dot{x}_k, \dot{x}_k) \le 2\|x_k\|_H^2 \le 4C^2 .$$

Hence, we can find a subsequence $(x_{\bar{k}})$ which converges weakly in H and strongly in $L^\infty(I, \mathbb{R}^n)$ and $L^2(I, \mathbb{R}^n)$ to some function $x \in H$. Because the subsequence is convergent in $L^\infty(I, \mathbb{R}^n)$ it is also Cauchy under this norm. As $\bar{j}, \bar{k} \to \infty$ we can apply (10.31) to show this subsequence satisfies (10.25),

$$|b(x_{\bar{j}} - x_{\bar{k}})| \le K_B\|x_{\bar{j}} - x_{\bar{k}}\|_{L^\infty(I,\mathbb{R}^n)}^2 \to 0 . \tag{10.32}$$

If in addition the subsequence $(x_{\bar{k}})$ converges to zero in $L^2(I, \mathbb{R}^n)$, the subsequence must also converge in $L^\infty(I, \mathbb{R}^n)$ to the zero function, and

$$b(x_{\bar{k}}) \to 0 .$$

Then the bilinear forms $a(\cdot, \cdot)$ and $b(\cdot, \cdot)$ satisfy the conditions of Theorem 10.17. The bilinear form $d(\cdot, \cdot)$ is the sum of these two bilinear forms as in (10.28). By this theorem, the operator $\mathcal{D}$ associated with this bilinear form has the same essential spectrum as the operator $\mathcal{A}$ associated with the bilinear form $a(\cdot, \cdot)$.

Now we show that for any constant $\epsilon > 0$ there exists a positive constant K_ϵ such that

$$|b(x)| \le \epsilon\|\dot{x}\|^2 + K_\epsilon\|x\|^2 \quad x \in D(\mathcal{A}). \tag{10.33}$$

We can find a constant K_B satisfying (10.31). For any $\epsilon > 0$, let $\xi = \epsilon/K_B$. Then there is a constant C_ξ which satisfies

$$|b(x)| \le K_B\|x\|_{L^\infty(I,\mathbb{R}^n)}^2$$

$$\leq K_B \left(\frac{\epsilon}{K_B} \|\dot{x}\|^2 + C_\xi \|x\|^2 \right)$$

$$\leq \epsilon \|\dot{x}\|^2 + \left(K_B \cdot C_\xi) \|x\|^2 \right) .$$

Setting $K_\epsilon = K_B \cdot C_\xi$ gives the stated inequality. To show $d(\cdot, \cdot)$ is closed, first apply this inequality with $\epsilon = 1/2$. Thus there is a constant C_0 such that

$$| b(u) | \leq \frac{1}{2} a(u) + C_0 \|u\|^2 . \tag{10.34}$$

Now suppose a sequence $(u_k) \subset D(d)$ satisfies

$$d(u_j - u_k) \to 0 , \tag{10.35}$$

and $(u_k) \to u$ in $L^2(I, \mathbb{R}^n)$. The sequence is Cauchy in $L^2(I, \mathbb{R}^n)$ and as j, k increase

$$\| u_j - u_k \|^2 \to 0 .$$

Suppose that $u \notin D(d)$. Because the domains of $d(\cdot, \cdot)$ and $a(\cdot, \cdot)$ are the same, $u \notin D(a)$. We have shown above that $a(\cdot, \cdot)$ is closed, so the sequence cannot be Cauchy and as j, k increase $a(u_j - u_k)$ does not approach zero. But by (10.35),

$$a(u_j - u_k) - b(u_j - u_k) \to 0 .$$

Applying the inequality in (10.34) bounds the magnitude of each $b(\cdot, \cdot)$ term, and since $a(u, u) \geq 0$, the following inequality is satisfied,

$$a(u_j - u_k) - b(u_j - u_k) \geq \tfrac{1}{2} a(u_j - u_k) - C_0 \|u_j - u_k\|^2 .$$

Adding the last term to both sides leaves only the positive bilinear form on the right side,

$$\begin{aligned} a(u_j - u_k) - b(u_j - u_k) + C_0 \|u_j - u_k\|^2 & \\ \geq \tfrac{1}{2} a(u_j - u_k) & \\ \geq 0 . & \end{aligned}$$

As j, k increase the left side of this equation approaches zero so the center term must also approach zero, a contradiction to the statement above. Therefore, $u \in D(a) = D(d)$, and $d(\cdot, \cdot)$ is also a closed bilinear form.

Next we show that there exists a positive constant N such that for any $x \in D(a)$,

$$d(x) + N\|x\|^2 \geq 0 . \tag{10.36}$$

For any positive constant $\epsilon > 0$ we can find K_ϵ which satisfies (10.33) and thereby find a lower bound for $b(x, x)$,

$$a(x) + b(x) + N\|x\|^2 \geq a(x) - \epsilon\|\dot{x}\|^2 - K_\epsilon\|x\|.$$

We have shown that $d(\cdot, \cdot)$ is closed, and as the sum of two Hermitian bilinear forms, $d(\cdot, \cdot)$ is clearly Hermitian. Its domain is dense in $L^2(I, \mathbb{R}^n)$ and the N in (10.36) satisfies the conditions of Theorem 10.16, so the operator $\mathcal{D}$ associated with this bilinear form is self-adjoint and has its spectrum bounded below by $-N$. We have shown that the essential spectrum of this operator is the null set, so the spectrum is discrete and we can number the eigenvalues in increasing order, and each eigenvalue is of finite multiplicity.

10.5 Proofs of the Theorems

We now give the proof of Theorem 10.1.

Proof. Let

$$I(x) = d(w), \quad J(x) = -d(v) + 2\int_I V(t, x(t))\,dt.$$

Thus,

$$G_\lambda(x) = \lambda I(x) - J(x), \quad x \in H. \tag{10.37}$$

By Hypothesis 1,

$$I(x) + J(x) \to \infty \text{ as } \|x\|_\infty \to \infty.$$

To see this, note that

$$d(w) - d(v) + 2\int_I V(t, x) \to \infty, \quad \|x\|_H \to \infty.$$

This follows from the fact that

$$2V(t, x) + \nu_l|x|^2 \to \infty, \quad |x| \to \infty,$$

where $\nu_l = \min[\lambda_l, |\lambda_{l-1}|]$. Hence,

$$d(w) - d(v) + 2\int_I V(t, x) \geq \nu_l\|x\|^2 + 2\int_I V(t, x)\,dt \to \infty, \quad \|x\|_H \to \infty.$$

We note that Hypothesis 1 implies

$$G_\lambda(v) \leq Q, \quad v \in N, \tag{10.38}$$

where

$$Q = \int_I W(t)\,dt.$$

In fact, we have

$$G_\lambda(x) = d(x) - 2\int_I V(t,x)\,dt \leq \int_I [\lambda_{l-1}|x|^2 - 2V(t,x)]\,dt \leq Q, \quad x \in N.$$

If $x \in M$, we have by Hypothesis 2 that

$$\begin{aligned} G_\lambda(x) &\geq \lambda d(x) - \int \mu |x(t)|^2\,dt - Q \\ &\geq (\lambda\lambda_l - \mu)\|x\|^2 - Q \geq -Q, \end{aligned} \tag{10.39}$$

provided $\lambda \geq \mu/\lambda_l$.

By Corollary 3.5, M and N form a sandwich pair. Then by Corollary 4.12, for almost every $\lambda \in [\mu/\lambda_l, \infty)$ there is a bounded sequence $\{x^{(k)}\} \subset H$ such that

$$G_\lambda(x^{(k)}) = d_\lambda(x^{(k)}) - 2\int_I V(t, x^{(k)}(t))\,dt \to c, \tag{10.40}$$

$$(G'_\lambda(x^{(k)}), z)/2 = d_\lambda(x^{(k)}, z) - \int_I \nabla_x V(t, x^{(k)}) \cdot z(t)\,dt \to 0, \quad z \in H \tag{10.41}$$

and

$$(G'_\lambda(x^{(k)}), x^{(k)})/2 = d_\lambda(x^{(k)}) - \int_I \nabla_x V(t, x^{(k)}) \cdot x^{(k)}\,dt \to 0, \tag{10.42}$$

where $-Q \leq c \leq Q$. Since

$$\rho_k = \|x^{(k)}\|_H \leq C,$$

there is a renamed subsequence such that $x^{(k)}$ converges to a limit $x \in H$ weakly in H and uniformly on I. From (10.42) we see that

$$(G'_\lambda(x), z)/2 = d_\lambda(x, z) - \int_I \nabla_x V(t, x(t)) \cdot z(t)\,dt = 0, \quad z \in H,$$

from which we conclude easily that x is a solution of (10.8). □

Proof of Theorem 10.2 We note that Hypothesis 1 implies

$$G_\lambda(v) \le 0, \quad v \in N. \tag{10.43}$$

In fact, we have

$$G_\lambda(x) = d(x) - 2\int_I V(t,x)\,dt \le \int_I [\lambda_{l-1}|x|^2 - 2V(t,x)]\,dt \le 0, \quad x \in N.$$

Note that there is a positive $\rho > 0$ such that

$$|x(t)| < m$$

when $\|x\|_H = \rho$. In fact, we have $|x(t)| \le c_0\|x\|_H$. If $\lambda > \mu/\lambda_l$ and $x \in M$, then

$$G_\lambda(x) = d_\lambda(x) - 2\int_I V(t,x)\,dt \ge d(x)[\lambda - \frac{\mu\|x\|^2}{d(x)}] > \varepsilon > 0.$$

Take

$$A = \partial\mathcal{B}_\rho \cap M,$$
$$B = N,$$

where

$$\mathcal{B}_\rho = \{x \in H : \|x\|_H < \rho\}.$$

By Example 2 of Chap. 3, A links B. Moreover,

$$\sup_A[-G_\lambda] \le 0 \le \inf_B[-G_\lambda]. \tag{10.44}$$

Hence, we may apply Corollary 4.13 to conclude that for a.e. $\lambda \in [\mu/\lambda_l, \infty)$ there is a bounded sequence $\{x^{(k)}\} \subset H$ such that

$$G_\lambda(x^{(k)}) = d_\lambda(x^{(k)}) - 2\int_I V(t, x^{(k)}(t))\,dt \to c \le 0, \tag{10.45}$$

$$(G'_\lambda(x^{(k)}), z)/2 = d_\lambda(x^{(k)}, z) - \int_I \nabla_x V(t, x^{(k)}) \cdot z(t)\,dt \to 0, \quad z \in H \tag{10.46}$$

and

$$(G'_\lambda(x^{(k)}), x^{(k)})/2 = d_\lambda(x^{(k)}) - \int_I \nabla_x V(t, x^{(k)}) \cdot x^{(k)}\,dt \to 0. \tag{10.47}$$

Since

$$\rho_k = \|x^{(k)}\|_H \le C,$$

there is a renamed subsequence such that $x^{(k)}$ converges to a limit $x \in H$ weakly in H and uniformly on I. From (10.41) we see that

$$(G'_\lambda(x), z)/2 = d_\lambda(x, z) - \int_I \nabla_x V(t, x(t)) \cdot z(t)\, dt = 0, \quad z \in H,$$

from which we conclude easily that x is a solution of (10.8). □

Proof of Theorem 10.3 Let

$$y(t) = v + sw_0,$$

where $v \in N$, $s \ge 0$, and $w_0 \in M$ is an eigenfunction of $\mathcal{D}$ corresponding to λ_l. Consequently,

$$\begin{aligned} G_\lambda(y) &= s^2\lambda d(w_0) + d(v) - 2\int_I V(t, y(t))\, dt \\ &\le \lambda\lambda_l s^2\|w_0\|^2 + \lambda_{l-1}\|v\|^2 - \gamma \int_I |y(t)|^2\, dt + Q \\ &\le (\lambda_{l-1} - \gamma)\|v\|^2 + (\lambda\lambda_l - \gamma)s^2\|w_0\|^2 + Q \\ &\to -\infty \text{ as } s^2 + |v|^2 \to \infty, \end{aligned}$$

provided $\lambda < \gamma/\lambda_l$, where

$$Q = \int_I W(t)\, dt.$$

Take

$$\begin{aligned} A &= \{v \in N : \|v\|_H \le R\} \cup \{sw_0 + v : v \in N, s \ge 0, \|sw_0 + v\|_H = R\}, \\ B &= \partial\mathcal{B}_\rho \cap M, \ 0 < \rho < R. \end{aligned}$$

By Example 3 of Chap. 3, A links B. Moreover, if R is sufficiently large,

$$\sup_A G_\lambda \le 0 < \varepsilon \le \inf_B G_\lambda. \tag{10.48}$$

We may now apply Corollary 4.13 to conclude that for almost all $\lambda \in [\mu/\lambda_l, \gamma/\lambda_l]$ there is a bounded sequence $\{x^{(k)}\} \subset H$ such that

$$G_\lambda(x^{(k)}) = d_\lambda(x^{(k)}) - 2\int_I V(t, x^{(k)}(t))\, dt \to c \ge \varepsilon > 0, \tag{10.49}$$

$$(G'_\lambda(x^{(k)}), z)/2 = d_\lambda(x^{(k)}, z) - \int_I \nabla_x V(t, x^{(k)}) \cdot z(t)\, dt \to 0, \quad z \in H \tag{10.50}$$

and

$$(G'_\lambda(x^{(k)}), x^{(k)})/2 = d_\lambda(x^{(k)}) - \int_I \nabla_x V(t, x^{(k)}) \cdot x^{(k)}\, dt \to 0. \tag{10.51}$$

Since

$$\rho_k = \|x^{(k)}\|_H \le C,$$

there is a renamed subsequence such that $x^{(k)}$ converges to a limit $x \in H$ weakly in H and uniformly on I. From (10.50) we see that

$$(G'_\lambda(x), z)/2 = d_\lambda(x, z) - \int_I \nabla_x V(t, x(t)) \cdot z(t)\, dt = 0, \quad z \in H,$$

from which we conclude easily that x is a solution of (10.8). Moreover, (10.51) implies

$$d_\lambda(x^{(k)}) \to \int_I \nabla_x V(t, x) \cdot x\, dt = d_\lambda(x). \tag{10.52}$$

Consequently,

$$x^{(k)} \to x \tag{10.53}$$

strongly in H. This means that

$$G_\lambda(x) = d_\lambda(x) - 2\int_I V(t, x)\, dt = c \ge \varepsilon > 0. \tag{10.54}$$

But

$$G_\lambda(0) = -2\int_I V(t, 0)\, dt \le 0.$$

Hence, $x(t) \not\equiv 0$.

The proofs of Theorems 10.8–10.10 are similar to those of Theorems 10.1–10.3 taking into account that $\lambda_0 > 0$ and $N = \{0\}$. □

10.6 Some Lemmas

Before giving the remaining proofs, we shall prove a few lemmas.

Lemma 10.19

$$\int_I [V(t,u) - V(t,rw) + (r^2 w - \frac{1+r^2}{2} u)V(t,u)] \leq C,$$

$$u \in D,\ w \in M,\ r \in [0,1],\ \|w\|_D \leq \|u\|_D,$$

where the constant C does not depend on u, w, r.

Proof. This follows from (10.11) if we take $t = u$, and $s = rw - u$. □

Lemma 10.20 *If u satisfies $G'_\lambda(u) = 0$ for some $\lambda > 0$, then there is a constant C independent of u, λ, r such that*

$$G_\lambda(rw) - r^2(\mathcal{D}v, v) - G_\lambda(u) \leq C \tag{10.55}$$

for all $r \in [0,1]$, where w,v are the projections of u onto M, N, respectively.

Proof. For such u, let $u = v + w$, where $v \in N,\ w \in M$. Then

$$(G'_\lambda(u), g)/2 = \lambda(\mathcal{D}w, g_1) + (\mathcal{D}v, g_2) - \int gV(t,u) = 0$$

for every $g \in D$, where g_1, g_2 are the projections of g onto M, N, respectively. Take

$$g = (r^2+1)v - (r^2-1)w = (r^2+1)u - 2r^2 w.$$

Then we have

$$\begin{aligned} G_\lambda(rw) - r^2(\mathcal{D}v, v) - G_\lambda(u) &= \lambda(r^2-1)(\mathcal{D}w, w) - (\mathcal{D}v, v) \\ &\quad + \lambda(\mathcal{D}w, g_1) + (\mathcal{D}v, g_2) - r^2(\mathcal{D}v, v) \\ &\quad + \int_I [2V(t,u) - 2V(t,rw) - gV(t,u)]\,dx \\ &= \int_I [2V(t,u) - 2V(t,rw) - ((r^2+1)u - 2r^2 w)V(t,u)]\,dx \\ &\leq C \end{aligned}$$

by Lemma 10.19. □

10.7 The Remaining Proofs

Proof of Theorem 10.4 Let $\tilde{\lambda} = \mu/\lambda_l$ and $\nu < \infty$. By Theorem 10.3, for a.e. $\lambda \in (\tilde{\lambda}, \nu)$, there exists u_λ such that $G'_\lambda(u_\lambda) = 0, \quad G_\lambda(u_\lambda) = a(\lambda) \geq a(\lambda_0)$. Let λ satisfy $\lambda_0 \leq \lambda < \nu$. Choose $\lambda_n \to \lambda,\ \lambda_n > \lambda$ such that there exists x_n satisfying

$$G'_{\lambda_n}(x_n) = 0, \quad G_{\lambda_n}(x_n) = a(\lambda_n) \geq a(\lambda_0).$$

Therefore,

$$\int_I \frac{2V(t, x_n)}{\|x_n\|_H^2} dt \leq C.$$

Now we prove that $\{x_n\}$ is bounded. If $\|x_n\|_H \to \infty$, let $\tilde{x}_n = x_n/\|x_n\|_H$. Then there is a renamed subsequence such that $\tilde{x}_n \to \tilde{x}$ weakly in H, strongly in $L^\infty(I)$, and a.e. in I. Let Ω_0 be the set where $\tilde{x} \neq 0$. Then $|x_n(t)| \to \infty$ for $t \in \Omega_0$. If Ω_0 had positive measure, then we would have

$$C \geq \int_I \frac{2V(t, x_n)}{\|x_n\|_H^2} dt = \int_I \frac{2V(t, x_n)}{|x_n|^2} |\tilde{x}_n|^2 dt$$

$$\geq \int_{\Omega_0} \frac{2V(t, x_n)}{|x_n|^2} |\tilde{x}_n|^2 dt + \lambda_{l-1} \int_{I \backslash \Omega_0} |\tilde{x}_n|^2 \, dt \to \infty,$$

showing that $\tilde{x} = 0$ a.e. in $L^\infty(I)$. Hence, $\tilde{x}_n \to 0$ in $L^\infty(I)$. Since

$$\|\tilde{x}_n\|_H^2 = d(\tilde{w}_n) - d(\tilde{v}_n) + \|\tilde{g}_n\|^2 = 1, \tag{10.56}$$

and $d(\tilde{v}_n) \to 0,\ \|\tilde{g}_n\|^2 \to 0$, we have $d(\tilde{w}_n) \to 1$. For any $s > 0$ and $h_n = s\tilde{x}_n$, we have

$$\int_I V(t, h_n) \, dt \to \int_I V(t, 0) \, dt = 0. \tag{10.57}$$

Take $r_n = s/R_n \to 0$. By Lemma 10.20

$$G_{\lambda_n}(r_n w_n) - r_n^2 (\mathcal{D} v_n, v_n) - G_{\lambda_n}(x_n) \leq C. \tag{10.58}$$

Hence,

$$G_{\lambda_n}(s\tilde{w}_n) - s^2 (\mathcal{D}\tilde{v}_n, \tilde{v}_n) \leq C'. \tag{10.59}$$

But

$$\begin{aligned} G_{\lambda_n}(s\tilde{w}_n) - s^2(\mathcal{D}\tilde{v}_n, \tilde{v}_n) =& \lambda_n s^2(\mathcal{D}\tilde{w}_n, \tilde{w}_n) - s^2(\mathcal{D}\tilde{v}_n, \tilde{v}_n) \\ & -2\int_I V(t, s\tilde{w}_n) \\ \geq & s^2(\lambda d(\tilde{w}_n) - d(\tilde{v}_n)) - 2\int_I V(t, s\tilde{w}_n) \\ & \to \lambda s^2 \end{aligned}$$

by (10.57). This implies

$$G_{\lambda_n}(s\tilde{w}_n) - s^2(\mathcal{D}\tilde{v}_n, \tilde{v}_n) \to \infty \quad as\ s \to \infty,$$

contrary to (10.59).

This contradiction shows that $\|x_n\|_H \leq C$. Then there is a renamed subsequence such that $x_n \to x$ weakly in H, strongly in $L^\infty(I)$, and a.e. in I. It now follows that for the bounded renamed subsequence,

$$G'_\lambda(x_n) \to 0, \quad G_\lambda(x_n) \to a(\lambda) \geq a(\lambda_0).$$

We can now apply Theorem 3.4.1 in [184, p. 64] to obtain the desired solution. □

Proof of Theorem 10.5 We follow the proof of Theorem 10.4 until (10.56) from which we conclude that $d(\tilde{w}_n) \to 1$. We define $\theta_n \in [0, 1]$ by

$$G_{\lambda_n}(\theta_n x_n) = \max_{\theta \in [0,1]} G_{\lambda_n}(\theta x_n).$$

For any $c > 0$ and $h_n = c\tilde{x}_n$, we have

$$\int_I V(t, h_n)\, dt \to \int_I V(t, 0)\, dt = 0.$$

Thus,

$$G_{\lambda_n}(\theta_n x_n) \geq G_{\lambda_n}(c\tilde{x}_n) = c^2\lambda_n d(\tilde{w}_n) - 2\int_I V(t, h_n)\, dt \to \lambda c^2, \quad n \to \infty.$$

Hence, $G_{\lambda_n}(\theta_n x_n) \geq \lambda c^2/2$ for n sufficiently large. That is, $\lim_{n\to\infty} G_{\lambda_n}(\theta_n x_n) = \infty$. If there is a renamed subsequence such that $\theta_n = 1$, then

$$G_{\lambda_n}(x_n) \to \infty. \tag{10.60}$$

If $0 \leq \theta_n < 1$ for all n, then we have $(G'_{\lambda_n}(\theta_n x_n), x_n) \leq 0$. Therefore,

$$\int_I H(t, \theta_n x_n)\,dt = \int_I \Big(\nabla_x V(t, \theta_n x_n)\theta_n x_n - 2V(t, \theta_n x_n)\Big)\,dt$$
$$= G_{\lambda_n}(\theta_n x_n) - (G'_{\lambda_n}(\theta_n x_n), \theta_n x_n)$$
$$\geq G_{\lambda_n}(\theta_n x_n) \to \infty.$$

By hypothesis,

$$G_{\lambda_n}(x_n) = \int_I H(t, x_n)\,dx$$
$$\geq \int_I H(t, \theta_n x_n)\,dt/C - \int_I W(t)\,dt \to \infty.$$

Thus, (10.60) holds in any case. But

$$G_{\lambda_n}(x_n) = a(\lambda_n) \leq a(\nu) < \infty.$$

This contradiction shows that $\|x_n\|_H \leq C$. It now follows that for a renamed subsequence,

$$G'_\lambda(x_n) \to 0, \quad G_\lambda(x_n) \to a(\lambda) \geq a(\lambda_0).$$

We can now apply Theorem 3.4.1 in [184, p. 64] to obtain the desired solution. □

Proof of Theorem 10.6 We may assume $\lambda = 1$. By the previous proof, $\mathcal{M}_\lambda \neq \phi$. Let

$$\alpha = \inf_{\mathcal{M}_\lambda} G(x).$$

There is a sequence $\{x^{(k)}\} \in \mathcal{M}_\lambda$ such that

$$G(x^{(k)}) = d(x^{(k)}) - 2\int_I V(t, x^{(k)}(t))\,dt \to \alpha, \tag{10.61}$$

$$(G'(x^{(k)}), z)/2 = d(x^{(k)}, z) - \int_I \nabla_x V(t, x^{(k)}) \cdot z(t)\,dt = 0, \quad z \in H \tag{10.62}$$

and

$$(G'(x^{(k)}), x^{(k)})/2 = d(x^{(k)}) - \int_I \nabla_x V(t, x^{(k)}) \cdot x^{(k)}\,dt = 0. \tag{10.63}$$

By the previous proof, there is a renamed subsequence such that

$$\rho_k = \|x^{(k)}\|_H \leq C.$$

Hence, there is a renamed subsequence such that $x^{(k)}$ converges to a limit $x \in H$ weakly in H and uniformly on I. Moreover,

$$\int_I H(t, x^{(k)}(t))\, dt = G(x^{(k)}) \to \alpha.$$

From (10.61) and (10.62) we see that

$$G(x) = d(x) - 2\int_I V(t, x(t))\, dt \leq \alpha, \tag{10.64}$$

and

$$(G'(x), z)/2 = d(x, z) - \int_I \nabla_x V(t, x(t)) \cdot z(t)\, dt = 0, \quad z \in H,$$

from which we conclude easily that x is a solution of (10.1). Hence, $x \in \mathcal{M}_\lambda$ and $G(x) = \alpha$. This completes the proof.

The proofs of Theorems 10.11–10.13 are similar to those of Theorems 10.4–10.6 taking into account that $\lambda_0 > 0$ and $N = \{0\}$. □

10.8 Notes and Remarks

In this chapter we followed [218]. In [178] the hypotheses are as follows:

(1) There exist functions $W_1, W_2 \in L^1(I)$ and consecutive eigenvalues λ_l, λ_{l+1} of $\mathcal{D}$ such that for any $t \in I$ and any $x \in \mathbb{R}^n$,

$$\lambda_l |x|^2 - W_1(t) \leq 2V(t, x) \leq \lambda_{l+1}|x|^2 + W_2(t). \tag{10.65}$$

(2) There exists a function $W_0(t) \in L^1(I)$ such that for any $t \in I$ and any $x \in \mathbb{R}^n$,

$$H(t, x) = 2V(t, x) - x \cdot \nabla_x V(t, x) \geq -W_0(t).$$

(3) $H(t, x) \to \infty$ uniformly in t as $|x| \to \infty$.

The periodic nonautonomous problem

$$\ddot{x}(t) = \nabla_x V(t, x(t)), \tag{10.66}$$

has an extensive history in the case of singular systems (cf., e.g., Ambrosetti-Coti Zelati [3]). The first to consider it for potentials satisfying (10.3) were Berger and the author [33] in 1977. We proved the existence of solutions to (10.8) under the condition that

$$V(t,x) \to \infty \quad \text{as} \quad |x| \to \infty$$

uniformly for a.e. $t \in I$. Subsequently, Willem [275], Mawhin [153], Mawhin-Willem [158], Tang [246, 263], Tang-Wu [250, 251], Wu-Tang [279], and others proved existence under various conditions (cf. the references given in these publications).

Most previous work considered the case when $B(t) = 0$. The first to study (10.1) with $B(t) \neq 0$ were Berger and the author [33]. We considered the case of (10.1) where $B(t)$ is a constant symmetric matrix that is positive definite, and showed existence of solutions when the magnitude of $\nabla_x V(t,x)$ is uniformly bounded, the potential is strictly convex, and if $y(t)$ is a T-periodic solutions of the linear system $-\ddot{y} = Ay$, then there exists a function $x(t)$ which is weakly differentiable with $\dot{x} \in L^2(\mathbb{R}, \mathbb{R}^n)$ and satisfies

$$\int_0^T \langle \nabla_x V(t, x(t)),\ y(t) \rangle_I \, dt \;=\; 0.$$

Ding and Girardi [74] considered the case of (10.1) when the potential oscillates in magnitude and sign,

$$-\ddot{x}(t) \;=\; B(t)x(t) \;+\; b(t)\nabla W(x(t)) \,, \tag{10.67}$$

and found conditions for solutions when the matrix $B(t)$ is symmetric and negative definite and the function $W(x)$ grows superquadratically and satisfies a homogeneity condition. Antonacci [6] gave conditions for existence of solutions with stronger constraints on the potential but without the homogeneity condition, and without the negative definite condition on the matrix. Generalizations of the above results are given by Antonacci and Magrone [7], Barletta and Livrea [10], Guo and Xu [113], Li and Zou [141], Faraci and Livrea [95], Bonanno and Livrea [37, 38], Jiang [121, 122], Shilgba [233, 234], Faraci and Iannizzotto [94], and Tang and Xiao [256].

Some authors considered the second order system (10.1) where the potential function $V(t,x)$ is quadratically bounded as $|x| \to \infty$. Han [114] gave conditions for existence of solutions when $B(t)$ was a multiple of the identity matrix, the system satisfies the resonance condition, and the potential has upper and lower subquadratic bounds. Li and Zou [141] considered the case where $B(t)$ is continuous and nonconstant and the system satisfies the resonance condition, and showed existence of solutions when the potential is even and grows no faster than linearly. Tang and Wu [251] required the function that satisfies the resonance condition to pass through the zero vector, and gave upper and lower conditions for subquadratic growth of the magnitude of $V(t,x)$ without the requirement that the potential be even. Faraci [93] considered the case where for each $t \in I$, $B(t)$ is negative definite with elements that are bounded but not necessarily continuous and the potential has an upper quadratic bound as $|x| \to \infty$, showing existence of a solution when the gradient of the potential is bounded near the origin and exceeds the matrix product in at least one direction.

The monotonicity trick was introduced by Struwe in [239, 240] for minimization problems. (This trick was also used by others to solve Landesman-Lazer type problems, for bifurcation problems, for Hamiltonian systems, and Schrödinger equations.)

The theory of sandwich pairs began in [235] and [186, 187] and was developed in subsequent publications such as [189, 193].

Chapter 11
Core Functions

11.1 Use of Core Functions

Let $\mathcal{A}$ be a self-adjoint operator on $L^2(\Omega)$, where Ω is a bounded domain in $\mathbb{R}^n$. Let $f(x,t)$ be a Carathéodory function on $\Omega \times \mathbb{R}$. A well-known semilinear problem is to solve

$$\mathcal{A}u = f(x,u), \quad u \in D(\mathcal{A}). \tag{11.1}$$

In particular, one searches for properties of $f(x,t)$ which guarantee the existence of solutions. This is not a trivial situation; there does not appear to be a criterion which tells us whether or not the problem is solvable.

The "asymptotically linear" (or "sublinear") case is when $f(x,t)$ satisfies an inequality of the form

$$|f(x,t)| \leq V(x)(|t|+1), \quad x \in \Omega,\ t \in \mathbb{R}. \tag{11.2}$$

If one examines the asymptotically linear problem, one will find (sometimes hidden) a role played by the "core" function

$$H(x,t) := tf(x,t) - 2F(x,t),$$

where

$$F(x,t) := \int_0^t f(x,s)\,ds.$$

In particular, if there is a number τ satisfying $0 < \tau < 2$ such that $H(x,t)$ satisfies an inequality of the form

M. Schechter, *Critical Point Theory*,
https://doi.org/10.1007/978-3-030-45603-0_11

$$|H(x,t)| \leq W(x)(|t|^{\tau}+1), \quad x \in \Omega,\ t \in \mathbb{R}, \tag{11.3}$$

we have the following.

Theorem 11.1 *If (11.3) holds, then $f(x,t)$ satisfies an inequality of the form*

$$|f(x,t)| \leq V(x)(|t|+1), \quad x \in \Omega,\ t \in \mathbb{R},$$

and there exist functions $\alpha_{\pm}(x)$ such that

$$f(x,t)/t \to \alpha_{\pm}(x) \text{ as } t \to \pm\infty.$$

Moreover, the function $f(x,t)$ is completely determined by the core function H(x,t) and the limits $\alpha_{\pm}(x)$.

We can then situate these limits within the framework of the spectrum of $\mathcal{A}$ and obtain sufficient conditions (depending on $H(x,t)$, the limits $\alpha_{\pm}(x)$, and the spectrum of $\mathcal{A}$) for the solvability of (11.1). For instance, we have the following.

Theorem 11.2 *Assume that the spectrum of $\mathcal{A}$ consists of isolated eigenvalues of finite multiplicity:*

$$0 < \lambda_0 < \lambda_1 < \cdots < \lambda_\ell < \cdots .$$

Let $\lambda_\ell < \lambda_p,\ \ell > 0$, be two of these eigenvalues, and define

$$N = \oplus_{\lambda<\lambda_\ell} N(\mathcal{A}-\lambda), \quad E = \oplus_{\lambda_\ell \leq \lambda < \lambda_p} N(\mathcal{A}-\lambda), \quad M = [N \oplus E]^{\perp}.$$

Assume that there are numbers a_1, a_2 such that $\alpha_\ell < a_1 < a_2$ and

$$a_1 < \alpha_-(x) < a_2, \quad \tilde{\gamma}_\ell(a_1) < \alpha_+(x) < \tilde{\Gamma}_p(a_2), \qquad x \in \Omega,\ t \in \mathbb{R}, \tag{11.4}$$

where

$$\alpha_\ell := \max\{(\mathcal{A}v, v) : v \in N, v \geq 0, \|v\| = 1\},$$

and the functions $\tilde{\gamma}_\ell(a),\ \tilde{\Gamma}_p(a)$ are defined by

$$\tilde{\gamma}_\ell(a) := \max\{(\mathcal{A}v, v) - a\|v^-\|^2 \ :\ v \in N \oplus E,\ \|v^+\| = 1\} \tag{11.5}$$

and

$$\tilde{\Gamma}_p(a) := \inf\{(\mathcal{A}w, w) - a\|w^-\|^2 :\ w \in M \oplus E,\ \|w^+\| = 1\}, \tag{11.6}$$

where $u^{\pm} = \max\{\pm u, 0\}$. Then (11.1) is solvable.

Note that the solvability of problem (11.1) depends only on the quantities $a_1, a_2, \tilde{\gamma}_\ell(a_1), \tilde{\Gamma}_p(a_2)$, but not on the specific details of the operator $\mathcal{A}$ or the function $f(x,t)$. We shall explore this fact in detail in the next section.

11.2 Boundary Value Problems

We assume that Ω is a bounded domain in $\mathbb{R}^n$ with boundary $\partial\Omega$ sufficiently regular so that the Sobolev inequalities hold and the embedding of $H^{m,2}(\Omega)$ in $L^2(\Omega)$ is compact for $m > 0$. Let $\mathcal{A}$ be a self-adjoint operator on $L^2(\Omega)$. We assume that $\mathcal{A} \geq \lambda_0 > 0$ and that

$$C_0^\infty(\Omega) \subset D := D(\mathcal{A}^{1/2}) \subset H^{m,2}(\Omega)$$

for some $m > 0$, where $C_0^\infty(\Omega)$ denotes the set of test functions in Ω (i.e., infinitely differentiable functions with compact supports in Ω) and $H^{m,2}(\Omega)$ denotes the Sobolev space. If m is an integer, the norm in $H^{m,2}(\Omega)$ is given by

$$\|u\|_{m,2} := \left(\sum_{|\mu| \leq m} \|D^\mu u\|^2 \right)^{1/2}. \tag{11.7}$$

Here D^μ represents the generic derivative of order $|\mu|$ and the norm on the right-hand side of (11.7) is that of $L^2(\Omega)$. We shall not assume that m is an integer. Note that

$$\|u\|_{m,2} \leq C\|u\|_D, \quad u \in D, \tag{11.8}$$

where

$$\|u\|_D := \|\mathcal{A}^{1/2} u\|. \tag{11.9}$$

With the norm (11.9), D becomes a Hilbert space.

Let $f(x,t)$ be a Carathéodory function on $\Omega \times \mathbb{R}$. This means that $f(x,t)$ is continuous in t for a.e. $x \in \Omega$ and measurable in x for every $t \in \mathbb{R}$. We assume the following.

(A) There is a number τ satisfying $0 < \tau < 2$ such that the core function

$$H(x,t) := tf(x,t) - 2F(x,t)$$

satisfies

$$|H(x,t)| \leq W(x)(|t|^\tau + 1), \quad x \in \Omega, \ t \in \mathbb{R}, \tag{11.10}$$

and

$$W(x),\ |F(x,\pm 1)| \in L^1(\Omega), \tag{11.11}$$

where

$$F(x,t) := \int_0^t f(x,s)\,ds.$$

We have the following.

Theorem 11.3 *If (11.10) and (11.11) hold, then $f(x,t)$ satisfies an inequality of the form*

$$|f(x,t)| \le V(x)(|t|+1), \quad x \in \Omega,\ t \in \mathbb{R},$$

and there exist $L^1(\Omega)$ functions $\alpha_\pm(x)$ such that

$$f(x,t)/t \to \alpha_\pm(x) \text{ as } t \to \pm\infty.$$

Moreover, the function $f(x,t)$ is completely determined by the core function H(x,t) and the limits $\alpha_\pm(x)$.

Since the embedding of D in $L^2(\Omega)$ is compact, the spectrum of $\mathcal{A}$ consists of isolated eigenvalues of finite multiplicity:

$$0 < \lambda_0 < \lambda_1 < \cdots < \lambda_\ell < \cdots.$$

Let $\lambda_\ell < \lambda_p,\ \ell > 0$, be two of these eigenvalues, and define

$$N = \oplus_{\lambda<\lambda_\ell} N(\mathcal{A}-\lambda), \quad E = \oplus_{\lambda_\ell \le \lambda < \lambda_p} N(\mathcal{A}-\lambda), \quad M = [N \oplus E]^\perp.$$

Let

$$\nu_\pm(x) := \limsup_{t\to\pm\infty} H(x,t)/|t|^\tau, \tag{11.12}$$

and

$$\mu_\pm(x) := \liminf_{t\to\pm\infty} H(x,t)/|t|^\tau. \tag{11.13}$$

We assume that either

$$\mu(y) := \int_\Omega \mu_+(x)|y(x)^+|^\tau + \mu_-(x)|y(x)^-|^\tau > 0, \quad y \in E\backslash\{0\}, \tag{11.14}$$

or

$$\nu(y) := \int_\Omega \nu_+(x)|y(x)^+|^\tau + \nu_-(x)|y(x)^-|^\tau < 0, \quad y \in E\backslash\{0\}. \tag{11.15}$$

We assume also that

$$a(u) \leq \alpha(u), \quad u \in \tilde{N} = N \oplus E, \tag{11.16}$$

$$a(u) < \alpha(u), \quad u \in \tilde{N}\backslash E, \tag{11.17}$$

$$\alpha(u) \leq a(u), \quad u \in \tilde{M} = M \oplus E, \tag{11.18}$$

$$\alpha(u) < a(u), \quad u \in \tilde{M}\backslash E, \tag{11.19}$$

where $a(u) = (\mathcal{A}u, u)$ and

$$\alpha(u) = \int_\Omega \alpha_+(x)[u(x)^+]^2 + \alpha_-(x)[u(x)^-]^2. \tag{11.20}$$

We wish to obtain a solution of

$$\mathcal{A}u = f(x, u), \quad u \in D. \tag{11.21}$$

By a solution of (11.21) we shall mean a function $u \in D$ such that

$$(u, v)_D = (f(\cdot, u), v), \quad v \in D. \tag{11.22}$$

If $f(x, u)$ is in $L^2(\Omega)$, then a solution of (11.22) is in $D(\mathcal{A})$ and solves (11.21) in the classical sense. Otherwise we call it a weak (or semi-strong) solution. We have

Theorem 11.4 *Under the above hypotheses,*

$$\mathcal{A}u = f(x, u), \quad u \in D \tag{11.23}$$

has at least one solution.

11.3 Examples

We now give some examples in which the hypotheses of Theorem 11.4 are satisfied.

Example 1 Assume that there are numbers a_1, a_2 such that $\alpha_\ell < a_1 < a_2$ and

$$a_1 < \alpha_-(x) < a_2, \quad \tilde{\gamma}_\ell(a_1) < \alpha_+(x) < \tilde{\Gamma}_p(a_2), \qquad x \in \Omega, \ t \in \mathbb{R}, \tag{11.24}$$

where

$$\alpha_\ell := \max\{(\mathcal{A}v, v) : v \in N, v \geq 0, \|v\| = 1\},$$

and the functions $\tilde{\gamma}_\ell(a)$, $\tilde{\Gamma}_p(a)$ are defined by

$$\tilde{\gamma}_\ell(a) := \max\{(\mathcal{A}v, v) - a\|v^-\|^2 \ : \ v \in \tilde{N}, \ \|v^+\| = 1\} \tag{11.25}$$

and

$$\tilde{\Gamma}_p(a) := \inf\{(\mathcal{A}w, w) - a\|w^-\|^2 : \ w \in \tilde{M}, \ \|w^+\| = 1\}, \tag{11.26}$$

where $u^\pm = \max\{\pm u, 0\}$. Then all of the hypotheses of Theorem 11.4 are satisfied. Note that

$$\int_\Omega [\alpha_-(x) - a_1](v^-)^2 + [\alpha_+(x) - \tilde{\gamma}_\ell(a_1)](v^+)^2\, dx > 0, \quad v \in \tilde{N}\backslash E,$$

with a similar inequality holding for $\tilde{M}\backslash E$.

Example 2 Let $\lambda_\ell \leq \lambda \leq \lambda_p$ and take $\alpha_\pm(x) \equiv \lambda$. Then all of the hypotheses of Theorem 11.4 are satisfied.

11.4 Lemmas

In proving Theorem 11.3, we shall make use of several lemmas. We assume all of the hypotheses of Theorem 11.3.

Lemma 11.5 *Let*

$$\alpha_+(x,t) := 2t^{-2}F(x,t) + 2\int_t^\infty s^{-3}H(x,s)ds, \quad t > 0, \tag{11.27}$$

and

$$\alpha_-(x,t) := 2t^{-2}F(x,t) - 2\int_{-\infty}^t s^{-3}H(x,s)ds, \quad t < 0. \tag{11.28}$$

Then $\alpha_\pm(x,t) \equiv \alpha_\pm(x)$ are independent of t.

Proof. We note that

$$\partial(Ft^{-2})/\partial t = t^{-3}H(x,t).$$

Thus,

$$t_2^{-2}F(x,t_2) - t_1^{-2}F(x,t_1) = \int_{t_1}^{t_2} s^{-3}H(x,s)ds. \tag{11.29}$$

Hence,

$$\frac{1}{2}\alpha_+(x,t_2) - \int_{t_2}^{\infty} s^{-3}H(x,s)ds - \frac{1}{2}\alpha_+(x,t_1) + \int_{t_1}^{\infty} s^{-3}H(x,s)ds$$
$$= \int_{t_1}^{t_2} s^{-3}H(x,s)ds.$$

Thus,

$$\alpha_+(x,t_2) = \alpha_+(x,t_1),$$

with a similar identity for $\alpha_-(x,t)$. (One could have differentiated $\alpha_\pm(x,t)$ with respect to t and shown that the derivatives vanish.) □

Lemma 11.6 *If $\alpha_\pm(x)$ are given by (11.27) and (11.28), then*

$$F(x,t) = \frac{1}{2}\alpha_\pm(x)(t^\pm)^2 - F_0(x,t), \quad t \in \mathbb{R}, \tag{11.30}$$

where

$$F_0(x,t) := t^2 \int_t^{\infty} s^{-3}H(x,s)\,ds, \quad t > 0 \tag{11.31}$$
$$:= -t^2 \int_{-\infty}^{t} s^{-3}H(x,s)ds, \quad t < 0.$$

Proof. Letting $t_2 \to +\infty$ in (11.29), we obtain (11.30) for $t > 0$, and letting $t_1 \to -\infty$, we obtain the formula for $t < 0$. □

Lemma 11.7 *If $F_0(x,t)$ is given by (11.31), then*

$$F_0(x,t)/t^2 \to 0 \text{ as } |t| \to \infty.$$

Proof. Apply (11.10) to (11.31). □

Lemma 11.8 *If $\rho_k = \|u_k\|_D \to \infty$, $\tilde{u}_k = u_k/\rho_k$ and $\tilde{u}_k \to \tilde{u}$ strongly in $L^2(\Omega)$ and a.e. in Ω, then*

$$\int_\Omega F_0(x,u_k)\,dx/\rho_k^2 \to 0.$$

Proof. Apply Lemma 11.7. We have

$$\int_\Omega F_0(x,u_k)\,dx/\rho_k^2 = \int_\Omega [F_0(x,\rho_k\tilde{u}_k)/\rho_k^2\tilde{u}_k^2]\tilde{u}_k^2\,dx \to 0.$$

□

Lemma 11.9 *We have*

$$f(x,t) = \alpha_+(x)t + 2t\int_t^\infty s^{-3}H(x,s)ds + t^{-1}H(x,t), \quad t > 0$$
$$= \alpha_-(x)t - 2t\int_{-\infty}^t s^{-3}H(x,s)ds - t^{-1}H(x,t), \quad t < 0.$$

Hence,

$$|f(x,t)| \leq CV(x)(|t|+1), \quad x \in \Omega,\ t \in \mathbb{R},$$

and

$$f(x,t)/t \to \alpha_\pm(x) \text{ as } t \to \pm\infty. \tag{11.32}$$

Proof. We differentiate (11.30) and use (11.31). □

Lemma 11.10 *If $\nu_\pm(x)$ and $\mu_\pm(x)$ are given by (11.12) and (11.13), then*

$$\nu_\pm(x) = (2-\tau)\limsup_{t\to\pm\infty} F_0(x,t)/|t|^\tau, \tag{11.33}$$

and

$$\mu_\pm(x) = (2-\tau)\liminf_{t\to\pm\infty} F_0(x,t)/|t|^\tau. \tag{11.34}$$

Proof. For $\varepsilon > 0$, take t so large that

$$H(x,t)/|t|^\tau \leq (\nu_+(x)+\varepsilon).$$

Hence,

$$F_0(x,t)/|t|^\tau \leq t^{2-\tau}\int_t^\infty s^{\tau-3}(\nu_+(x)+\varepsilon)\,dx \leq (\nu_+(x)+\varepsilon)/(2-\tau).$$

The other cases are considered in the same way. □

Lemma 11.11 *All solutions of*

$$Au = \alpha_+u^+ - \alpha_-u^-$$

are in E.

Proof. We have

$$a(u) = (Au,u) = \int[\alpha_+(u^+)^2 + \alpha_-(u^-)^2]\,dx = \alpha(u).$$

By hypothesis, u is not in $\tilde{N}\backslash E$ and not in $\tilde{M}\backslash E$. Thus, it must be in E. □

Lemma 11.12 *If* $\rho_k = \|u_k\|_D \to \infty$, $\tilde{u}_k = u_k/\rho_k$ *and* $\tilde{u}_k \to \tilde{u}$ *strongly in* $L^2(\Omega)$ *and a.e. in* Ω, *then*

$$\limsup \int_\Omega F_0(x, u_k)\, dx/\rho_k^\tau \le \nu(\tilde{u})/(2-\tau),$$

and

$$\liminf \int_\Omega F_0(x, u_k)\, dx/\rho_k^\tau \ge \mu(\tilde{u})/(2-\tau).$$

Proof. Apply Lemma 11.10. □

Define

$$G(u) := \|u\|_D^2 - 2\int_\Omega F(x, u)\, dx. \tag{11.35}$$

It is readily shown that G is a continuously differentiable functional on the whole of D (cf., e.g., [193]).

Lemma 11.13 *If (11.14) holds, then*

$$G(u) \to \infty, \quad \|u\|_D \to \infty,\ u \in \tilde{M}. \tag{11.36}$$

Proof. If $u_k \in \tilde{M}$ and $\rho_k = \|u_k\|_D \to \infty$, let $\tilde{u}_k = u_k/\rho_k$. Then for a subsequence, $\tilde{u}_k \to \tilde{u}$ weakly in D, strongly in $L^2(\Omega)$, and a.e. in Ω. Then

$$G(u_k)/\rho_k^2 = \|\tilde{u}_k\|_D^2 - \alpha(\tilde{u}_k) - 2\int_\Omega F_0(x, u)\, dx/\rho_k^2 \to [1 - \|\tilde{u}\|_D^2] + [\|\tilde{u}\|_D^2 - \alpha(\tilde{u})]$$

This is positive unless $\tilde{u}_k \to \tilde{u}$ strongly in D and $\tilde{u} \in E$. Then $\tilde{u} \in E\backslash\{0\}$, and

$$G(u_k)/\rho_k^\tau \ge -2\int_\Omega F_0(x, u_k)\, dx/\rho_k^\tau \ge -\frac{2}{2-\tau}\nu(\tilde{u}) > 0.$$

Thus, (11.36) holds. □

Corollary 11.14 *Without (11.14) we have*

$$G(u) \to \infty, \quad \|u\|_D \to \infty,\ u \in \tilde{M}\backslash E. \tag{11.37}$$

Proof. In this case $\tilde{u} \notin E$. □

Lemma 11.15 *If (11.15) holds, then*

$$G(u) \to -\infty, \quad \|u\|_D \to \infty,\ u \in \tilde{N}. \tag{11.38}$$

Proof. If $u_k \in \tilde{N}$ and $\rho_k = \|u_k\|_D \to \infty$, let $\tilde{u}_k = u_k/\rho_k$. Because of the finite dimensionality of $\tilde{N}$, we have for a subsequence, $\tilde{u}_k \to \tilde{u}$ strongly in D and $L^2(\Omega)$ and a.e. in Ω. Then

$$G(u_k)/\rho_k^2 = \|\tilde{u}_k\|_D^2 - \alpha(\tilde{u}_k) - 2\int_\Omega F_0(x,u)\,dx/\rho_k^2 \to \|\tilde{u}\|_D^2 - \alpha(\tilde{u}).$$

This is negative unless $\tilde{u} \in E$. Then $\tilde{u} \in E\backslash\{0\}$, and

$$G(u_k)/\rho_k^\tau \le -2\int_\Omega F_0(x,u_k)\,dx/\rho_k^\tau \le -\frac{2}{2-\tau}\mu(\tilde{u}) < 0.$$

Thus, (11.38) holds. □

Corollary 11.16 *Without (11.15) we have*

$$G(u) \to -\infty, \quad \|u\|_D \to \infty,\ u \in \tilde{N}\backslash E. \tag{11.39}$$

Proof. In this case $\tilde{u} \notin E$. □

11.5 Nontriviality

If $f(x,0) \equiv 0$, then $u(x) \equiv 0$ is a solution of (11.23). In this case, one desires to find conditions which will guarantee the existence of another solution. For this purpose, we add the following assumptions.

(B) There is a $\delta > 0$ such that

$$2F(x,t) \le \lambda_0 t^2, \quad |t| \le \delta. \tag{11.40}$$

(C) Eigenfunctions of $\mathcal{A}$ corresponding to λ_0 are bounded. One eigenfunction $\phi(x)$ of $\mathcal{A}$ corresponding to λ_0 does not change sign in Ω, and

$$\alpha_+(x) > \lambda_0\ a.e., \quad x \in \Omega.$$

(D) Let $\nu_\pm(x)$ and $\mu_\pm(x)$ be given by (11.12) and (11.13), respectively. We assume that either

$$\mu(y) := \int_\Omega \mu_+(x)|y(x)^+|^\tau + \mu_-(x)|y(x)^-|^\tau > 0, \quad y \in D\backslash\{0\}, \tag{11.41}$$

or

$$\nu(y) := \int_\Omega \nu_+(x)|y(x)^+|^\tau + \nu_-(x)|y(x)^-|^\tau < 0, \quad y \in D\backslash\{0\}. \tag{11.42}$$

Theorem 11.17 *Under hypotheses* **(A)**, **(B)**, **(C)**, *and* **(D)**, *Eq.* (11.23) *has at least two solutions.*

Remark 11.18 Note that we do not need to assume any hypothesis of Theorem 11.4 other than **(A)**.

The following theorem is well known (cf., e.g., [193], Theorem 3.4.1).

Theorem 11.19 *Let $\mathcal{A}$ and $f(x,t)$ satisfy the hypotheses stated above, and assume that there is a sequence $\{u_k\} \subset D$ satisfying*

$$G(u_k) \to c, \ (1 + \|u_k\|_D)^\beta G'(u_k) \to 0,$$

with

$$-\infty \le c \le \infty, \quad -\infty < \beta < \infty.$$

If

$$\|u_k\|_D \le C,$$

then c is finite and there is a $u \in D$ such that

$$G(u) = c, \ G'(u) = 0.$$

The following is a special case of Example 1a of Chap. 3.

Theorem 11.20 *Assume that $G \in C^1(X, \mathbb{R})$ and there is a $\delta > 0$ such that*

$$G(0) \le \alpha \le G(u), \quad u \in \partial B_\delta \tag{11.43}$$

and that there is a $\varphi_0 \in \partial B_1$ such that

$$G(R\varphi_0) \le \gamma < \infty, \quad R > R_0. \tag{11.44}$$

Then there is a sequence $\{u_k\} \subset X$ such that

$$G(u_k) \to c, \ \alpha \le c \le \gamma, \ (1 + \|u_k\|)G'(u_k) \to 0. \tag{11.45}$$

11.6 Proofs of the Theorems

Theorem 11.3 follows from Lemmas 11.5, 11.6, and 11.9. In proving Theorem 11.4 we shall make use of the following theorem which is a special case of Corollary 3.5.

Theorem 11.21 *Let N be a closed subspace of a Hilbert space X and let $M = N^{\perp}$. Assume that at least one of the subspaces M, N is finite dimensional. Let G be a C^1-functional on X such that*

$$m_0 := \inf_{w \in M} G(w) \neq -\infty \tag{11.46}$$

and

$$m_1 := \sup_{v \in N} G(v) \neq \infty. \tag{11.47}$$

Then for any sequence $\{R_k\} \subset \mathbb{R}^+$, such that $R_k \to \infty$, there are a constant $c \in \mathbb{R}$ and a sequence $\{u_k\} \subset E$ such that

$$G(u_k) \to c, \ m_0 \leq c \leq m_1, \ (R_k + \|u_k\|)\|G'(u_k)\| \leq \frac{m_1 - m_0}{\ln(4/3)}. \tag{11.48}$$

Proof of Theorem 11.4 By Lemmas 11.13, 11.14 and their corollaries, we have either

$$G(u) \to \infty, \ \|u\|_D \to \infty, \ u \in \tilde{M}; \quad G(u) \to -\infty, \ \|u\|_D \to \infty, \ u \in N. \tag{11.49}$$

or

$$G(u) \to \infty, \ \|u\|_D \to \infty, \ u \in M; \quad G(u) \to -\infty, \ \|u\|_D \to \infty, \ u \in \tilde{N}. \tag{11.50}$$

In either case we can apply Theorem 11.21 to conclude that for any sequence $\{R_k\} \subset \mathbb{R}^+$, such that $R_k \to \infty$, there are a constant $c \in \mathbb{R}$ and a sequence $\{u_k\} \subset D$ such that

$$G(u_k) = \rho_k^2 - 2\int_\Omega F(x, u_k)\,dx \to c \tag{11.51}$$

and

$$|(G'(u_k), u_k)|/2 = |\rho_k^2 - \int_\Omega f(x, u_k)u_k\,dx| \leq C.$$

Hence,

$$|\int_\Omega H(x, u_k)\,dx| \leq C'. \tag{11.52}$$

Let $\tilde{u_k} = u_k/\rho_k$. If $\rho_k \to \infty$, then there is a renamed subsequence such that $\tilde{u}_k \to \tilde{u}$ weakly in D, strongly in $L^2(\Omega)$, and a.e. in Ω. Moreover, for $v \in D$

$$(G'(u_k), v)/2\rho_k = (\tilde{u}_k, v)_D - (f(u_k)/\rho_k, v) \to 0.$$

This implies

$$(\tilde{u}, v)_D = \alpha(\tilde{u}, v), \quad v \in D$$

or

$$\mathcal{A}\tilde{u} = \alpha_+ \tilde{u}^+ - \alpha_- \tilde{u}^-.$$

Thus, $\tilde{u} \in E\backslash\{0\}$ by Lemma 11.11. By (11.12) and (11.13),

$$\limsup \int_\Omega H(x, u_k)\, dx/\rho_k^\tau \leq \int_\Omega \limsup[H(x, u_k)/|u_k|^\tau]|\tilde{u}_k|^\tau\, dx \leq \nu(\tilde{u}) < 0 \tag{11.53}$$

or

$$\liminf \int_\Omega H(x, u_k)\, dx/\rho_k^\tau \geq \int_\Omega \liminf[H(x, u_k)/|u_k|^\tau]|\tilde{u}_k|^\tau\, dx \geq \mu(\tilde{u}) > 0. \tag{11.54}$$

Either of these contradicts (11.52). This contradiction tells us that the ρ_k must be bounded. We can now apply Theorem 11.19 to conclude that there is a $u \in D$ satisfying

$$G(u) = c, \quad G'(u) = 0, \tag{11.55}$$

and the proof is complete. □

Before proving Theorem 11.17, we note the following.

Theorem 11.22 *Under assumptions (B), the following alternative holds:*
Either

(a) *there is an infinite number of $y(x) \in D(\mathcal{A}) \setminus \{0\}$ such that*

$$\mathcal{A}y = f(x, y) = \lambda_0 y \tag{11.56}$$

or

(b) *for each $\rho > 0$ sufficiently small, there is an $\epsilon > 0$ such that*

$$G(u) \geq \epsilon, \ \|u\|_D = \rho. \tag{11.57}$$

Proof. Let $\lambda_1 > \lambda_0$ be the next point in the spectrum of $\mathcal{A}$, and let N_0 denote the eigenspace of λ_0. We take $M = N_0^\perp \cap D$. By hypothesis, the functions in N_0 are bounded. Thus, there is a $\rho > 0$ such that

$$\|y\|_D \leq \rho \Rightarrow |y(x)| \leq \delta/2, \quad y \in N_0.$$

Now suppose $u \in D$ satisfies

$$\|u\|_D \leq \rho \text{ and } |u(x)| \geq \delta \tag{11.58}$$

for some $x \in \Omega$. We write

$$u = w + y, \ \ w \in M, \ \ y \in N_0. \tag{11.59}$$

Then for those $x \in \Omega$ satisfying (11.58) we have

$$\delta \leq |u(x)| \leq |w(x)| + |y(x)| \leq |w(x)| + (\delta/2).$$

Hence

$$|y(x)| \leq \delta/2 \leq |w(x)|, \tag{11.60}$$

and consequently,

$$|u(x)| \leq 2|w(x)| \tag{11.61}$$

for all such x. Now we have by (11.40)

$$\begin{aligned} G(u) &\geq \|u\|_D^2 - \lambda_0 \int_{|u|<\delta} u^2 dx - C \int_{|u|>\delta} |u|^2 \, dx \\ &\geq \|u\|_D^2 - \lambda_0 \|u\|^2 - C' \int_{|u|>\delta} |u|^2 dx \\ &\geq \|w\|_D^2 - \lambda_0 \|w\|^2 - C'' \int_{2|w|>\delta} |w|^2 dx \end{aligned} \tag{11.62}$$

in view of the fact that $\|y\|_D^2 = \lambda_0 \|y\|^2$ and (11.61) holds. We shall show that

$$\int_{2|w|>\delta} |w|^2 dx / \|w\|_D^2 \to 0 \text{ as } \|w\|_D \to 0. \tag{11.63}$$

Assuming this for the moment, we see that

$$G(u) \geq \left(1 - \frac{\lambda_0}{\lambda_1} - o(1)\right) \|w\|_D^2, \quad \|u\|_D \leq \rho. \tag{11.64}$$

Now suppose alternative (b) of the theorem did not hold. Then there would be a sequence such that

$$G(u_k) \to 0, \ \ \|u_k\|_D = \rho. \tag{11.65}$$

If ρ is taken sufficiently small, (11.64) implies that $\|w_k\|_D \to 0$. Consequently, $\|y_k\|_D \to \rho$. Since N_0 is finite dimensional, there is a renamed subsequence such that $y_k \to y_0$ in N_0. Thus we have

$$\|y_0\| = \rho, \ G(y_0) = 0, \ |y_0(x)| \leq \delta/2, \quad x \in \Omega.$$

Consequently,

$$2F(x, y_0(x)) \leq \lambda_0 y_0(x)^2, \quad x \in \Omega. \tag{11.66}$$

Since

$$\int_\Omega \{\lambda_0 y_0(x)^2 - 2F(x, y_0(x))\}\, dx = G(y_0) = 0$$

and the integrand is ≥ 0 a.e. by (11.66), we see that

$$2F(x, y_0(x)) \equiv \lambda_0 y_0(x)^2, \quad x \in \Omega.$$

Let $\varphi(x)$ be any function in $C_0^\infty(\Omega)$. Then for $t > 0$ sufficiently small

$$t^{-1}[2F(x, y_0 + t\varphi) - \lambda_0(y_0 + t\varphi)^2 - 2F(x, y_0) + \lambda_0 y_0^2] \leq 0.$$

Taking the limit as $t \to 0$, we have

$$(f(x, y_0) - \lambda_0 y_0)\varphi(x) \leq 0, \quad x \in \Omega.$$

Since this is true for every $\varphi \in C_0^\infty(\Omega)$, we see that

$$f(x, y_0(x)) \equiv \lambda_0 y_0(x), \quad x \in \Omega.$$

Since $y_0 \in N_0$, it follows that (11.56) holds.

Thus, the theorem will be proved once we have established (11.63). Assume that there is a sequence $\{w_k\}$ such that

$$\int_{2|w_k|>\delta} |w_k|^2 dx/\|w_k\|_D^2 \geq \epsilon > 0 \tag{11.67}$$

while $\rho_k = \|w_k\|_D \to 0$. Let $\tilde{w}_k = w_k/\rho_k$. Then $\|\tilde{w}_k\|_D = 1$. Let $\mu_k(x)$ be the characteristic function of the set of those $x \in \Omega$ such that $2|w_k(x)| \geq \delta$. Then (11.67) becomes

$$\int_\Omega |\tilde{w}_k|^2 \mu_k(x)\, dx \geq \epsilon. \tag{11.68}$$

Since $\rho_k = \|w_k\|_D \to 0$, we have $w_k \to 0$ in $L^2(\Omega)$. Thus, for a renamed subsequence, $w_k \to 0$ a.e. in Ω. Consequently, $w_k \to 0$ in measure, and this implies that the measure of the set where $\mu_k(x) = 1$ converges to 0. Thus, $\mu_k(x) \to 0$ in measure as well. Hence, for a renamed subsequence, $\mu_k(x) \to 0$ a.e. We know that there is a renamed subsequence such that $\tilde{w}_k$ converges in $L^2(\Omega)$ and a.e. in Ω. But

$$\tilde{w}_k(x)^2 \mu_k(x) \le \tilde{w}_k(x)^2,$$

and the right-hand side converges in $L^1(\Omega)$. Since the left-hand side converges a.e. to 0, we see that

$$\int |\tilde{w}_k(x)|^2 \mu_k(x)\, dx \to 0.$$

Hence there cannot exist an $\epsilon > 0$ such that

$$\int_{2|w_k|>\delta} |\tilde{w}_k|^2 dx \ge \epsilon.$$

This completes the proof of (11.63) and of the theorem. □

Proof of Theorem 11.17 By Theorem 11.22, we may assume that for each $\rho > 0$ sufficiently small, there is an $\varepsilon > 0$ such that

$$G(u) \ge \varepsilon, \ \ \|u\|_D = \rho. \tag{11.69}$$

Otherwise, Theorem 11.17 is proved. We may assume that $\phi(x) \ge 0$ a.e. in Ω. Then

$$G(R\phi_0)/R^2 = \int [\lambda_0 - \alpha_+(x)]\phi_0^2\, dx - \int F_0(x, R\phi_0)/R^2\, dx.$$

Thus

$$\limsup G(R\phi_0)/R^2 = \int [\lambda_0 - \alpha_+(x)]\phi_0^2\, dx < 0.$$

Consequently, there is a function $v \in D$ such that $\|v\|_D > \rho$ and $G(v) \le 0$. Hence, we have linking. By Theorem 11.20, we know that there is a constant $c \in \mathbb{R}$ and a Cerami sequence $\{u_k\} \subset D$ such that

$$G(u_k) = \rho_k^2 - 2\int_\Omega F(x, u_k)\, dx \to c \tag{11.70}$$

and

$$|(G'(u_k), u_k)|/2 = |\rho_k^2 - \int_\Omega f(x, u_k) u_k\, dx| \to 0.$$

If $\rho_k = \|u_k\|_D \to \infty$, let $\tilde{u}_k = u_k/\rho_k$. Since $G(0) = 0$, we see that (11.70) holds with $c \geq \varepsilon > 0$. Again, this implies (11.52). By Hypothesis (D),

$$\limsup \int_\Omega H(x, u_k)\,dx/\rho_k^\tau \leq \int_\Omega \limsup[H(x, u_k)/|u_k|^\tau]|\tilde{u}_k|^\tau\,dx \leq \nu(\tilde{u}) < 0 \tag{11.71}$$

or

$$\liminf \int_\Omega H(x, u_k)\,dx/\rho_k^\tau \geq \int_\Omega \liminf[H(x, u_k)/|u_k|^\tau]|\tilde{u}_k|^\tau\,dx \geq \mu(\tilde{u}) > 0. \tag{11.72}$$

Either of these contradicts (11.52). This contradiction shows that there is a solution satisfying (11.55). □

11.7 Superlinear Problems

In studying superlinear problems, we make the following assumptions: Let q be a number satisfying

$$\begin{aligned} &2 < q \leq 2n/(n-2m), && 2m < n, \\ &2 < q < \infty, && n \leq 2m. \end{aligned} \tag{11.73}$$

We assume

($\tilde{\mathbf{A}}$) The function $f(x,t)$ satisfies

$$|f(x,t)| \leq V(x)^q(|t|^{q-1} + 1) \tag{11.74}$$

and

$$f(x,t)/V(x)^q = o(|t|^{q-1}) \text{ as } |t| \to \infty, \tag{11.75}$$

where $V(x) > 0$ is a function in $L^q(\Omega)$ such that

$$\|Vu\|_q \leq C\|u\|_D, \quad u \in D. \tag{11.76}$$

Here

$$\|u\|_q := \left(\int_\Omega |u(x)|^q dx\right)^{1/q}, \tag{11.77}$$

$$\|u\|_D := \|A^{1/2}u\|, \tag{11.78}$$

and $q' = q/(q-1)$.

($\tilde{\mathbf{B}}$) The point λ_0 is an isolated simple eigenvalue with a bounded eigenfunction $\varphi_0(x) \neq 0$ a.e. in Ω.

($\tilde{\mathbf{C}}$) There is a $\delta > 0$ such that

$$2F(x,t) \leq \lambda_0 t^2, \quad |t| \leq \delta, \quad x \in \Omega,$$

where

$$F(x,t) := \int_0^t f(x,s)ds. \tag{11.79}$$

($\tilde{\mathbf{D}}$) There is a function $W(x) \in L^1(\Omega)$ such that either

$$-W(x) \leq F(x,t)/t^2 \to \infty \text{ as } t \to \infty, \quad x \in \Omega.$$

or

$$-W(x) \leq F(x,t)/t^2 \to \infty \text{ as } t \to -\infty, \quad x \in \Omega.$$

We shall prove the following:

Theorem 11.23 *Assume that* ($\tilde{\mathbf{A}}$)–($\tilde{\mathbf{D}}$) *hold. Then for almost every* $\beta \in [0,1]$, *the equation*

$$Au = \beta f(x,u) \tag{11.80}$$

has a nontrivial solution

(cf. Theorem 12.1.)

Theorem 11.24 *Let* λ *be a parameter satisfying* $1 < \lambda \leq K < \infty$. *Under hypotheses* ($\tilde{\mathbf{A}}$)–($\tilde{\mathbf{D}}$), *for each* $\rho > 0$ *sufficiently small (not depending on* λ), *we have*

$$G_\lambda(u) := \lambda \|u\|_D^2 - 2\int_\Omega F(x,u)dx \geq (\lambda - 1)\rho^2, \quad \|u\|_D = \rho \tag{11.81}$$

(cf. (12.37)).

In order to pinpoint a solution for $\beta = 1$, we add the hypothesis:

($\tilde{\mathbf{E}}$) The core function

$$H(x,t) := tf(x,t) - 2F(x,t) \tag{11.82}$$

satisfies:

$$H(x,st) \leq C[H(x,t) + W(x)], \quad s \in [0,1], \ t \in \mathbb{R}, \ x \in \Omega,$$

for some constant C and function $W(x) \in L^1(\Omega)$. We shall prove

Theorem 11.25 *Under hypotheses* $(\tilde{\mathbf{A}})$–$(\tilde{\mathbf{E}})$, *the problem*

$$Au = f(x,u), \ u \in D \tag{11.83}$$

has at least one nontrivial solution.

Proof. We take $\lambda = 1/\beta$. By Theorems 11.23 and 11.24, for each arbitrary $K > 1$, and a.e. $\lambda \in (1, K)$, there exists u_λ such that $G'_\lambda(u_\lambda) = 0, \quad G_\lambda(u_\lambda) = a(\lambda) \geq (\lambda - 1)\rho^2$. Choose $\lambda_n \to 1, \ \lambda_n > 1$. Then there exists u_n such that

$$G'_{\lambda_n}(u_n) = 0, \quad G_{\lambda_n}(u_n) = a(\lambda_n) \geq a(1) \geq \varepsilon > 0.$$

Therefore,

$$\int_\Omega \frac{2F(x,u_n)}{\|u_n\|_D^2} dx \leq c.$$

Now we prove that $\{u_n\}$ is bounded. If $\|u_n\|_D \to \infty$, let $w_n = u_n/\|u_n\|_D$. Then there is a renamed subsequence such that $w_n \to w$ weakly in D, strongly in $L^2(\Omega)$, and a.e. in Ω.

Let Ω_0 be the set where $w \neq 0$. Then $|u_n(x)| \to \infty$ for $x \in \Omega_0$. If Ω_0 had positive measure, then we would have

$$c \geq \int_\Omega \frac{2F(x,u_n)}{\|u_n\|_D^2} dx = \int_\Omega \frac{2F(x,u_n)}{u_n^2} |w_n|^2 dx$$

$$\geq \int_{w \neq 0} \frac{2F(x,u_n)}{u_n^2} |w_n|^2 dx - \int_{w=0} W(x)\, dx \to \infty,$$

showing that $w = 0$ a.e. in Ω. We define $t_n \in [0,1]$ by

$$G_{\lambda_n}(t_n u_n) = \max_{t \in [0,1]} G_{\lambda_n}(t u_n).$$

Note that by (11.81), we cannot have $t_n = 0$. For any $c > 0$ and $\bar{w}_n = c w_n$, we have

$$\int_\Omega F(x, \bar{w}_n)\, dx \to 0$$

(cf., e.g., [193, p. 64]). Thus,

$$G_{\lambda_n}(t_n u_n) \geq G_{\lambda_n}(c w_n) = c^2 \lambda_n - 2 \int_\Omega F(x, \bar{w}_n)\, dx \geq c^2/2$$

for n large enough. That is, $\lim_{n\to\infty} G_{\lambda_n}(t_n u_n) = \infty$. If there is a renamed subsequence such that $t_n = 1$, then

$$G_{\lambda_n}(u_n) \to \infty. \tag{11.84}$$

If $0 < t_n < 1$ for all n, then we have $(G'_{\lambda_n}(t_n u_n), u_n) = 0$. Therefore,

$$\begin{aligned} G_{\lambda_n}(t_n u_n) &= \int_\Omega \Big(f(x, t_n u_n) t_n u_n - 2F(x, t_n u_n) \Big)\, dx \\ &= \int_\Omega H(x, t_n u_n)\, dx \to \infty. \end{aligned}$$

By hypothesis ($\tilde{\mathbf{E}}$),

$$G_{\lambda_n}(u_n) = \int_\Omega H(x, u_n)\, dx \geq \int_\Omega H(x, t_n u_n)\, dx / C - \int_\Omega W(x)\, dx \to \infty.$$

Thus, (11.84) holds in any case. But

$$\begin{aligned} G_{\lambda_n}(u_n) = a(\lambda_n) &\leq \sup_{s\in[0,1],\, u\in A} G_{\lambda_n}((1-s)u) \\ &\leq \sup_{s\in[0,1],\, u\in A} G_K((1-s)u) \\ &< c, \end{aligned}$$

a contradiction, where $A = \{u \in D : \|u\|_D = \rho\}$. Thus, $\|u_n\|_D \leq C$. It now follows that

$$G'(u_n) \to 0, \quad G(u_n) \to a(1) \geq \varepsilon.$$

We can now apply Theorem 11.19 to obtain the desired solution. □

11.8 Notes and Remarks

This chapter follows [187].

Costa-Magalhães [66, 67] solved (18.2) under the following assumptions:

$$|f(x,t)| \leq a_0 |t|^{p-1} + b_0, \quad x \in \Omega,\ t \in \mathbb{R},$$

$$\limsup_{|t|\to\infty} \frac{F(x,t)|}{|t|^q} \leq b < \infty \quad uniformly\ for\ a.e.\ x \in \Omega,$$

$$\liminf_{|t|\to\infty} \frac{H(x,t)}{|t|^\mu} \geq a > 0 \quad uniformly\ for\ a.e.\ x \in \Omega,$$

$$\limsup_{t\to 0}\ 2F(x,t)/t^2 \leq \alpha < \lambda_0 \quad uniformly\ for\ a.e.\ x\ \in \Omega,$$

$$F(x,t)/t^2 \to \infty \text{ as } |t| \to \infty,$$

where $1 \leq p < 2n/(n-2),\ \mu > n(q-2)/2$.

Willem-Zou [276] proved a weaker form of Theorem 11.25 for a special case. They do not require superquadraticity, but they do assume

$$tf(x,t) \geq 0, \quad t \in \mathbb{R}$$

and

$$tf(x,t) \geq c_0|t|^\mu, \quad |t| \geq r$$

for some constants $c_0 > 0,\ r \geq 0,\ \mu > 2$. Schechter-Zou [230] proved Theorem 11.25 under the hypothesis that $\tilde{H}(x,t)$ is convex. Using our argument, Miyagaki-Souto [160] noticed that this assumption can be weakened to

$$\tilde{H}(x,t) \leq \tilde{H}(x,s) + C_*$$

for all $0 < t < s$ or $s < t < 0$. In Theorem 11.25 we weaken this further.

Chapter 12
Custom Monotonicity Methods

12.1 Introduction

Consider the problem

$$-\Delta u = f(x,u),\ x \in \Omega; \quad u = 0 \text{ on } \partial\Omega, \tag{12.1}$$

where $\Omega \subset \mathbb{R}^n$ is a bounded domain whose boundary is a smooth manifold, and $f(x,t)$ is a continuous function on $\bar{\Omega}\times\mathbb{R}$. The following theorem will be a corollary of the results of this chapter.

Theorem 12.1 *Assume*

(a_1) *There are constants* $c_1,\ c_2 \geq 0$ *such that*

$$|f(x,t)| \leq c_1 + c_2|t|^s,$$

where $0 \leq s < (n+2)/(n-2)$ *if* $n > 2$.

(a_2)

$$f(x,t) = o(t) \text{ as } t \to 0, \quad \textit{uniformly} \quad \textit{in } x.$$

(a_3) *Either*

$$F(x,t)/t^2 \to \infty \text{ as } t \to \infty$$

or

$$F(x,t)/t^2 \to \infty \text{ as } t \to -\infty,$$

M. Schechter, *Critical Point Theory*,
https://doi.org/10.1007/978-3-030-45603-0_12

where

$$F(x,t)=\int_0^t f(x,s)\,ds.$$

Then the boundary value problem

$$-\Delta u=\beta f(x,u),\ x\in\Omega;\quad u=0 \text{ on } \partial\Omega, \tag{12.2}$$

has a nontrivial solution for almost every positive β.

In this chapter we shall prove

Theorem 12.2 *In place of hypothesis (a_2) assume that there is a positive $\tilde{\lambda}\le\lambda_0$ such that*

$$2F(x,t)\le\tilde{\lambda}t^2,\quad |t|\le\delta \tag{12.3}$$

for some $\delta>0$, and in place of hypothesis (a_3) assume that there is a $\lambda>\tilde{\lambda}$ such that

$$\sup_{r>0}\int_\Omega [\nu r^2\varphi^2-2F(x,r\varphi)]\,dx<\infty \tag{12.4}$$

holds for all ν satisfying $\tilde{\lambda}<\nu<\lambda$, where φ is an eigenfunction corresponding to the first eigenvalue λ_0 of $-\Delta u$. Then the boundary value problem

$$-\Delta u=\beta f(x,u),\ x\in\Omega;\quad u=0 \text{ on } \partial\Omega, \tag{12.5}$$

has a nontrivial solution for almost every β in the interval $[\lambda_0/\lambda,\lambda_0/\tilde{\lambda}]$.

Corollary 12.3 *The conclusion of Theorem 12.2 holds if there is a $\tilde{\lambda}\le\lambda_0$ such that*

$$2F(x,t)\le\tilde{\lambda}t^2,\quad |t|\le\delta \tag{12.6}$$

for some $\delta>0$ and either

$$2F(x,t)\ge\lambda t^2-W_+(x,t),\quad t>0, \tag{12.7}$$

or

$$2F(x,t)\ge\lambda t^2-W_-(x,t),\quad t<0, \tag{12.8}$$

for some $\lambda>\tilde{\lambda}$, where the $W_\pm(x,t)/t^2\le W(x)\in L^1(\Omega)$ satisfy

$$W_\pm(x,t)/t^2\to 0 \text{ a.e. as } t\to\pm\infty,$$

as the case may be.

Corollary 12.4 *Assume hypotheses* (a_1) *and*

$$\limsup_{t\to 0} F(x,t)/t^2 \le 0, \quad x \in \Omega.$$

If (12.4) holds for all positive v*, then the boundary value problem*

$$-\Delta u = \beta f(x,u), \ x \in \Omega; \quad u = 0 \text{ on } \partial\Omega, \tag{12.9}$$

has a nontrivial solution for almost every positive β.

Note that Corollaries 12.3 and 12.4 allow a wide range of functions $f(x,t)$ both superlinear and sublinear for the conclusion of Theorem 12.1 to hold for a β interval. Also note that they imply Theorem 12.1 in the superlinear case. Our method centers about the construction of a collection $\mathcal{K}_0$ of subsets K such that

$$A \in \mathcal{K}_0, \quad B \cap K \neq \phi, \quad K \in \mathcal{K}_0 \tag{12.10}$$

together with

$$a_0 := \sup_A G < \infty, \qquad b_0 := \inf_B G > G(0), \tag{12.11}$$

implies the existence of a PS sequence.

12.2 Custom Flows

Let E be a Banach space, and let Σ_0 be the set of all continuous maps $\sigma = \sigma(t)$ from $E \times [0,1]$ to E such that

1. $\sigma(0)$ is the identity map,
2. for each $t \in [0,1]$, $\sigma(t)$ is a homeomorphism of E onto E,
3. $\sigma'(t)$ is piecewise continuous on [0,1] and satisfies

$$\|\sigma'(t)u\| \le const., \quad u \in E, \ t \in [0,1], \tag{12.12}$$

4.

$$\sigma(t)0 = 0, \quad t \in [0,1]. \tag{12.13}$$

The mappings in Σ_0 are called *custom flows*. We have customized them for our purposes. The property (12.13) is not usually assumed. We make use of this property in our applications. We note the following.

Remark 12.5 If $\sigma_1,\ \sigma_2$ are in Σ_0, define $\sigma_3 = \sigma_1 \circ \sigma_2$ by

$$\sigma_3(s) = \begin{cases} \sigma_1(2s), & 0 \le s \le \frac{1}{2}, \\ \sigma_2(2s-1)\sigma_1(1), & \frac{1}{2} < s \le 1. \end{cases}$$

Then $\sigma_1 \circ \sigma_2 \in \Sigma_0$.

12.3 Custom Sandwich Systems

Let E be a Banach space. We define a nonempty collection $\mathcal{K}_0$ of nonempty subsets $K \subset E$ to be a **custom sandwich system** if $\mathcal{K}_0$ has the following property:

$$\sigma(1)K \in \mathcal{K}_0, \quad \sigma \in \Sigma_0,\ K \in \mathcal{K}_0.$$

This property of $\mathcal{K}_0$ takes into account the special nature of Σ_0. We have

Theorem 12.6 *Let $\mathcal{K}_0$ be a custom sandwich system, and let $G(u)$ be a C^1 functional on E. Define*

$$a := \inf_{K\in\mathcal{K}_0} \sup_K G, \tag{12.14}$$

and assume that a is finite and $G(0) < a$. Assume, in addition, that there is a constant C_0 such that for each $\delta > 0$ there is a $K \in \mathcal{K}_0$ satisfying

$$\sup_K G \le a + \delta, \tag{12.15}$$

such that the inequality

$$G(u) \ge a - \delta, \quad u \in K, \tag{12.16}$$

implies $\|u\| \le C_0$. Then there is a bounded sequence $\{u_k\} \subset E$ such that

$$G(u_k) \to a, \quad \|G'(u_k)\| \to 0. \tag{12.17}$$

The advantage of this theorem is the fact that for most applications there is need to add appropriate hypotheses to obtain a convergent subsequence. This is usually achieved by hypotheses that cause the PS sequence to be bounded. Theorem 12.6 obviates this requirement.

Theorem 12.7 *Let $\mathcal{K}_0$ be a custom sandwich system, and let G(u) be a C^1 functional on E. Assume that there are subsets A,B of E such that*

$$a_0 := \sup_A G < \infty, \qquad b_0 := \inf_B G > G(0), \tag{12.18}$$

$A \in \mathcal{K}_0$ and

$$B \cap K \neq \phi, \quad K \in \mathcal{K}_0. \tag{12.19}$$

Then a given by (12.14) satisfies $b_0 \leq a \leq a_0$. Assume, in addition, that there is a constant C_0 such that for each $\delta > 0$ there is a $K \in \mathcal{K}_0$ satisfying (12.15) such that the inequality (12.16) implies $\|u\| \leq C_0$. Then there is a bounded sequence $\{u_k\} \subset E$ satisfying (12.17).

Definition 12.8 We shall say that sets A,B in E form a **custom sandwich pair** if A is a member of a custom sandwich system $\mathcal{K}_0$ and B satisfies (12.19).

We have the following.

Theorem 12.9 *Let A be a continuous curve in E connecting 0 and ∞, and let B the boundary of a bounded open set in E containing 0. Then A,B form a custom sandwich pair.*

12.4 The Custom Parameter Problem

Let E be a reflexive Banach space with norm $\|\cdot\|$. Suppose that $G \in C^1(E, \mathbb{R})$ is of the form: $G(u) := I(u) - J(u), u \in E$, where $I, J \in C^1(E, \mathbb{R})$ map bounded sets to bounded sets. Define

$$G_\lambda(u) = \lambda I(u) - J(u), \quad \lambda \in \Lambda,$$

where Λ is an open interval contained in $(0, +\infty)$. Assume one of the following alternatives holds.

(H$_1$) $I(u) \geq 0$ for all $u \in E$ and $I(u) + |J(u)| \to \infty$ as $\|u\| \to \infty$
or
(H$_2$) $I(u) \leq 0$ for all $u \in E$ and $|I(u)| + |J(u)| \to \infty$ as $\|u\| \to \infty$.

Furthermore, we suppose that $\mathcal{K}_0$ is a custom sandwich system satisfying

(H$_3$) $a(\lambda) := \inf_{K \in \mathcal{K}_0} \sup_K G_\lambda < \infty$ for any $\lambda \in \Lambda$.

Theorem 12.10 *Assume that* **(H$_1$)** *(or* **(H$_2$)***) and* **(H$_3$)** *hold. Then we have*

(1) For almost all $\lambda \in \Lambda$ there exists a constant $k_0(\lambda) := k_0$ (depending only on λ) such that for each $\delta > 0$ there exists a $K \in \mathcal{K}_0$ such that $\sup_K G_\lambda \leq a(\lambda) + \delta$ and

$$\|u\| \leq k_0 \quad \text{whenever} \quad u \in K \text{ and } G_\lambda(u) \geq a(\lambda) - \delta. \tag{12.20}$$

(2) For almost all $\lambda \in \Lambda$ there exists a bounded sequence $u_k(\lambda) \in E$ such that

$$\|G'_\lambda(u_k)\| \to 0, \quad G_\lambda(u_k) \to a(\lambda) := \inf_{K\in\mathcal{K}_0} \sup_K G_\lambda, \qquad k \to \infty.$$

Corollary 12.11 *The conclusions of Theorem 12.10 hold if we replace hypothesis* ($\mathbf{H}_3$) *with*

($\mathbf{H}'_3$) *There is a custom sandwich pair A,B such that*

$$a_0 := \sup_A G_\mu < \infty, \qquad G_\mu(0) < b_0 := \inf_B G_\mu \tag{12.21}$$

for each $\mu \in \Lambda$.

To prove Theorem 12.10 and Corollary 12.11, see the Proof of Theorem 4.8.

12.5 Some Applications

We now show how Theorem 12.10 and Corollary 12.11 can be used to prove the results stated earlier in the chapter. As before, we let Ω be a domain in $\mathbb{R}^n$, and let $\mathcal{A}$ be a self-adjoint operator on $L^2(\Omega)$. We assume that $\mathcal{A} \geq \lambda_0 > 0$ and that

$$C_0^\infty(\Omega) \subset D := D(\mathcal{A}^{1/2}) \subset H^{m,2}(\Omega) \tag{12.22}$$

for some $m > 0$, where $C_0^\infty(\Omega)$ denotes the set of test functions in Ω (i.e., infinitely differentiable functions with compact supports in Ω), and $H^{m,2}(\Omega)$ denotes the Sobolev space. If m is an integer, the norm in $H^{m,2}(\Omega)$ is given by

$$\|u\|_{m,2} := \left(\sum_{|\mu|\leq m} \|D^\mu u\|^2 \right)^{1/2}. \tag{12.23}$$

Here D^μ represents the generic derivative of order $|\mu|$ and the norm on the right-hand side of (12.23) is that of $L^2(\Omega)$. We shall not assume that m is an integer.

Let q be any number satisfying

$$\begin{aligned} &2 < q < 2^* = 2n/(n-2m), && 2m < n \\ &2 < q < \infty, && n \leq 2m, \end{aligned}$$

and let $f(x,t)$ be a Carathéodory function on $\Omega \times \mathbb{R}$. This means that $f(x,t)$ is continuous in t for a.e. $x \in \Omega$ and measurable in x for every $t \in \mathbb{R}$. Through this section we make the following assumptions:

(**A**) The function $f(x,t)$ satisfies

$$|f(x,t)| \leq V_0(x)^q |t|^{q-1} + V_0(x)^q W_0(x) \tag{12.24}$$

and

$$f(x,t)/V_0(x)^q = o(|t|^{q-1}) \text{ as } |t| \to \infty, \tag{12.25}$$

where $V_0(x) > 0$ is a function such that

$$\|V_0 u\|_q \leq C\|u\|_D, \quad u \in D \tag{12.26}$$

and W_0 is a function in $L^\infty(\Omega)$. Here

$$\|u\|_q := (\int_\Omega |u(x)|^q dx)^{1/q}, \tag{12.27}$$

and

$$\|u\|_D := \|\mathcal{A}^{1/2}u\|. \tag{12.28}$$

If Ω and $V_0(x)$ are bounded, then (12.26) will hold automatically by the Sobolev inequality. However, there are functions $V_0(x)$ which are unbounded and such that (12.26) holds even on unbounded regions Ω. With the norm (12.28), D becomes a Hilbert space. Define

$$F(x,t) := \int_0^t f(x,s)ds \tag{12.29}$$

and

$$G(u) := \|u\|_D^2 - 2\int_\Omega F(x,u)\,dx. \tag{12.30}$$

It follows that G is a continuously differentiable functional on the whole of D (cf., e.g., [184]).

For $\mu > 0$, we let

$$G_\mu(u) := \mu\|u\|_D^2 - 2\int_\Omega F(x,u)\,dx. \tag{12.31}$$

We wish to obtain a solution of

$$\mu\mathcal{A}u = f(x,u), \; u \in D. \tag{12.32}$$

By a solution of (12.32) we shall mean a function $u \in D$ such that

$$\mu(u,v)_D = (f(\cdot,u),v), \quad v \in D. \tag{12.33}$$

If $f(x,u)$ is in $L^2(\Omega)$, then a solution of (12.33) is in $D(\mathcal{A})$ and solves (12.32) in the classical sense. Otherwise we call it a weak (or semi-strong) solution.

We assume that λ_0 is a simple isolated eigenvalue of $\mathcal{A}$ having a bounded eigenfunction $\varphi(x)$. In addition, we assume that there is a positive number $\tilde{\lambda} \leq \lambda_0$ such that

$$2F(x,t) \leq \tilde{\lambda}t^2, \quad |t| < \delta \tag{12.34}$$

for some positive constant δ. Moreover, we assume that

$$\sup_{r>0} \int_\Omega [\nu r^2\varphi^2 - 2F(x, r\varphi)]\,dx < \infty \tag{12.35}$$

for each ν satisfying $\tilde{\lambda} < \nu < \lambda$ for some $\lambda > \tilde{\lambda}$. We have the following.

Theorem 12.12 *Under the above hypotheses, the equation*

$$\mathcal{A}u = \beta f(x,u), \ u \in D \tag{12.36}$$

has at least one nontrivial solution for almost all $\beta \in [\lambda_0/\lambda, \lambda_0/\tilde{\lambda}]$.

Proof. We apply Corollary 12.11. We let N be the eigenspace $E(\lambda_0)$, and we take $M = N^\perp$. We note that (12.34) implies

$$G_\mu(u) \geq (\mu - \eta)\rho^2, \quad \|u\|_D = \rho \tag{12.37}$$

for $\rho > 0$ sufficiently small, where $\eta = \tilde{\lambda}/\lambda_0$. To see this, let $u = w + y$, where $w \in M$ and $y \in N$. Note that there is a $\rho > 0$ such that

$$\|y\|_D \leq \rho \Rightarrow |y(x)| \leq \delta/2, \quad y \in E(\lambda_0).$$

Now suppose u satisfies

$$\|u\|_D \leq \rho \text{ and } |u(x)| \geq \delta \tag{12.38}$$

for some $x \in \Omega$. Then for those $x \in \Omega$ satisfying (12.38) we have

$$\delta \leq |u(x)| \leq |w(x)| + |y(x)| \leq |w(x)| + (\delta/2).$$

Hence

$$|y(x)| \leq \delta/2 \leq |w(x)|,$$

and consequently,

$$|u(x)| \leq 2|w(x)| \tag{12.39}$$

for all such x. Now we have by hypothesis (A) and (12.34)

$$\begin{aligned} G_\mu(u) &\geq \mu\|u\|_D^2 - \tilde{\lambda}\int_{|u|<\delta} u^2 dx - C\int_{|u|>\delta} (|Vu|^q + V^q|u|)dx \\ &\geq \mu\|u\|_D^2 - \tilde{\lambda}\|u\|^2 - C'\int_{|u|>\delta} |Vu|^q dx \\ &\geq (\mu-\eta)\|y\|_D^2 + \mu\|w\|_D^2 - \tilde{\lambda}\|w\|^2 - C''\int_{2|w|>\delta} |Vw|^q dx \end{aligned}$$

in view of the fact that $\|y\|_D^2 = \lambda_0\|y\|^2$ and (12.39) holds. Thus, by (12.26),

$$G_\mu(u) \geq (\mu-\eta)\|y\|_D^2 + \left(\mu - \frac{\tilde{\lambda}}{\lambda_1} - C'''\|w\|_D^{q-2}\right)\|w\|_D^2, \quad \|u\|_D \leq \rho, \tag{12.40}$$

where λ_1 is the next point in the spectrum of $\mathcal{A}$. We take $\rho > 0$ to satisfy

$$\eta - \frac{\tilde{\lambda}}{\lambda_1} > C'''\rho^{q-2}$$

. Consequently,

$$G_\mu(u) \geq (\mu-\eta)\rho^2 + \left(\mu - \frac{\tilde{\lambda}}{\lambda_1} - C'''\rho^{q-2} - \mu + \eta\right)\|w\|_D^2 \geq (\mu-\eta)\rho^2, \quad \|u\|_D = \rho.$$

Thus, (12.37) holds.

We let $A = \{r\varphi : r \geq 0\}$ and $B = \partial\mathcal{B}_\rho$. By Theorem 12.9, they form a custom sandwich pair. Note that

$$G_\mu(r\varphi) = \int_\Omega [\mu\lambda_0 r^2\varphi^2 - 2F(x, r\varphi)]\,dx.$$

By (12.35) and (12.37), we see that (12.21) holds for each G_μ, for $\tilde{\lambda}/\lambda_0 < \mu < \lambda/\lambda_0$ with $b_0 > 0$. Apply Corollary 12.11, and take $\beta = 1/\mu$. □

Proof of Theorem 12.2 Since the Dirichlet problem (12.2) is a special case of problem (12.36), Theorem 12.12 implies Theorem 12.2. □

Proof of Corollary 12.3 We know that φ does not change sign in Ω. We take it to be positive and satisfy $\|\varphi\|_D = 1$. Since

$$G_\mu(\pm r\varphi) = \int_\Omega [\mu\lambda_0 r^2\varphi^2 - 2F(x, \pm r\varphi)]\,dx,$$

we have by (12.7) or (12.8)

$$G_\mu(\pm r\varphi)/r^2 \leq \mu\lambda_0 - \lambda + \int_\Omega [W_\pm(x, \pm r\varphi)/r^2\varphi^2]\varphi^2\,dx \to \mu\lambda_0 - \lambda < 0 \quad \text{as } r \to \infty,$$

as the case may be. This shows that (12.21) holds. Apply Corollary 12.11. □

12.6 Finding the Sequences

We proceed to the proof of Theorem 12.6. Let $M = C_0 + 1$. Then

$$\|\sigma(1)v\| \leq M$$

whenever $\sigma \in \Sigma_0$ satisfies $\|\sigma'(t)\| \leq 1$ and $v \in E$ satisfies $\|v\| \leq C_0$. If the theorem was false, then there would be a $\delta > 0$ such that

$$\|G'(u)\| \geq 3\delta \tag{12.41}$$

when

$$u \in \{u \in E : \|u\| \leq M + 1, |G(u) - a| \leq 3\delta\}. \tag{12.42}$$

Take $\delta < 1/3$ so small that $G(0) < b_0 - 2\delta$. Since $G \in C^1(E, \mathbb{R})$, for each $\theta < 1$ there is a locally Lipschitz continuous mapping $Y(u)$ of $\hat{E} = \{u \in E : G'(u) \neq 0\}$ into E such that

$$\|Y(u)\| \leq 1, \ \theta\|G'(u)\| \leq (G'(u), Y(u)), \quad u \in \hat{E} \tag{12.43}$$

(cf., e.g., [184]). Take $\theta > 2/3$. Let

$$\begin{aligned} Q_0 &= \{u \in E : \|u\| \leq M + 1, |G(u) - a| \leq 2\delta\}, \\ Q_1 &= \{u \in E : \|u\| \leq M, |G(u) - a| \leq \delta\}, \\ Q_2 &= E \setminus Q_0, \\ \eta(u) &= d(u, Q_2)/[d(u, Q_1) + d(u, Q_2)]. \end{aligned}$$

It is easily checked that $\eta(u)$ is locally Lipschitz continuous on E and satisfies

$$\begin{cases} \eta(u) = 1, & u \in Q_1, \\ \eta(u) = 0, & u \in \bar{Q}_2, \\ \eta(u) \in (0, 1), & \text{otherwise}. \end{cases} \tag{12.44}$$

Let

$$W(u) = -\eta(u)Y(u).$$

Then

$$\|W(u)\| \le 1, \quad u \in E.$$

By Theorem 4.5 of [189], for each $v \in E$ there is a unique solution $\sigma(t)v$ of

$$\sigma'(t) = W(\sigma(t)), \ t \in \mathbb{R}^+, \quad \sigma(0) = v. \tag{12.45}$$

We have

$$\begin{aligned} dG(\sigma(t)v)/dt &= -\eta(\sigma(t)v)(G'(\sigma(t)v), Y(\sigma(t)v)) \\ &\le -\theta\eta(\sigma)\|G'(\sigma)\| \\ &\le -3\theta\delta\eta(\sigma). \end{aligned} \tag{12.46}$$

Let $K \in \mathcal{K}_0$ satisfy the hypotheses of the theorem. Let v be any element of $K \cap Q_1$. Then $\|v\| \le C_0$. If there is a $t_1 \le 1$ such that $\sigma(t_1)v \notin Q_1$, then

$$G(\sigma(1)v) < a - \delta, \tag{12.47}$$

since $\|\sigma(1)v\| \le M$,

$$G(\sigma(1)v) \le G(\sigma(t_1)v),$$

and the right-hand side cannot be greater than $a + \delta$ by (12.46). On the other hand, if $\sigma(t)v \in Q_1$ for all $t \in [0, 1]$, then we have by (12.46)

$$G(\sigma(1)v) \le a + \delta - 3\delta\theta < a - \delta.$$

If $v \in K\backslash Q_1$, then we must have

$$G(\sigma(1)v) \le G(v) < a - \delta,$$

since $G(v) \ge a - \delta$ would put v into Q_1. Hence

$$G(\sigma(1)v) < a - \delta, \quad v \in K. \tag{12.48}$$

Next, we note that $\sigma \in \Sigma_0$. To see this, note that $\eta(u) = 0$ when $G(u) < a - 2\delta$. Since $G(0) < a - 2\delta$ and $G(\sigma(t)0)$ decreases as t increases, we see that $\sigma'(t)0 = 0$ for $t \in [0, 1]$. By hypothesis, $\tilde{K} = \sigma(1)K \in \mathcal{K}_0$. This means that

$$G(w) < a - \delta, \quad w \in \tilde{K}. \tag{12.49}$$

But this contradicts the definition (12.14) of a. Hence (12.41) cannot hold for u satisfying (12.42). This proves the theorem.

Proof of Theorem 12.7 Since $A \in \mathcal{K}_0$, clearly $a \le a_0$. Moreover, for any $K \in \mathcal{K}_0$, we have

$$b_0 = \inf_B G \le \inf_{B \cap K} G \le \sup_{B \cap K} G \le \sup_K G.$$

Hence, $b_0 \le a$. Apply Theorem 12.6. □

Proof of Theorem 12.9 Take

$$\mathcal{K}_0 = \{\sigma(1)A : \sigma \in \Sigma_0\}.$$

If $\sigma \in \Sigma_0$ and $v \in A \cap \partial B_R$, then

$$\|\sigma(1)v - v\| = \|\int_0^1 \sigma'(t)v\,dt\| \le 1.$$

Consequently,

$$\|\sigma(1)v\| \ge \|v\| - \|\sigma(1)v - v\| \ge R - 1 \to \infty \text{ as } R \to \infty.$$

Since $\sigma(1)0 = 0$, we see that $\sigma(1)A \cap B \ne \phi$. □

12.7 Notes and Remarks

Here we follow [227, 230], and [210].

Chapter 13
Elliptic Systems

13.1 Introduction

In this chapter we show how monotonicity methods combined with infinite dimensional sandwich pairs can be used to solve very general systems of equations whether or not they are semibounded.

The purpose of this investigation is to solve systems of equations of the form

$$\mathcal{A}v = f(x, v, w) \tag{13.1}$$

$$\mathcal{B}w = g(x, v, w), \tag{13.2}$$

where $\mathcal{A}, \mathcal{B}$ are linear partial differential operators. We assume that there is a Carathéodory function $F(x, v, w)$ on $\Omega \times \mathbb{R}^2$ such that

$$f(x, v, w) = \partial F/\partial v, \quad g(x, v, w) = \partial F/\partial w. \tag{13.3}$$

If $\mathcal{A}, \mathcal{B}$ are positive, then the problem (13.1), (13.2) is called cooperative. This situation has the advantage that it leads to a coercive problem.

On the other hand, if one operator is positive and the other is negative, i.e., the system is of the form

$$-\mathcal{A}v = f(x, v, w) \tag{13.4}$$

$$\mathcal{B}w = g(x, v, w), \tag{13.5}$$

then the system is called noncooperative. In this case serious problems arise. The corresponding functional

M. Schechter, *Critical Point Theory*,
https://doi.org/10.1007/978-3-030-45603-0_13

$$G(u) = (\mathcal{B}w, w) - (\mathcal{A}v, v) - 2\int_{\Omega} F(x, v, w)dx, \quad u \in D \tag{13.6}$$

is unbounded from above and below on infinite dimensional subspaces.

The purpose of the present chapter is to weaken the hypotheses in both cases considerably. In particular, we do not assume that the functions $f(x, v, w)$, $g(x, v, w)$ are differentiable in any sense, nor do we assume any asymptotic limits for them. Moreover, we do not assume that the systems satisfy any Palais-Smale or Cerami conditions.

13.2 The Cooperative System

Let $\mathcal{A}, \mathcal{B}$ be positive, self-adjoint operators on $L^2(\Omega)$ with compact resolvents, where $\Omega \subset \mathbb{R}^n$. Let $F(x, v, w)$ be a Carathéodory function on $\Omega \times \mathbb{R}^2$ such that

$$f(x, v, w) = \partial F/\partial v, \quad g(x, v, w) = \partial F/\partial w \tag{13.7}$$

are also Carathéodory functions satisfying

$$|f(x, v, w)| + |g(x, v, w)| \leq C_0(|v| + |w| + 1), \quad v, w \in \mathbb{R}. \tag{13.8}$$

We wish to solve the system

$$\mathcal{A}v = f(x, v, w) \tag{13.9}$$

$$\mathcal{B}w = g(x, v, w). \tag{13.10}$$

Let $\lambda_0(\mu_0)$ be the lowest eigenvalue of $\mathcal{A}(\mathcal{B})$. These are assumed positive. There are corresponding eigenfunctions $v_0, \ w_0$ which do not vanish in Ω.

Our first result is the following.

Theorem 13.1 *Assume*

$$2F(x, s, t) \leq \lambda(x)s^2 + \mu(x)t^2 + W(x), \quad x \in \Omega, \ s, t \in \mathbb{R}, \tag{13.11}$$

where $W(x) \in L^1(\Omega)$ and

$$\lambda(x) \leq\not\equiv \lambda_0, \ \mu(x) \leq\not\equiv \mu_0, \quad x \in \Omega. \tag{13.12}$$

Then the system (13.9) and (13.10) has a solution. If, in addition,

$$\int_{\Omega} [\lambda_0 v_0(x)^2 + \mu_0 w_0(x)^2 - 2F(x, v_0(x), w_0(x))]dx < 0, \tag{13.13}$$

then the system (13.9) and (13.10) has a nontrivial solution.

Let $D = D(\mathcal{A}^{1/2}) \times D(\mathcal{B}^{1/2})$. Then D becomes a Hilbert space with norm given by

$$\|u\|_D^2 = (\mathcal{A}v, v) + (\mathcal{B}w, w), \quad u = (v, w) \in D. \tag{13.14}$$

We let N be the set of those $(v, 0) \in D$ and M the set of those $(0, w) \in D$. Then M, N are orthogonal closed subspaces such that

$$D = M \oplus N. \tag{13.15}$$

If u_k is a bounded sequence in D, then there is a renamed subsequence and a $u \in D$ such that $u_k \to u$ weakly in D, strongly in $L^2(\Omega)$, and a.e. in Ω.

We define

$$G(u) = b(w) + a(v) - 2\int_\Omega F(x, v, w)dx, \quad u \in D, \tag{13.16}$$

where

$$a(v) = (\mathcal{A}v, v), \quad b(w) = (\mathcal{B}w, w). \tag{13.17}$$

Then $G \in C^1(D, \mathbb{R})$ and

$$(G'(u), h)/2 = b(w, h_2) + a(v, h_1) - (f(u), h_1) - (g(u), h_2), \tag{13.18}$$

where we write $f(u), g(u)$ in place of $f(x, v, w), g(x, v, w)$, respectively. It is readily seen that the system (13.9) and (13.10) is equivalent to

$$G'(u) = 0. \tag{13.19}$$

13.3 Some Lemmas

Lemma 13.2 *If $b_k^2 = b(w_k) \to \infty$, then*

$$b(w_k) - \int_\Omega \mu(x)w_k^2 \to \infty.$$

If $a(v_k) \to \infty$, then

$$a(v_k) - \int_\Omega \lambda(x)v_k^2 \to \infty.$$

Proof. Let $D = D(\mathcal{A}^{1/2}) \times D(\mathcal{B}^{1/2})$. Then D becomes a Hilbert space with norm given by

$$\|u\|_D^2 = (\mathcal{A}v, v) + (\mathcal{B}w, w), \quad u = (v, w) \in D. \tag{13.20}$$

We let N be the set of those $(v, 0) \in D$ and M the set of those $(0, w) \in D$. Then M, N are orthogonal closed subspaces such that

$$D = M \oplus N. \tag{13.21}$$

If u_k is a bounded sequence in D, then there is a renamed subsequence and a $u \in D$ such that $u_k \to u$ weakly in D, strongly in $L^2(\Omega)$, and a.e. in Ω. Let N' be the orthogonal complement of $N_0 = E(\lambda_0)$ in N. Then $N = N' \oplus N_0$. Let $M_0 = E(\mu_0)$, and let M' be its orthogonal complement in M.

Assume the lemma is not true. Then there would be a sequence $\{v_k, w_k\}$ such that $a(v_k) + b(w_k) \to \infty$ and $b(w_k) - \int_\Omega \mu(x) w_k^2 + a(v_k) - \int_\Omega \lambda(x) v_k^2$ is bounded. Write $v_k = v_k' + y_k,\ v_k' \in N',\ y_k \in N_0$ and $w_k = w_k' + h_k,\ w_k' \in M',\ h_k \in M_0$. If $b_k^2 = b(w_k) \to \infty$, let $\tilde{w}_k = w_k/b_k$. Then $b(\tilde{w}_k) = 1$, and there is a renamed subsequence such that $\tilde{w}_k \to \tilde{w},\ \tilde{w}_k' \to \tilde{w}',\ \tilde{h}_k \to \tilde{h}$ weakly in D, strongly in $L^2(\Omega)$, and a.e. in Ω. Then

$$\begin{aligned}
[b(w_k) - \int_\Omega \mu(x) w_k^2]/b_k^2 &= 1 - \int_\Omega \mu(x) \tilde{w}_k(x)^2 \\
&\to 1 - \int_\Omega \mu(x) \tilde{w}(x)^2 \\
&= [1 - b(\tilde{w})] \\
&\quad + [b(\tilde{w}) - \mu_0 \|\tilde{w}\|^2] \\
&\quad + \int_\Omega [\mu_0 - \mu(x)] \tilde{w}^2 \\
&= A + B + C,
\end{aligned}$$

with A, B, C nonnegative. If $A = 0$, then $b(\tilde{w}) = 1$. If $B = 0$ also, then $\tilde{w}' = 0,\ \tilde{w} = \tilde{h}$. If, in addition, $C = 0$, then

$$\int_\Omega [\mu_0 - \mu(x)] \tilde{h}^2 = 0.$$

Consequently, $\tilde{h} = 0$ on a set of positive measure. By hypothesis, $\tilde{h} \equiv 0$. This means that $\tilde{w} \equiv 0$, contradicting the fact that $b(\tilde{w}) = 1$. Thus, $b(w_k) \to \infty$ implies

$$b(w_k) - \int_\Omega \mu(x) w_k^2 \to \infty.$$

Similarly, $a(v_k) \to \infty$ implies

$$a(v_k) - \int_\Omega \lambda(x)v_k^2 \to \infty.$$

□

Lemma 13.3

$$G(u) \to \infty \text{ as } \|u\|_D \to \infty. \tag{13.22}$$

Lemma 13.4

$$b_0 = \inf_D G > -\infty. \tag{13.23}$$

Proof. By (13.11) and (13.12),

$$\begin{aligned} G(u) &= b(w) + a(v) - 2\int_\Omega F(x, v, w)dx \\ &\geq b(w) - \int_\Omega \mu(x)w^2 + a(v) - \int_\Omega \lambda(x)v^2 - \int_\Omega W(x) \\ &\geq \int_\Omega [\mu_0 - \mu(x)]w(x)^2 + \int_\Omega [\lambda_0 - \lambda(x)]v(x)^2 - \int_\Omega W(x) \\ &> -\int_\Omega W(x), \quad u = (v, w) \in D. \end{aligned}$$

Moreover, by Lemma 13.2

$$G(u) \geq b(w) - \int_\Omega \mu(x)w^2 + a(v) - \int_\Omega \lambda(x)v^2 - \int_\Omega W(x) \to \infty$$

as $\|u\|_D \to \infty$. □

13.4 Proof of Theorem 13.1

We can now apply Corollary 3.22 of [189] to conclude that there is a sequence $\{u_k\} \subset D$ such that

$$G(u_k) \to b_0, \quad (1 + \|u_k\|_D)\|G'(u_k)\| \to 0. \tag{13.24}$$

Once this is known we can use the usual procedures to show that there is a renamed subsequence such that $u_k \to u$ in D, and u satisfies (13.19). In fact, since the u_k are bounded in D, there is a renamed subsequence converging to a limit $u \in D$ weakly in D, strongly in $L^2(\Omega)$, and a.e. in Ω. For any $\varphi = (\varphi_1, \varphi_2) \in C_0^\infty(\Omega)$, we have

$$\begin{aligned}(G'(u_k), \varphi)/2 &= b(w_k, \varphi_2) + a(v_k, \varphi_1)\\ &\quad - \int_\Omega [f(x, u_k(x))\varphi_1(x) + g(x, u_k(x))\varphi_1(x)]dx \to 0.\end{aligned}$$

Hence,

$$(G'(u), \varphi)/2 = b(w, \varphi) + a(v, \varphi) - \int_\Omega [f(x, u(x))\varphi_1(x) + g(x, u(x)\varphi_2(x)] = 0,$$

showing that $G'(u) = 0$. Moreover,

$$\begin{aligned}&\int_\Omega [f(x, u_k(x))v_k(x) + g(x, u_k(x))w_k(x)]dx\\ &\qquad \to \int_\Omega [f(x, u(x))v(x) + g(x, u(x))w(x)]dx\end{aligned}$$

since the integrands converge a.e. and in $L^1(\Omega)$. Consequently,

$$\begin{aligned}(G'(u_k), u_k)/2 &= b(w_k) + a(v_k)\\ &\quad - \int_\Omega [f(x, u_k(x))v_k(x) + g(x, u_k(x))w_k(x)]dx\\ &\to 0 = (G'(u), u)/2 = b(w) + a(v)\\ &\quad - \int_\Omega [f(x, u(x))v(x) + g(x, u(x))w(x)] = 0.\end{aligned}$$

Thus,

$$b(w_k) + a(v_k) \to b(w) + a(v).$$

Since

$$\int_\Omega F(x, v_k, w_k)dx \to \int_\Omega F(x, v, w)dx,$$

we see that $G(u_k) \to G(u)$. But $G(u_k) \to b_0$. Hence, $G(u) = b_0$ and $G'(u) = 0$. If (13.13) holds as well, then we know that $b_0 < 0$. This shows that the solution obtained from (13.24) is not 0.

13.5 The Noncooperating System

Let $\mathcal{A}, \mathcal{B}$ be positive, self-adjoint operators on $L^2(\Omega)$ with compact resolvents, where $\Omega \subset \mathbb{R}^n$. Let $F(x, v, w)$ be a Carathéodory function on $\Omega \times \mathbb{R}^2$ such that

$$f(x, v, w) = \partial F/\partial v, \quad g(x, v, w) = \partial F/\partial w \tag{13.25}$$

are also Carathéodory functions satisfying

$$|f(x, v, w)| + |g(x, v, w)| \leq C_0(|v| + |w| + 1), \quad v, w \in \mathbb{R}. \tag{13.26}$$

We wish to solve the system

$$-\mathcal{A}v = f(x, v, w) \tag{13.27}$$

$$\lambda \mathcal{B}w = g(x, v, w). \tag{13.28}$$

We want to solve this system for $\lambda = 1$. But we discovered that we can achieve this under weaker hypotheses if we introduce a parameter λ and make use of monotonicity methods to solve the system (13.27), (13.28) for almost all λ in some interval containing $\lambda = 1$ and then zeroing in on $\lambda = 1$. Let $\lambda_0(\mu_0)$ be the lowest eigenvalue of $\mathcal{A}(\mathcal{B})$. These are assumed positive.

Our first result is the following.

Theorem 13.5 *Assume*

$$2F(x, 0, t) \leq \mu_0 t^2 + W_1(x), \quad x \in \Omega, \ s \in \mathbb{R}, \tag{13.29}$$

and

$$2F(x, s, t) \geq -\lambda(x)s^2 - \mu(x)t^2 - W_2(x), \quad x \in \Omega, \ s, t \in \mathbb{R}, \tag{13.30}$$

where $W_i(x) \in L^1(\Omega)$ and

$$\lambda(x) \leq\not\equiv \lambda_0, \ \mu(x) \leq\not\equiv \mu_0, \quad x \in \Omega. \tag{13.31}$$

Then the system (13.27) and (13.28) has a solution for a.e. $\lambda \geq \beta/\mu_0$, where $\beta =$ $\sup \mu(x)$.

Proof. Let $D = D(\mathcal{A}^{1/2}) \times D(\mathcal{B}^{1/2})$. Then D becomes a Hilbert space with norm given by

$$\|u\|_D^2 = (\mathcal{A}v, v) + (\mathcal{B}w, w), \quad u = (v, w) \in D. \tag{13.32}$$

We define

$$G_\lambda(u) = \lambda b(w) - a(v) - 2\int_\Omega F(x, v, w)dx, \quad u \in D, \tag{13.33}$$

where

$$a(v) = (\mathcal{A}v, v), \quad b(w) = (\mathcal{B}w, w). \tag{13.34}$$

Then $G_\lambda \in C^1(D, \mathbb{R})$ and

$$(G'_\lambda(u), h)/2 = \lambda b(w, h_2) - a(v, h_1) - (f(u), h_1) - (g(u), h_2) \tag{13.35}$$

where we write $f(u), g(u)$ in place of $f(x, v, w), g(x, v, w)$, respectively. It is readily seen that the system (13.27) and (13.28) is equivalent to

$$G'_\lambda(u) = 0. \tag{13.36}$$

We let N be the set of those $(v, 0) \in D$ and M the set of those $(0, w) \in D$. Then M, N are orthogonal closed subspaces such that

$$D = M \oplus N. \tag{13.37}$$

If we define

$$L_\lambda u = 2(-v, \lambda w), \quad u = (v, w) \in D, \tag{13.38}$$

then L_λ is a self-adjoint bounded operator on D. Also

$$G'_\lambda(u) = L_\lambda u + c_0(u), \tag{13.39}$$

where

$$c_0(u) = -(\mathcal{A}^{-1} f(u), \mathcal{B}^{-1} g(u)) \tag{13.40}$$

is compact on D. This follows from (13.26) and the fact that $\mathcal{A}$ and $\mathcal{B}$ have compact resolvents. It also follows that G'_λ has weak-to-weak continuity. For if $u_k \to u$ weakly, then $L_\lambda u_k \to L_\lambda u$ weakly and $c_0(u_k)$ has a convergent subsequence. Now by (13.29)

$$G_\lambda(0, w) \geq \lambda b(w) - \int_\Omega [\mu_0 w^2 + W_1(x)]\, dx, \quad (0, w) \in M. \tag{13.41}$$

Thus

$$\inf_M G_\lambda \geq -\int_\Omega W_1(x)\, dx \equiv b_0. \tag{13.42}$$

On the other hand, (13.30) implies

$$G_\lambda(v,0) \le -a(v) + \lambda_0\|v\|^2 + \int_\Omega W_2(x)\,dx, \quad (v,0) \in N. \tag{13.43}$$

Thus

$$\sup_N G_\lambda \le \int_\Omega W_2(x)\,dx \equiv a_0. \tag{13.44}$$

Define

$$I(u) = b(w), \quad J(u) = a(v) + 2\int_\Omega F(x,v,w)dx, \quad u \in D. \tag{13.45}$$

Then, $G_1(u) = I(u) - J(u)$ and

$$I(u) + J(u) = b(w) + a(v) + 2\int_\Omega F(x,v,w)dx, \quad u \in D.$$

Then,

$$I(u) + J(u) \to \infty \text{ as } \|u\|_D \to \infty. \tag{13.46}$$

To see this, let N' be the orthogonal complement of $N_0 = E(\lambda_0)$ in N. Then $N = N' \oplus N_0$. Let $M_0 = E(\mu_0)$, and let M' be its orthogonal complement in M. Assume (13.46) is not true. Then there would be a sequence $\{v_k, w_k\}$ such that $a(v_k) + b(w_k) \to \infty$ and $b(w_k) - \int_\Omega \mu(x)w_k^2 + a(v_k) - \int_\Omega \lambda(x)v_k^2$ is bounded. Write $v_k = v_k' + y_k,\ v_k' \in N',\ y_k \in N_0$ and $w_k = w_k' + h_k,\ w_k' \in M',\ h_k \in M_0$. If $b_k^2 = b(w_k) \to \infty$, let $\tilde{w}_k = w_k/b_k$. Then $b(\tilde{w}_k) = 1$, and there is a renamed subsequence such that $\tilde{w}_k \to \tilde{w},\ \tilde{w}_k' \to \tilde{w}',\ \tilde{h}_k \to \tilde{h}$ weakly in D, strongly in $L^2(\Omega)$, and a.e. in Ω. Then

$$\begin{aligned}
[b(w_k) - \int_\Omega \mu(x)w_k^2]/b_k^2 &= 1 - \int_\Omega \mu(x)\tilde{w}_k(x)^2 \\
&\to 1 - \int_\Omega \mu(x)\tilde{w}(x)^2 \\
&= [1 - b(\tilde{w})] \\
&\quad + [b(\tilde{w}) - \mu_0\|\tilde{w}\|^2] \\
&\quad + \int_\Omega [\mu_0 - \mu(x)]\tilde{w}^2 \\
&= A + B + C,
\end{aligned}$$

with A, B, C nonnegative. If $A = 0$, then $b(\tilde{w}) = 1$. If $B = 0$ also, then $\tilde{w}' = 0$, $\tilde{w} = \tilde{h}$. If, in addition, $C = 0$, then

$$\int_\Omega [\mu_0 - \mu(x)]\tilde{h}^2 = 0.$$

Consequently, $\tilde{h} = 0$ on a set of positive measure. By hypothesis, $\tilde{h} \equiv 0$. This means that $\tilde{w} \equiv 0$, contradicting the fact that $b(\tilde{w}) = 1$. Thus, $b(w_k) \to \infty$ implies

$$b(w_k) - \int_\Omega \mu(x) w_k^2 \to \infty.$$

Similarly, $a(v_k) \to \infty$ implies

$$a(v_k) - \int_\Omega \lambda(x) v_k^2 \to \infty.$$

Thus, (13.46) holds.

We can now apply Corollary 5.21 to conclude that for a.e. $\lambda \geq \beta/\mu_0$ there is a bounded sequence $\{u_k\} \subset D$ such that

$$G_\lambda(u_k) \to a, \quad \|G'_\lambda(u_k)\| \to 0. \tag{13.47}$$

Once this is known we can use the usual procedures to show that there is a renamed subsequence such that $u_k \to u$ in D, and u satisfies (13.36). □

Theorem 13.6 *In addition, assume that the eigenfunctions of* λ_0 *and* μ_0 *are bounded and* $\neq 0$ *a.e. in* Ω, *and there is a* $q > 2$ *such that*

$$\|w\|_q^2 \leq Cb(w), \quad w \in M. \tag{13.48}$$

Assume that for some $\delta > 0$,

$$2F(x, s, t) \leq \beta t^2 - \lambda_0 s^2, \quad |t| + |s| \leq \delta, \tag{13.49}$$

where $\beta = \sup \mu(x)$. *Then the system (13.27) and (13.28) has a nontrivial solution for a.e.* $\lambda \geq \beta/\mu_0$.

Theorem 13.7 *Under the same hypotheses, if*

$$H(x, s, t) \leq W(x) \tag{13.50}$$

and

$$H(x, s, t) \to -\infty \text{ as } |s| + |t| \to \infty, \tag{13.51}$$

where $W(x) \in L^2(\mathbb{R}^n)$ *and*

$$H(x,s,t) = f(x,s,t)s + g(x,s,t)t - 2F(x,s,t), \tag{13.52}$$

then the system (13.27) and (13.28) has a nontrivial solution for all $\lambda \geq \beta/\mu_0$. *In particular, this is true for* $\lambda = 1$.

Proof of Theorem 13.6 Let N' be the orthogonal complement of $N_0 = \{\varphi_0\}$ in N, where φ_0 is the eigenfunction of $\mathcal{A}$ corresponding to λ_0. Then $N = N' \oplus N_0$. Let M_0 be the subspace of M spanned by the eigenfunctions of $\mathcal{B}$ corresponding to μ_0, and let M' be its orthogonal complement in M. Since N_0 and M_0 are contained in $L^\infty(\Omega)$, there is a positive constant ρ such that

$$a(y) \leq \rho^2 \Rightarrow \|y\|_\infty \leq \delta/4, \quad y \in N_0 \tag{13.53}$$

$$b(h) \leq \rho^2 \Rightarrow \|h\|_\infty \leq \delta/4, \quad h \in M_0, \tag{13.54}$$

where δ is the number given in (13.49). If

$$a(y) \leq \rho^2, b(w) \leq \rho^2, |y(x)| + |w(x)| \geq \delta \tag{13.55}$$

we write $w = h + w', h \in M_0, w' \in M'$ and

$$\delta \leq |y(x)| + |w(x)| \leq |y(x)| + |h(x)| + |w'(x)| \leq (\delta/2) + |w'(x)|. \tag{13.56}$$

Thus

$$|y(x)| + |h(x)| \leq \delta/2 \leq |w'(x)| \tag{13.57}$$

and

$$|y(x)| + |w(x)| \leq 2|w'(x)|. \tag{13.58}$$

Now by (13.49) and (13.58)

$$G_\lambda(y,w) = \lambda b(w) - a(y) - 2\int_\Omega F(x,y,w)\,dx \tag{13.59}$$

$$\geq \lambda b(w) - a(y) - \int_{|y|+|w|<\delta} \{\beta w^2 - \lambda_0 y^2\} dx \tag{13.60}$$

$$- c_0 \int_{|y|+|w|>\delta} (|y| + |w| + 1)^2\,dx \tag{13.61}$$

$$\geq \lambda b(w) - a(y) - \beta\|w\|^2 + \lambda_0\|y\|^2 - c_1 \int\limits_{2|w'|>\delta} |w'|^q dx \tag{13.62}$$

$$\geq \lambda b(w') - \beta\|w'\|^2 - c_2 b(w')^{q/2} \tag{13.63}$$

$$\geq \left(\lambda - \frac{\beta}{\mu_1} - c_2 b(w')^{(q/2)-1}\right) b(w'), \quad a(y) \leq \rho^2,\ b(w) \leq \rho^2, \tag{13.64}$$

where μ_1 is the next eigenvalue of $\mathcal{B}$ after μ_0. If we reduce ρ accordingly, we can find a positive constant ν such that

$$G_\lambda(y, w) \geq \nu b(w'), \quad a(y) \leq \rho^2, b(w) \leq \rho^2. \tag{13.65}$$

I claim that either system (13.27), (13.28) has a nontrivial solution or there is an $\epsilon > 0$ such that

$$G_\lambda(y, w) \geq \epsilon, \quad a(y) + b(w) = \rho^2. \tag{13.66}$$

For suppose (13.66) did not hold. Then there would be a sequence $\{y_k, w_k\}$ such that $a(y_k) + b(w_k) = \rho^2$ and $G_\lambda(y_k, w_k) \to 0$. If we write $w_k = w_k' + h_k$, $w_k' \in M'$, $h_k \in M_0$, then (13.65) tells us that $b(w_k') \to 0$. Thus $a(y_k)+b(h_k) \to \rho^2$. Since N_0, M_0 are finite dimensional, there is a renamed subsequence such that $y_k \to y$ in N_0 and $h_k \to h$ in M_0. By (13.53) and (13.54), $\|y\|_\infty \leq \delta/4$ and $\|h\|_\infty \leq \delta/4$. Consequently (13.49) implies

$$2F(x, y, h) \leq \lambda\mu_0 h^2 - \lambda_0 y^2. \tag{13.67}$$

Since

$$G_\lambda(y, h) = \lambda b(h) - a(y) - 2\int_\Omega F(x, y, h)dx = 0, \tag{13.68}$$

we have

$$\int_\Omega \{2F(x, y, h) + \lambda_0 y^2 - \lambda\mu_0 h^2\}dx = 0. \tag{13.69}$$

In view of (13.67), this implies

$$2F(x, y, h) \equiv \lambda\mu_0 h^2 - \lambda_0 y^2. \tag{13.70}$$

For $\zeta \in C_0^\infty(\Omega)$ and $t > 0$ small we have

$$2[F(x, y + t\zeta, h) - F(x, y, h)]/t \leq -\lambda_0[(y + t\zeta)^2 - y^2]/t. \tag{13.71}$$

Taking $t \to 0$, we have

$$f(x, y, h)\zeta \le -\lambda_0 y\zeta. \tag{13.72}$$

Since this is true for all $\zeta \in C_0^\infty(\Omega)$, we have

$$f(x, y, h) = -\lambda_0 y = -\mathcal{A}y. \tag{13.73}$$

Similarly,

$$2[F(x, y, h + t\zeta) - F(x, y, h)]/t \le \lambda\mu_0[(h + t\zeta)^2 - h^2]/t \tag{13.74}$$

and consequently

$$g(x, y, h)\zeta \le \lambda\mu_0 h\zeta \tag{13.75}$$

and

$$g(x, y, h) = \lambda\mu_0 h = \lambda\mathcal{B}h. \tag{13.76}$$

We see from (13.73) and (13.76) that (13.27) and (13.28) has a nontrivial solution. Thus, we may assume that (13.66) holds.

Next, we note that there is an $\varepsilon > 0$ depending on ρ and β/μ_0 such that

$$G_\lambda(0, w) \ge \varepsilon, \quad b(w) \ge \rho > 0, \ w \in M.$$

To see this, suppose that $\{w_k\} \subset M$ is a sequence such that

$$G_\lambda(0, w_k) \to 0, \quad b(w_k) \ge \rho.$$

If

$$b_k^2 = b(w_k) \le C,$$

this implies

$$b(w_k) - \mu_0\|w_k\|^2 \to 0$$

and

$$\int [\lambda\mu_0 - \mu(x)]w_k^2 dx \to 0,$$

since

$$G_\lambda(0,w) \geq \lambda[b(w) - \mu_0\|w\|^2] + \int [\lambda\mu_0 - \mu(x)]w^2 dx, \quad w \in M.$$

If we write $w_k = w_k' + h_k, w_k' \in M', h_k \in M_0$ as before, then this tells us that $b(w_k') \to 0$. Since M_0 is finite dimensional, there is a renamed subsequence such that $h_k \to h$. But the two conclusions above tell us that $h = 0$. Since $b(h) \geq \rho$, we see that $\varepsilon > 0$ exists for any constant C. If the sequence $\{b_k\}$ is not bounded, we take $\tilde{w}_k = w_k/b_k$. Then

$$G_\lambda(0,w_k)/b_k^2 \geq \lambda[b(\tilde{w}_k) - \mu_0\|\tilde{w}_k\|^2] + \int [\lambda\mu_0 - \mu(x)]\tilde{w}_k^2 dx,$$

and the inequality is true in this case as well.

Next we note that there is a $\nu > 0$ such that

$$G_\lambda(0,w) \geq \nu b(w), \quad w \in M. \tag{13.77}$$

Assuming this for the moment, we see that

$$\inf_B G_\lambda \geq \varepsilon_1 > 0, \tag{13.78}$$

where

$$B = \{w \in M : b(w) \geq \rho^2\} \cup \{u = (s\varphi_0, w) : s \geq 0, w \in M, \|u\|_D = \rho\}, \tag{13.79}$$

and $\varepsilon_1 = \min\{\varepsilon, \nu\rho^2\}$. By (13.44) there is an $R > \rho$ such that

$$\sup_A G_\lambda = a_\lambda < \infty, \tag{13.80}$$

where $A = N \cap \partial B_R$. By Corollary 5.12, A, B form a weak sandwich pair. Hence, for a.e. $\lambda > \beta/\mu_0$ there is a bounded sequence $\{u_k\} \subset D$ such that

$$G_\lambda(u_k) \to c_\lambda, \quad \|G_\lambda'(u_k)\| \to 0 \tag{13.81}$$

holds with $c_\lambda \geq \varepsilon_1$. Arguing as in the proof of Theorem 13.5, we see that there is a $u \in D$ such that $G_\lambda(u) = c_\lambda \geq \varepsilon_1 > 0$, $G_\lambda'(u) = 0$. Since $c_\lambda \neq 0$ and $G_\lambda(0) = 0$, we see that $u \neq 0$, and we have a nontrivial solution of the system (13.27) and (13.28).

It therefore remains only to prove (13.77). Clearly $\nu \geq 0$. If $\nu = 0$, then there is a sequence $\{w_k\} \subset M$ such that

$$G_\lambda(0,w_k) \to 0, \quad b(w_k) = 1. \tag{13.82}$$

Thus there is a renamed subsequence such that $w_k \to w$ weakly in M, strongly in $L^2(\Omega)$, and a.e. in Ω. Consequently

$$\int_\Omega [\lambda\mu_0 - \mu(x)]w_k^2\,dx \le \lambda - \int_\Omega \mu(x)w_k^2\,dx \le G_\lambda(0, w_k) \to 0 \tag{13.83}$$

and

$$\lambda = \int_\Omega \mu(x)w^2dx \le \lambda\mu_0\|w\|^2 \le \lambda b(w) \le \lambda, \tag{13.84}$$

which means that we have equality throughout. It follows that we must have $w \in E(\mu_0)$, the eigenspace of μ_0. Since $w \not\equiv 0$, we have $w \neq 0$ a.e. But

$$\int_\Omega [\lambda\mu_0 - \mu(x)]w^2dx = 0 \tag{13.85}$$

implies that the integrand vanishes identically on Ω, and consequently $\beta = \lambda\mu_0$, violating the hypothesis of the theorem. This establishes (13.77) and completes the proof. □

Proof of Theorem 13.7 By Theorem 13.6, for each $\lambda \ge \beta/\mu_0$ there is a sequence $\lambda_k > \lambda$ such that $\lambda_k \to \lambda$ and the system (13.27) and (13.28) has a nontrivial solution $u_k = (v_k, w_k)$ for λ_k. Hence,

$$G_{\lambda_k}(u_k) = a_k \ge \varepsilon > 0, \quad G'_{\lambda_k}(u_k) = 0, \tag{13.86}$$

where ε depends on β/μ_0 but not on λ_k. This means that

$$G_{\lambda_k}(u_k) = \lambda_k b(w_k) - a(v_k) - 2\int_\Omega F(x, v_k, w_k)dx = a_k \ge \varepsilon, \tag{13.87}$$

and

$$\begin{aligned}(G'_{\lambda_k}(u_k), h)/2 =& \lambda_k b(w_k, h_2) - a(v_k, h_1) \\ & -(f(u_k), h_1) - (g(u_k), h_2) = 0, \quad h = (h_1, h_2) \in D.\end{aligned} \tag{13.88}$$

Consequently, we have

$$\lambda_k b(w_k) = (g(u_k), w_k), \quad a(v_k) + (f(u_k), v_k) = 0, \tag{13.89}$$

and

$$\int_\Omega H(x, v_k, w_k)dx = G_{\lambda_k}(u_k) \to a \ge \varepsilon. \tag{13.90}$$

If u_k is a bounded sequence in D, then there is a renamed subsequence and a $u \in D$ such that $u_k \to u$ weakly in D, strongly in $L^2(\Omega)$, and a.e. in Ω. In that case we obtain

$$\begin{aligned}(G'_\lambda(u), h)/2 =& \lambda b(w, h_2) - a(v, h_1) \\ & -(f(u), h_1) - (g(u), h_2) = 0, \quad h = (h_1, h_2) \in D.\end{aligned} \tag{13.91}$$

This implies that u is a nontrivial solution of the system (13.27) and (13.28).

To complete the proof, we show that the sequence u_k is bounded in D. To see this, assume that $r_k = \|u_k\|_D \to \infty$, and let $\tilde{u}_k = u_k/r_k$. Then $\|\tilde{u}_k\|_D = 1$, and there is a renamed subsequence such that $\tilde{u}_k \to \tilde{u}$, weakly in D, strongly in $L^2(\Omega)$, and a.e. in Ω. Since,

$$\lambda_k b(w_k) + a(v_k) = (g(u_k), w_k) - (f(u_k), v_k),$$

we have

$$\lambda_k b(\tilde{w}_k) + a(\tilde{v}_k) \leq C\|\tilde{u}_k\|^2.$$

Hence,

$$1 = \|\tilde{u}_k\|_D^2 \leq C\|\tilde{u}_k\|^2 \to C\|\tilde{u}\|^2.$$

Consequently, $\tilde{u} \not\equiv 0$.

Let Ω_0 be the subset of Ω on which $\tilde{u} \neq 0$. Then

$$|u_k(x)| = r_k|\tilde{u}_k(x)| \to \infty, \quad x \in \Omega_0. \tag{13.92}$$

If $\Omega_1 = \Omega \setminus \Omega_0$, then we have

$$\int_\Omega H(x, u_k)\, dx = \int_{\Omega_0} + \int_{\Omega_1} \leq \int_{\Omega_0} H(x, u_k)\, dx + \int_{\Omega_1} W_1(x) dx \to -\infty. \tag{13.93}$$

This contradicts (13.90), and we see that $r_k = \|u_k\|_D$ is bounded. □

13.6 Contrast

We now exhibit theorems corresponding to Theorems 13.5 and 13.6 which do not use monotonicity and require the solvability of system (13.27) and (13.28) for $\lambda = 1$.

Theorem 13.8 *In addition to (13.29) and (13.30), assume*

$$f(x, ty, tz)/t \to \alpha_+(x)v^+ - \alpha_-(x)v^- + \beta_+(x)w^+ - \beta_-(x)w^- \tag{13.94}$$

$$g(x, ty, tz)/t \to \gamma_+(x)v^+ - \gamma_-(x)v^- + \delta_+(x)w^+ - \delta_-(x)w^- \tag{13.95}$$

as $t \to +\infty$, $y \to v, z \to w$, *where* $a^{\pm} = \max(\pm a, 0)$. *We also assume that the only solution of*

$$-\mathcal{A}v = \alpha_+v^+ - \alpha_-v^- + \beta_+w^+ - \beta_-w^- \tag{13.96}$$

$$\mathcal{B}w = \gamma_+v^+ - \gamma_-v^- + \delta_+w^+ - \delta_-w^- \tag{13.97}$$

is $v = w = 0$. *Then the system (13.27) and (13.28) has a solution.*

Proof. We follow the proof of Theorem 13.5 for the case $\lambda = 1$. We conclude that there is a sequence $\{u_k\} \subset D$ such that

$$G_1(u_k) \to c_1, \quad \|G_1'(u_k)\| \to 0. \tag{13.98}$$

Let $u_k = (v_k, w_k)$. I claim that

$$\rho_k^2 = a(v_k) + b(w_k) \leq C. \tag{13.99}$$

To see this, assume that $\rho_k \to \infty$, and let $\tilde{u}_k = u_k/\rho_k$. Then there is a renamed subsequence such that $\tilde{u}_k \to \tilde{u}$ weakly in D, strongly in $L^2(\Omega)$, and a.e. in Ω. If $h = (h_1, h_2) \in D$, then

$$(G'(u_k), h)/\rho_k = 2b(\tilde{w}_k, h_2) - 2a(\tilde{v}_k, h_1) - 2(f(u_k), h_1)/\rho_k - 2(g(u_k), h_2)/\rho_k. \tag{13.100}$$

Taking the limit and applying (13.26), (13.94), and (13.95) we see that $\tilde{u} = (\tilde{v}, \tilde{w})$ is a solution of (13.96) and (13.97). Hence $\tilde{u} = 0$ by hypothesis. On the other hand, since $a(\tilde{v}_k) + b(\tilde{w}_k) = 1$, there is a renamed subsequence such that $a(\tilde{v}_k) \to \tilde{a}, b(\tilde{w}_k) \to \tilde{b}$ with $\tilde{a} + \tilde{b} = 1$. Thus by (13.94), (13.95), and (13.35)

$$\begin{aligned}(G'(u_k), (\tilde{v}_k, 0))/2\rho_k &= -a(\tilde{v}_k) - (f(u_k), \tilde{v}_k)/\rho_k \\ &\to -\tilde{a} - \int_\Omega (\alpha_+\tilde{v}^+ - \alpha_-\tilde{v}^- + \beta_+\tilde{w}^+ - \beta_-\tilde{w}^-)\tilde{v}\, dx\end{aligned}$$

and

$$\begin{aligned}(G'(u_k), (0, \tilde{w}_k))/2\rho_k &= b(\tilde{w}_k) - (g(u_k), \tilde{w}_k)/\rho_k \\ &\to \tilde{b} - \int_\Omega (\gamma_+\tilde{v}^+ - \gamma_-\tilde{v}^- + \delta_+\tilde{w}^+ - \delta_-\tilde{w}^-)\tilde{w}\, dx.\end{aligned}$$

Thus by (13.98),

$$\tilde{a} = -\int_\Omega (\alpha_+ \tilde{v}^+ - \alpha_- \tilde{v}^- + \beta_+ \tilde{w}^+ - \beta_- \tilde{w}^-)\tilde{v} dx \tag{13.101}$$

and

$$\tilde{b} = \int_\Omega (\gamma_+ \tilde{v}^+ - \gamma_- \tilde{v}^- + \delta_+ \tilde{w}^+ - \delta_- \tilde{w}^-)\tilde{w}\, dx. \tag{13.102}$$

Since one of the two numbers $\tilde{a}, \tilde{b}$ is not zero, we see that we cannot have $\tilde{u} \equiv 0$. This contradiction proves (13.99). Once this is known we can use the usual procedures to show that there is a renamed subsequence such that $u_k \to u$ in D, and u satisfies (13.36). □

13.7 Notes and Remarks

These problems have been attacked before under quite strong hypotheses (cf., e.g., [48, 67, 68, 71, 72, 90, 91, 106, 116, 126, 133, 189, 200, 218, 230, 262, 298, 299], and the references contained there).

Chapter 14
Flows and Critical Points

14.1 Introduction

In this chapter we study equations of the form

$$\begin{Bmatrix} -\Delta_p u = f(x,u) & \text{in } \Omega \\ u = 0 & \text{on } \partial\Omega \end{Bmatrix}, \tag{14.1}$$

where Ω is a bounded domain in $\mathbb{R}^n$, $n \geq 1$, $\Delta_p u = \operatorname{div}\left(|\nabla u|^{p-2}\nabla u\right)$ is the p-Laplacian of u, $1 < p < \infty$, and f is a Carathéodory function on $\Omega \times \mathbb{R}$ with subcritical growth. We show that sandwich pairs can be used in solving such problems.

14.2 Flows

Let E be a Banach space, and let Σ be the set of all flows as described in Chap. 2. We now give a sufficient condition for two subsets of E to form a strong sandwich pair.

The following is a corollary of Theorem 2.1.

Theorem 14.1 *Let G be a C^1-functional on E, and let A be a subset of E. Assume that*

$$c := \inf_{\sigma\in\Sigma}\sup_{u\in A} G(\sigma(1)u) \tag{14.2}$$

M. Schechter, *Critical Point Theory*,
https://doi.org/10.1007/978-3-030-45603-0_14

is finite. Let $\rho(t)$ be any function in $\mathcal{Q}$. Then there is a sequence $\{u_k\} \subset E$ such that

$$G(u_k) \to c, \quad \rho(\|u_k\|)G'(u_k) \to 0. \tag{14.3}$$

Corollary 14.2 *Under the hypotheses of Theorem 14.1 there is a sequence $\{u_k\} \subset E$ such that*

$$G(u_k) \to c, \quad (1 + \|u_k\|)G'(u_k) \to 0. \tag{14.4}$$

Proof. We merely take $\rho(u) = 1 + \|u\|$ in Theorem 14.1. □

We now give some consequences of Theorem 14.1.

Theorem 14.3 *Let A,B be subsets of E such that*

$$\sigma(1)A \cap B \neq \phi, \quad \sigma \in \Sigma. \tag{14.5}$$

Assume that

$$a_0 := \sup_A G < \infty, \quad b_0 := \inf_B G > -\infty. \tag{14.6}$$

Then the conclusions of Theorem 14.1 hold.

Proof. Clearly, the quantity c given by (14.2) satisfies

$$b_0 \le c \le a_0 \tag{14.7}$$

by (14.5), (14.6) and the fact that $\sigma(t)u \equiv u$ is in Σ. □

Next, we adjust the hypotheses of Theorem 14.1.

Theorem 14.4 *Let $\rho(t)$ be any function in $\mathcal{Q}$. For subsets A,B of E let $\tilde{\Sigma}$ be the set of those $\sigma \in \Sigma$ which satisfy*

$$\min_{t\in[0,1]} d(\sigma(t)(Ru), B) \to \infty \quad as\ R \to \infty, \quad u \in A \cap \partial\mathcal{B}_1.$$

Assume that $\tilde{\Sigma} \neq \phi$ and

$$\tilde{c} := \inf_{\sigma\in\tilde{\Sigma}} \sup_{u\in A} G(\sigma(1)u) \tag{14.8}$$

is finite. Then there is a sequence $\{u_k\} \subset E$ such that

$$G(u_k) \to \tilde{c}, \quad \rho(d(u_k, B))G'(u_k) \to 0. \tag{14.9}$$

14.3 Linking

We now apply Theorem 14.4 to our specific situation.

Theorem 14.5 *Let A_1 link $B_1 \subset B$. Assume*

$$d(\sigma(t)(RA_1), B) \to \infty, \quad R \to \infty, \ \sigma \in \Sigma, \tag{14.10}$$

uniformly in t, where $RA_1 = \{Ru : u \in A_1\}$. Assume also that $A \cap B = \phi$, where $A = \{\lambda u : u \in A_1, \ \lambda \geq 1\}$. Then {A,B} satisfies the hypotheses of Theorem 14.3.

Proof. Suppose they do not. Then (14.5) does not hold for all $\sigma \in \Sigma$. Hence, there is a $\sigma \in \Sigma$ such that

$$\sigma(1)A \cap B = \phi. \tag{14.11}$$

Now

$$d(\sigma(t)(RA_1), B) \to \infty, \quad R \to \infty$$

uniformly in t. Hence, for R sufficiently large, the set

$$\{\sigma(t)(Ru) : u \in A_1\}$$

does not intersect B. Define $\Gamma \in \Phi$ by

$$\Gamma(t)u = \begin{cases} \sigma(1 - 3t + 3Rt)\,u, & 0 \leq t \leq 1/3 \\ \sigma(3t - 1)Ru, & 1/3 < t \leq 2/3 \\ \sigma(1)(3(1 - t)\,Ru), & 2/3 < t \leq 1. \end{cases} \tag{14.12}$$

Then $\Gamma(t) \in \Phi$ since

$$\Gamma(1)u = \sigma(1)(0), \quad u \in E.$$

But

$$\Gamma(t)A_1 \cap B = \phi, \quad t \in [0, 1].$$

Since $B_1 \subset B$, we see that A_1 does not link B_1, contradicting the hypotheses of the theorem. □

The following is a corollary to Theorem 14.5.

Proposition 14.6 *Let A_1, B_1 be a pair of nonempty subsets of the unit sphere S in a Banach space E. If A_1 links B_1 in S, then*

$$A = \pi^{-1}(A_1) \cup \{0\}, \quad B = \pi^{-1}(B_1) \cup \{0\} \tag{14.13}$$

forms a sandwich pair, where $\pi : E \setminus \{0\} \to S$, $u \mapsto u/\|u\|$ is the radial projection onto S.

14.4 Cerami Sequences

We can now prove Theorem 14.4.

Proof. If the theorem were not true, there would be a $\delta > 0$ such that

$$\rho(d(u, B))\|G'(u)\| \geq \delta \tag{14.14}$$

would hold for all u in the set

$$Q = \{u \in E : \tilde{c} - 3\delta \leq G(u) \leq \tilde{c} + 3\delta\}. \tag{14.15}$$

Let

$$Q_0 = \{u \in Q : \tilde{c} - 2\delta \leq G(u) \leq \tilde{c} + 2\delta\} \tag{14.16}$$

$$Q_1 = \{u \in Q : \tilde{c} - \delta \leq G(u) \leq \tilde{c} + \delta\} \tag{14.17}$$

and

$$Q_2 = E \setminus Q_0, \; \eta(u) = d(u, Q_2)/[d(u, Q_1) + d(u, Q_2)]. \tag{14.18}$$

As before we note that it is easily checked that $\eta(u)$ is locally Lipschitz continuous on E and satisfies

$$\begin{cases} \eta(u) = 1, & u \in Q_1, \\ \eta(u) = 0, & u \in \bar{Q}_2, \\ \eta(u) \in (0, 1), & \text{otherwise.} \end{cases} \tag{14.19}$$

For some $\theta < 1$ there is a locally Lipschitz continuous map $Y(u)$ of $\hat{E} = \{u \in E : G'(u) \neq 0\}$ into itself such that

$$\|Y(u)\| \leq 1, \ \theta\|G'(u)\| \leq (G'(u), Y(u)), \quad u \in \hat{E} \tag{14.20}$$

(cf., e.g., [184]). Let $\sigma(t)$ be the flow generated by

$$W(u) = -\eta(u)Y(u)\rho(d(u, B)). \tag{14.21}$$

Since $\|W(u)\| \leq \rho(d(u, B))$ and is locally Lipschitz continuous, $\sigma(t)$ exists for all $t \in \mathbb{R}^+$ in view of Theorem 6.10. We also have

$$\begin{aligned} dG(\sigma(t)u)/dt &= (G'(\sigma), \sigma') \\ &= -\eta(\sigma)(G'(\sigma), Y(\sigma))\rho(d(\sigma, B)) \\ &\leq -\theta\eta(\sigma)\|G'(\sigma)\|\rho(d(\sigma, B)) \\ &\leq -\theta\delta\eta(\sigma) \end{aligned} \tag{14.22}$$

in view of (14.14) and (14.20). Let $T > 2\delta/\theta$ and suppose $u \in Q_1$ is such that there is a $t_1 \in [0, T]$ for which $\sigma(t_1)u \notin Q_1$. Then

$$G(\sigma(t_1)u) < \tilde{c} - \delta,$$

since we cannot have $G(\sigma(t_1)u) > \tilde{c} + \delta$ for $u \in Q_1$ by (14.22). But this implies

$$G(\sigma(T)u) < \tilde{c} - \delta. \tag{14.23}$$

On the other hand, if $\sigma(t)u \in Q_1$ for all $t \in [0, T]$, then

$$G(\sigma(T)u) \leq G(u) - \theta \int_0^T dt \leq \tilde{c} + \delta - \theta T < \tilde{c} - \delta$$

by (14.22). Thus, (14.23) holds for $u \in Q_1$.

By the definition of $\tilde{c}$, there is a $\tilde{\sigma} \in \tilde{\Sigma}$ such that

$$G(\tilde{\sigma}(1)u) < \tilde{c} + \delta, \quad u \in A. \tag{14.24}$$

Let $v = \tilde{\sigma}(1)u$, where $u \in A$. Then $v \in Q_1$. Hence

$$G(\sigma(T)\tilde{\sigma}(1)u) < \tilde{c} - \delta, \quad u \in A. \tag{14.25}$$

Let

$$\sigma_1(s) = \begin{cases} \sigma(2sT), & 0 \leq s \leq \frac{1}{2}, \\ \sigma(T)\tilde{\sigma}(2s - 1), & \frac{1}{2} < s \leq 1. \end{cases} \tag{14.26}$$

Then $\sigma_1 \in \tilde{\Sigma}$. To see this we first note that since

$$\sigma(t)u - u = \int_0^t W(\sigma(\tau)u)d\tau, \tag{14.27}$$

we have

$$\|\sigma(t)u - \sigma(s)u\| \leq \int_s^t \rho(d(\sigma(r)u, B))\,dr.$$

If $v \in B$, we have

$$h(s) = d(\sigma(s)u, B) \leq \|\sigma(s)u - v\| \leq \|\sigma(t)u - v\| + \int_s^t \rho(d(\sigma(r)u, B))\,dr.$$

This implies,

$$h(s) \leq h(t) + \int_s^t \rho(h(r))\,dr. \tag{14.28}$$

Moreover, by Lemma 6.3 and Theorem 6.8, $h(s)$ satisfies

$$m(R) \leq h(s) \leq R, \quad 0 \leq s \leq T,$$

where $m(R)$ is given by

$$\int_{m(R)}^R \frac{d\tau}{\rho(\tau)} = T.$$

Note that $m(R) \to \infty\ as\ R \to \infty$ by (1.9). Thus

$$\min_{t\in[0,1]} d(\sigma(t)(Ru), B) \to \infty \quad as\ R \to \infty$$

when $u \in Q_1$ satisfies $\|u\| \geq 1$. Hence, $\sigma_1 \in \tilde{\Sigma}$ and

$$G(\sigma_1(1)u) < \tilde{c} - \delta, \quad u \in A. \tag{14.29}$$

But this contradicts the definition (14.8) of $\tilde{c}$. Hence (14.14) cannot hold for u satisfying (14.15). The proof is complete. □

14.5 Some Applications

We shall study equations of the form

$$\left\{\begin{aligned} -\Delta_p u &= f(x,u) && \text{in } \Omega \\ u &= 0 && \text{on } \partial\Omega \end{aligned}\right\}, \tag{14.30}$$

where Ω is a bounded domain in $\mathbb{R}^n$, $n \geq 1$, $\Delta_p u = \operatorname{div}\left(|\nabla u|^{p-2} \nabla u\right)$ is the p-Laplacian of u, $1 < p < \infty$, and f is a Carathéodory function on $\Omega \times \mathbb{R}$ with subcritical growth. Solutions of (14.30) coincide with the critical points of the C^1 functional

$$G(u) = \int_\Omega |\nabla u|^p - p\,F(x,u), \tag{14.31}$$

where $F(x,t) = \displaystyle\int_0^t f(x,s)\,ds$, defined on the Sobolev space $W_0^{1,\,p}(\Omega)$. Consider the nonlinear eigenvalue problem

$$\left\{\begin{aligned} -\Delta_p u &= \lambda\,|u|^{p-2}\,u && \text{in } \Omega \\ u &= 0 && \text{on } \partial\Omega. \end{aligned}\right. \tag{14.32}$$

Its eigenvalues coincide with the critical values of the C^1 functional

$$I(u) = \frac{1}{\displaystyle\int_\Omega |u|^p} \tag{14.33}$$

on the unit sphere S in $W_0^{1,\,p}(\Omega)$. Let $\tilde{\Gamma}_l$ be the class of odd continuous maps γ from the unit sphere S^{l-1} in R^l to S and set

$$\lambda_l := \inf_{\gamma \in \tilde{\Gamma}_l}\ \max_{u \in \tilde{\Gamma}(S^{l-1})} I(u). \tag{14.34}$$

Then $0 < \lambda_1 < \lambda_2 \leq \cdots \to \infty$ are eigenvalues of (14.32) (cf. Drábek and Robinson [82]).

Setting $H(x,t) = p\,F(x,t) - tf(x,t)$, we shall prove the following.

Theorem 14.7 *If* $\lambda_l < \lambda_{l+1}$ *and*

$$(\lambda_l + \varepsilon)\,|t|^p - W(x) \leq p\,F(x,t) \leq \lambda_{l+1}\,|t|^p + W(x), \quad t \in \mathbb{R} \tag{14.35}$$

for some l, $\varepsilon > 0$ *and* $W \in L^1(\Omega)$, *and*

$$H(x,t) \le C\,(|t|^{\alpha}+1), \qquad \overline{H}(x) := \overline{\lim_{|t|\to\infty}} \frac{H(x,t)}{|t|^{\alpha}} < 0 \ \ a.e., \tag{14.36}$$

for some α *satisfying* $0 < \alpha \le p$, *then* (14.30) *has a solution.*

Theorem 14.8 *If*

$$\lambda_l\,|t|^p - W(x) \le p\,F(x,t) \le (\lambda_{l+1} - \varepsilon)\,|t|^p + W(x), \qquad t \in \mathbb{R} \tag{14.37}$$

for some $l, \varepsilon > 0$ *and* $W \in L^1(\Omega)$, *and*

$$H(x,t) \ge -C\,(|t|^{\alpha}+1), \qquad \underline{H}(x) := \underline{\lim_{|t|\to\infty}} \frac{H(x,t)}{|t|^{\alpha}} > 0 \ \ a.e., \tag{14.38}$$

then (14.30) *has a solution.*

Theorem 14.9 *If* $\lambda_l < \lambda_{l+1}$ *and (14.35) holds for some* $l, \varepsilon > 0$ *and* $W \in L^1(\Omega)$, *and*

$$H(x,t) \le W_1(x) \in L^1(\Omega), \qquad H(x,t) \to -\infty \text{ as } |t| \to \infty, \tag{14.39}$$

then (14.30) *has a solution.*

Theorem 14.10 *If (14.37) holds for some* $l, \varepsilon > 0$ *and* $W \in L^1(\Omega)$, *and*

$$H(x,t) \ge -W_1(x) \in L^1(\Omega), \qquad H(x,t) \to \infty \text{ as } |t| \to \infty, \tag{14.40}$$

then (14.30) *has a solution.*

Proof of Theorem 14.7 Let ε satisfy $\lambda_l + 3\varepsilon < \lambda_{l+1}$. By (14.34), there is a $\gamma \in \tilde{\Gamma}_l$ such that $I \le \lambda_l + \varepsilon$ on $A_1 = \gamma(S^{l-1})$. Let $B_1 = \{u \in S : I(u) \ge \lambda_{l+1}\}$. Since $\lambda_l + \varepsilon < \lambda_{l+1}$, A_1 and B_1 are disjoint. Since A_1 is compact and B_1 is closed, it follows that $\operatorname{dist}(A_1, B_1) > 0$. We claim that A_1 links B_1 in W. Given $\varphi \in C(CA_1, S)$ such that $\varphi(\cdot, 0) = \mathrm{id}_{A_1}$, writing $x \in S^l$ as $(x', x_{l+1}) \in \mathbb{R}^l \oplus \mathbb{R}$, define $\overline{\gamma} \in \tilde{\Gamma}_{l+1}$ by

$$\overline{\gamma}(x) = \begin{cases} \varphi(\tilde{\gamma}(x'/|x'|), x_{l+1}), & 0 \le x_{l+1} < 1 \\ \varphi(A_1 \times \{1\}), & x_{l+1} = 1 \\ -\overline{\gamma}(x', -x_{l+1}), & x_{l+1} < 0. \end{cases} \tag{14.41}$$

Then $\overline{\gamma}(S^l) \cap B_1 \ne \phi$ by the definition of λ_{l+1}, so (14.10) holds since B_1 is symmetric. Consequently, A, B given by Proposition 14.6 forms a sandwich pair. Let G be given by (14.31). Since

$$\int_\Omega |\nabla u|^p \geq \lambda_{l+1} \int_\Omega |u|^p, \quad u \in B \tag{14.42}$$

and

$$\int_\Omega |\nabla u|^p \leq (\lambda_l + \varepsilon) \int_\Omega |u|^p, \quad u \in A, \tag{14.43}$$

(14.35) implies

$$-\int_\Omega W \leq \inf_B G \leq \sup_A G \leq \int_\Omega W, \tag{14.44}$$

so there is a sequence $\{u_j\} \subset W_0^{1,p}(\Omega)$ satisfying (14.35) with $\psi(t) \equiv 1$.

We claim that $\{u_j\}$ is bounded and hence has a convergent subsequence by a standard argument. If $\rho_j = \|u_j\| \to \infty$, a subsequence of $\widetilde{u}_j = u_j/\rho_j$ converges to some $\widetilde{u}$ weakly in $W_0^{1,p}(\Omega)$, strongly in $L^p(\Omega)$, and a.e. in Ω. Then

$$\int_\Omega \frac{H(x, u_j)}{\rho_j^\alpha} = \frac{G'(u_j)\, u_j/p - G(u_j)}{\rho_j^\alpha} \to 0 \tag{14.45}$$

by (14.35) and

$$\overline{\lim} \int_\Omega \frac{H(x, u_j)}{\rho_j^\alpha} \leq \int_\Omega \overline{\lim}\, \frac{H(x, u_j)}{|u_j|^\alpha}\, |\widetilde{u}_j|^\alpha \leq \int_\Omega \overline{H}(x)\, |\widetilde{u}|^\alpha \leq 0 \tag{14.46}$$

by (14.36). Since $\overline{H} < 0$ a.e., it follows that $\widetilde{u} = 0$ a.e. Now passing to the limit in

$$1 - \frac{G(u_j)}{\rho_j^p} = \int_\Omega \frac{p\, F(x, u_j)}{\rho_j^p} \leq \int_\Omega \lambda_{l+1}\, |\widetilde{u}_j|^p + \frac{W}{\rho_j^p} \tag{14.47}$$

gives

$$1 \leq \int_\Omega \lambda_{l+1}\, |\widetilde{u}|^p,$$

showing that $\tilde{u} \neq 0$ a.e. and providing a contradiction. □

Proof of Theorem 14.8 Take a sequence $\{\varepsilon_j\} \subset (0, \varepsilon]$ decreasing to 0 and let

$$G_j(u) = G(u) - \varepsilon_j \int_\Omega |u|^p. \tag{14.48}$$

Then

$$(\lambda_l + \varepsilon_j)\, |t|^p - W(x) \leq p\, F(x, t) + \varepsilon_j\, |t|^p \leq \lambda_{l+1}\, |t|^p + W(x) \tag{14.49}$$

by (14.37), so there is a sequence $\left(u_j\right) \subset W_0^{1,p}(\Omega)$ such that

$$G_j(u_j) \text{ is bounded}, \quad G'_j(u_j) \to 0 \tag{14.50}$$

as in the proof of Theorem 14.7.

We claim that $\left\{u_j\right\}$ is bounded and hence a subsequence converges to a critical point of G. If $\rho_j = \left\|u_j\right\| \to \infty$, a subsequence of $\widetilde{u}_j = u_j/\rho_j$ converges to some $\widetilde{u}$ weakly in $W_0^{1,p}(\Omega)$, strongly in $L^p(\Omega)$, and a.e. in Ω. Then

$$\int_\Omega \frac{H(x, u_j)}{\rho_j^\alpha} = \frac{G'_j(u_j)\, u_j/p - G_j(u_j)}{\rho_j^\alpha} \to 0 \tag{14.51}$$

by (14.50) and

$$\underline{\lim} \int_\Omega \frac{H(x, u_j)}{\rho_j^\alpha} \geq \int_\Omega \underline{\lim}\, \frac{H(x, u_j)}{|u_j|^\alpha}\, |\widetilde{u}_j|^\alpha \geq \int_\Omega \underline{H}(x)\, |\widetilde{u}|^\alpha \geq 0 \tag{14.52}$$

by (14.38). Since $\underline{H} > 0$ a.e., it follows that $\widetilde{u} = 0$ a.e. Now passing to the limit in

$$1 - \frac{G_j(u_j)}{\rho_j^p} = \int_\Omega \frac{p\, F(x, u_j)}{\rho_j^p} + \varepsilon_j\, |\widetilde{u}_j|^p \leq \int_\Omega \lambda_{l+1}\, |\widetilde{u}_j|^p + \frac{W}{\rho_j^p} \tag{14.53}$$

again gives

$$1 \leq \int_\Omega \lambda_{l+1}\, |\widetilde{u}|^p,$$

showing that $\tilde{u} \neq 0$ a.e. and providing a contradiction. □

Proof of Theorem 14.9 We follow the proof of Theorem 14.7, but this time we take $\psi(r) = 1/(r+1)$. Now we can conclude that there is a sequence $\{u_k\} \subset W_0^{1,p}(\Omega)$ satisfying (14.3). Again we claim that $\left\{u_j\right\}$ is bounded, and hence a subsequence converges to a critical point of G. If $\rho_j = \left\|u_j\right\| \to \infty$, a subsequence of $\widetilde{u}_j = u_j/\rho_j$ converges to some $\widetilde{u}$ weakly in $W_0^{1,p}(\Omega)$, strongly in $L^p(\Omega)$, and a.e. in Ω. Now

$$\int_\Omega H(x, u_k)\, dx = \int_\Omega \{G'(u_k)\, u_k/p - G(u_k)\}\, dx \to -c. \tag{14.54}$$

This implies

$$\left|\int_\Omega H(x, u_k)\, dx\right| \leq K. \tag{14.55}$$

As before, we show that $\tilde{u}(x) \not\equiv 0$. Let Ω_0 be the subset of Ω on which $\tilde{u} \neq 0$. Then

$$|u_k(x)| = \rho_k|\tilde{u}_k(x)| \to \infty, \quad x \in \Omega_0. \tag{14.56}$$

If $\Omega_1 = \Omega \setminus \Omega_0$, then we have

$$\int_\Omega H(x, u_k)\,dx = \int_{\Omega_0} + \int_{\Omega_1} \leq \int_{\Omega_0} H(x, u_k)dx + \int_{\Omega_1} W_1(x)dx \to -\infty. \tag{14.57}$$

This contradicts (14.55), and we see that $\rho_k = \|u_k\|$ is bounded.

□

Proof of Theorem 14.10 In this case we have

$$\int_\Omega H(x, u_k)\,dx = \int_{\Omega_0} + \int_{\Omega_1} \geq \int_{\Omega_0} H(x, u_k)\,dx - \int_{\Omega_1} W_1(x)\,dx \to \infty \tag{14.58}$$

which contradicts (14.55) as well. □

14.6 Notes and Remarks

Here we follow [175–177].

Similar resonance problems have been studied by Perera [168] when $f(x,t)/|t|^{p-2}t \to \alpha_\pm(x) \in L^\infty(\Omega)$ as $t \to \pm\infty$ and by Arcoya and Orsina [8], Bouchala and Drábek [41], and Drábek and Robinson [82] for the special case $\alpha_\pm(x) \equiv \lambda_l$.

Chapter 15
The Semilinear Wave Equation

15.1 Introduction

In this chapter we study periodic solutions of the Dirichlet problem for the semilinear wave equation:

$$\Box u - \mu u := u_{tt} - u_{rr} - \mu u = p(t, r, u), \quad t \in \mathbb{R}, \quad 0 < r < R, \tag{15.1}$$

$$u(t, R) = u(t, 0) = 0, \quad t \in \mathbb{R}, \tag{15.2}$$

$$u(t + T, r) = u(t, r), \quad t \in \mathbb{R}, \quad 0 \leq r \leq R. \tag{15.3}$$

Our basic assumption is that the ratio R/T is rational. Thus, we can write

$$2R/T = a/b, \tag{15.4}$$

where a, b are relatively prime positive integers. We also assume

$$|p(t, r, s)| \leq C(|s| + 1), \quad s \in \mathbb{R}. \tag{15.5}$$

We have the following.

Theorem 15.1 *Under assumptions (15.2)–(15.4), the operator $\Box$ has a self-adjoint extension L having discrete spectrum except for the point $\{0\}$. Assume that*

$$\alpha s^2 - W_1(t, r) \leq 2P(t, r, s) \leq \beta s^2 + W_2(t, r), \tag{15.6}$$

M. Schechter, *Critical Point Theory*,
https://doi.org/10.1007/978-3-030-45603-0_15

where

$$P(t, r, s) = \int_0^s p(t, r, \sigma)\, d\sigma, \tag{15.7}$$

$\mu \in (m_-, m_+)$ is contained in the resolvent set of L, $\alpha > \max[\mu - m_+, m_- - \mu]$, $\beta > 0$, and the functions W_1, W_2 are in $L^1(Q)$ with $Q = [0, T] \times [0, R]$. We also assume that $p(t, r, s)$ is nondecreasing in s if $m_- \geq 0$, and nonincreasing in s if $m_+ \leq 0$. We let M be the subspace of $E = D(|L|^{1/2})$ on which $L > \mu$, and N the subspace on which $L < \mu$. For $\lambda > 0$, we define

$$L_\lambda u = \lambda[L - \mu]w + [L - \mu]v, \quad u = v + w, \; w \in M, \; v \in N.$$

Then

$$L_1 = L - \mu.$$

We imbed (15.1) in the problem

$$L_\lambda u = p(t, r, u), \quad u \in D(L), \; \lambda > 0. \tag{15.8}$$

Then (15.8) has at least one weak solution for a.e. $\lambda > \beta/(m_+ - \mu)$.

An important aspect of this theorem is that both subspaces M and N are permitted to be infinite dimensional. Moreover, all rational values of R/T are allowed and the hypotheses are given in terms of the primative (15.7) of $p(t, r, s)$.

15.2 The Spectrum of the Linear Operator

In considering problem (15.1)–(15.3), we shall need to calculate the spectrum of the linear operator $\Box$.

Theorem 15.2 *Consider the operator*

$$\Box u = u_{tt} - u_{rr} \tag{15.9}$$

applied to functions $u(t, r)$ in $C^\infty(\bar{Q})$ satisfying

$$u(t + T, r) = u(t, r), \quad t \in \mathbb{R}, \quad 0 \leq r \leq R \tag{15.10}$$

$$u(t, R) = u(t, 0) = 0, \quad t \in \mathbb{R}, \tag{15.11}$$

where $Q = [0, T] \times [0, R]$. Then $\Box$ is symmetric on $L^2(Q)$. Assume that $2R/T = a/b$, where a, b are relatively prime integers (i.e., $(a, b) = 1$). Then $\Box$ has a self-adjoint extension L having no essential spectrum other than $\{0\}$.

Remark 15.3 The essential spectrum of a self-adjoint operator consists of those points of the spectrum that are not isolated eigenvalues of finite multiplicity.

Proof. If

$$\psi_{jk}(t,r) = \sin(j\pi r/R)e^{2\pi ikt/T}, \tag{15.12}$$

then

$$\Box\psi_{jk} = [(j\pi/R)^2 - (2\pi k/T)^2]\psi_{jk}. \tag{15.13}$$

Thus $\psi_{jk}(t,r)$ is an eigenfunction of $\Box$ with eigenvalue

$$\lambda_{jk} = (j\pi/R)^2 - (2\pi k/T)^2. \tag{15.14}$$

It is easily checked that the functions ψ_{jk}, when normalized, form a complete orthonormal sequence in $L^2(Q)$. We shall show that the corresponding eigenvalues (15.14) are not dense in $\mathbb{R}$. It will then follow that $\Box$ has a self-adjoint extension L with spectrum equal to the closure of the set $\{\lambda_{jk}\}$ (cf., e.g., [195]). Now

$$\lambda_{jk} = \frac{\pi^2}{R^2b^2}(bj - ak)(bj + ak). \tag{15.15}$$

Hence

$$|\lambda_{jk}| \geq \frac{\pi^2}{R^2b^2}|bj + ak| \tag{15.16}$$

when $bj \neq ak$, and

$$\lambda_{jk} = 0, \quad bj = ak. \tag{15.17}$$

Thus

$$\lim_{\substack{j,|k|\to\infty \\ bj\neq ak}} |\lambda_{jk}| = \infty \tag{15.18}$$

and

$$\lim_{\substack{j,|k|\to\infty \\ bj=ak}} |\lambda_{jk}| = 0. \tag{15.19}$$

Hence, the point 0 is the only limit point of eigenvalues. Consequently, it is in $\sigma_e(L)$. This completes the proof. □

15.3 Proof of Theorem 15.1

Proof. We define

$$I(u) = ([L-\mu]w, w), \quad u = v + w, \ w \in M, \ v \in N,$$

$$J(u) = -([L-\mu]v, v) + 2\int_Q\int P(t, r, u)\,dtdr, \quad u = v + w, \ w \in M, \ v \in N,$$

$$G(u) = I(u) - J(u), \quad u \in E.$$

Then

$$\begin{aligned} I(u) + J(u) &\geq (m_+ - \mu)\|w\|^2 + (\mu - m_-)\|v\|^2 + \alpha\|v + w\|^2 - B_1 \\ &= (m_+ - \mu + \alpha)\|w\|^2 + (\mu - m_- + \alpha)\|v\|^2 - B_1 \\ &\to \infty \text{ as } \|u\| \to \infty, \end{aligned}$$

where

$$B_j = \int_Q\int W_j(t, r)\,dtdr.$$

Define

$$G_\lambda(u) = \lambda I(u) - J(u), \quad u \in E.$$

Then

$$(G_\lambda'(u), g)/2 = \lambda(w, g)_E - (v, g)_E - (p(u), g), \quad g \in E, \tag{15.20}$$

where $u = v+w, \ v \in N, \ w \in M$, and u is a weak solution of (15.8) iff $G_\lambda'(u) = 0$. Moreover,

$$\begin{aligned} G_\lambda(v) = G(v) &= ([L-\mu]v, v) - 2\int_Q\int P(t, r, v)\,dtdr \\ &\leq (m_- - \mu)\|v\|^2 - \alpha\|v\|^2 + B_1 \\ &\leq B_1, \quad v \in N, \end{aligned}$$

and

$$G_\lambda(w) \geq \lambda(m_+ - \mu)\|w\|^2 - \beta\|w\|^2 - B_2 \geq -B_2, \quad w \in M \tag{15.21}$$

by (15.6), provided $\lambda > \beta/(m_+ - \mu)$.

Assume that $\mu < 0$ with $p(s)$ nonincreasing. Then $N \cap N_0 = \{0\}$, where $N_0 = \{u \in E : Lu = 0\}$. If $\{u_k\} \subset E$ is a sequence such that $v_k = Pu_k \to v = Pu$ weakly in N and $w_k = (I - P)u_k \to w = (I - P)u$ strongly in M, where P is the projection of E onto N, then $\{u_k\}$ has a renamed subsequence which converges strongly in $L^2(Q)$ since the embedding of $E \ominus N_0$ in $L^2(Q)$ is compact. Now,

$$(G'_\lambda(u_k), v)/2 = \lambda(w_k, v)_E - (v_k, v)_E - (p(u_k), v), \quad v \in E, \tag{15.22}$$

where $u_k = v_k + w_k,\ v_k \in N,\ w_k \in M$. Clearly, the first two terms on the right converge. Concerning the last, note that

$$\begin{aligned} |[p(u_k) - p(u)]v| &\le C(|u_k| + |u| + 1)|v| \\ &\le C|u_k v| + C(|u| + 1)|v| \\ &\le C(u_k^2 + v^2 + (|u| + 1)|v|), \end{aligned}$$

which converges in $L^1(Q)$. Thus $G'_\lambda(u_k) \to G'_\lambda(u)$ weakly in E. Hence all of the hypotheses of Corollary 4.13 are satisfied, and we can conclude that for a.e. $\lambda > \beta/(m_+ - \mu)$, there is a bounded sequence $\{u_k\}$ satisfying

$$\|G'_\lambda(u_k)\| \to 0, \quad G_\lambda(u_k) \to a(\lambda). \tag{15.23}$$

Write $u_k = v_k + w_k + y_k$, where $v_k \in N,\ w_k \in M \ominus N_0,\ y_k \in N_0$. Thus $\|u_k\|_E$ has a renamed subsequence which converges weakly in E and such that $u'_k = v_k + w_k$ converges strongly in $L^2(Q)$. Now

$$\begin{aligned} (G'_\lambda(u_k), y_k - y)/2 &= \lambda([L - \mu](y_k - y), y_k - y) \\ &\quad - (p(u_k) - p(u'_k + y), y_k - y) \\ &\quad - (p(u'_k + y) - p(u), y_k - y) \\ &\quad + (\lambda[L - \mu]y - p(u), y_k - y), \end{aligned} \tag{15.24}$$

where $y_k \to y$ weakly in E and $L^2(Q)$ and $u'_k \to u'$ weakly in E and strongly in $L^2(Q)$. By hypothesis

$$(p(u_k) - p(u'_k + y), y_k - y) \le 0 \tag{15.25}$$

if $\mu < 0$. Moreover,

$$\begin{aligned} (G'_\lambda(u_k), y_k - y) &\to 0 \\ (p(u'_k + y) - p(u), y_k - y) &\to 0 \end{aligned}$$

and

$$(\lambda[L-\mu]y - p(u), y_k - y) \to 0. \tag{15.26}$$

Hence

$$\lambda\|y_k - y\|_E^2 = \lambda([L-\mu](y_k - y), y_k - y) \le o(1), \quad k \to \infty. \tag{15.27}$$

This shows that $y_k \to y$ in E. Consequently, there is a renamed subsequence which converges weakly in E, a.e. in Ω and strongly in $L^2(Q)$. Taking the limit in

$$(G'_\lambda(u_k), v)/2 = (L_\lambda u_k, v) - (p(u_k), v), \tag{15.28}$$

we obtain a weak solution of (15.8). If $\mu > 0$, we apply Corollary 4.13 to $-G_\lambda(u)$ and come to the same conclusion. In this case, the inequality in (15.25) is reversed. This completes the proof. □

15.4 Notes and Remarks

The one-dimensional semilinear wave equation has been studied by many researchers over a long period of time; cf., e.g., [24, 25, 42, 103, 104, 150–152, 154–157] for results, references, and surveys. In particular, we call attention to the surveys [42, 151, 154].

Our approach differs in that all rational values of R/T are allowed and the hypotheses are given in terms of the primative (15.7) of $p(t, r, s)$ and the "core function" (23).

Chapter 16
Nonlinear Optics

16.1 Introduction

Light waves propagating in a photo refractive crystal are governed by a nonlinear Schrödinger equation of the form

$$i\frac{\partial u}{\partial z} + D\Delta u = g(x, |u|^2)u,$$

where $D > 0$ is the beam diffraction coefficient and the functions are periodic with respect to the variables $x = (x_1, x_2) \in \Omega \subset \mathbb{R}^2$. Here,

$$g(x, |u|^2) = \frac{K}{1 + V(x) + |u|^2},$$

where $V(x)$ is a continuous, nonnegative function periodic in $\overline{\Omega}$. Steady state solutions satisfy the following equation over a periodic domain $\Omega \subset \mathbb{R}^2$:

$$\Delta u = \frac{Pu}{1 + V(x) + |u|^2} + \lambda u, \tag{16.1}$$

where P, λ are parameters. The solutions u are to be periodic in Ω with the same periods as those of Ω. This equation has the trivial solution $u = 0$.

We shall show that there are many intervals of the parameters in which nontrivial solutions exist. Our results are true in any dimension.

In stating our results, we shall make use of the following considerations. Let Ω be a bounded periodic domain in $\mathbb{R}^n$, $n \geq 1$. Consider the operator $-\Delta$ on functions in $L^2(\Omega)$ having the same periods as Ω. The spectrum of $-\Delta$ consists of isolated eigenvalues of finite multiplicity:

M. Schechter, *Critical Point Theory*,
https://doi.org/10.1007/978-3-030-45603-0_16

$$0 = \lambda_0 < \lambda_1 < \cdots < \lambda_\ell < \cdots ,$$

with eigenfunctions in $L^\infty(\Omega)$. Let $\lambda_\ell,\ \ell \geq 0$, be one of these eigenvalues, and define

$$N = \bigoplus_{\lambda \leq \lambda_\ell} E(\lambda), \quad M = N^\perp,$$

where $E(\lambda)$ is the eigenspace corresponding to λ. Let $a(x) = 1/(1 + V(x))$, and assume that $a(x)$ is positive and bounded: $0 < m_0 \leq a(x) \leq m_1 \leq 1$. Then (16.1) becomes

$$\Delta u = \frac{Pau}{1 + a|u|^2} + \lambda u. \tag{16.2}$$

We shall prove the following:

Theorem 16.1

1. *If $P > 0,\ \lambda < 0$ and there is an $\ell \geq 0$ such that $\lambda_\ell + m_1 P \leq |\lambda| < \lambda_{\ell+1} + m_0 P,\ |\lambda| > \lambda_{\ell+1}$, then (16.1) has a nontrivial solution.*
2. *If $P < 0,\ \lambda < 0$ and there is an $\ell \geq 0$ such that $\lambda_\ell + m_0 P < |\lambda| \leq \lambda_{\ell+1} + m_1 P,\ |\lambda| < \lambda_\ell$, then (16.1) has a nontrivial solution.*
3. *If $P > 0,\ \lambda < 0$ and $0 < |\lambda| < m_0 P$, then (16.1) has a nontrivial solution.*
4. *If $P < 0,\ \lambda > 0$ and $0 < \lambda < m_0|P|$, then (16.1) has a nontrivial solution.*

16.2 Some Lemmas

In proving our results we shall make use of the following lemmas.

We let E be the subspace of $H^{1,2}(\Omega)$ consisting of those functions having the same periodicity as Ω with norm given by

$$\|w\|_E^2 = \|\nabla w\|^2 + \|w\|^2.$$

Define

$$I(u) = \frac{1}{P}\|\nabla u\|^2 + \frac{\lambda}{P}\|u\|^2 + \int_\Omega \ln\{1 + a(x)|u|^2\}\,dx. \tag{16.3}$$

Then,

$$(I'(u), v)/2 = \frac{1}{P}(\nabla u, \nabla v) + \frac{\lambda}{P}(u, v) + \int_\Omega \frac{au}{1 + au^2} v\,dx. \tag{16.4}$$

We have the following.

Lemma 16.2 *If I(u) is given by (16.3), then every sequence satisfying*

$$I(u_k) \to c > -\infty, \quad (1 + \|u_k\|_E)\|I'(u_k)\| \to 0. \tag{16.5}$$

has a subsequence converging in E. Consequently, there is a u ∈ E such that I(u)=c and $I'(u) = 0$.

Proof. The sequence satisfies

$$I(u_k) = \frac{1}{P}\|\nabla u_k\|^2 + \frac{\lambda}{P}\|u_k\|^2 + \int_\Omega \ln\{1 + a|u_k|^2\}\, dx \to c, \tag{16.6}$$

$$(I'(u_k), v)/2 = \frac{1}{P}(\nabla u_k, \nabla v) + \frac{\lambda}{P}(u_k, v) + \int_\Omega \frac{a u_k}{1 + a u_k^2} v\, dx \to 0 \tag{16.7}$$

and

$$(I'(u_k), u_k)/2 = \frac{1}{P}(\nabla u_k, \nabla u_k) + \frac{\lambda}{P}(u_k, u_k) + \int_\Omega \frac{a u_k^2}{1 + a u_k^2}\, dx \to 0. \tag{16.8}$$

Thus,

$$\int_\Omega H(x, u_k)\, dx \to c, \tag{16.9}$$

where

$$H(x,t) = \ln(1 + at^2) - \frac{at^2}{1 + at^2}. \tag{16.10}$$

Let $\rho_k = \|u_k\|_E$. Assume first that $\rho_k \to \infty$. Let $\tilde{u}_k = u_k/\rho_k$. Then $\|\tilde{u}_k\|_E = 1$. Hence, there is a renamed subsequence such that $\tilde{u}_k \rightharpoonup \tilde{u}$ in E, and $\tilde{u}_k \to \tilde{u}$ in $L^2(\Omega)$ and a.e. By (16.8),

$$P(I'(u_k), u_k)/2 = \|\nabla u_k\|^2 + \lambda\|u_k\|^2 + P\int_\Omega \frac{a u_k^2}{1 + a u_k^2}\, dx \to 0. \tag{16.11}$$

Hence,

$$\begin{aligned} 1 = \|\tilde{u}_k\|_E^2 &\le |P(I'(u_k), u_k)/2\rho_k^2| \\ &\quad + |1 - \lambda| \cdot \|\tilde{u}_k\|^2 + |P| \int_\Omega \frac{a\tilde{u}_k^2}{1 + a u_k^2}\, dx \\ &\le [|1 - \lambda| + m_1|P|] \cdot \|\tilde{u}_k\|^2 + o(1). \end{aligned}$$

In the limit we have,

$$1 \leq [|1-\lambda| + m_1|P|] \cdot \|\tilde{u}\|^2.$$

This shows that $\tilde{u} \not\equiv 0$. Let Ω_0 be the subset of Ω where $\tilde{u}(x) \neq 0$. Then $|\Omega_0| \neq 0$ and $|u_k(x)| \to \infty$ when $x \in \Omega_0$. Consequently,

$$0 \leq H(x, u_k) \to \infty, \quad x \in \Omega_0.$$

Thus,

$$\begin{aligned}\int_\Omega H(x, u_k)\,dx &= \int_{\Omega_0} H(x, u_k)\,dx + \int_{\Omega\setminus\Omega_0} H(x, u_k)\,dx \\ &\geq \int_{\Omega_0} H(x, u_k)\,dx \to \infty.\end{aligned}$$

This contradicts (16.9). Thus, the sequence satisfying (16.5) is bounded in E. Hence, there is a renamed subsequence such that $u_k \rightharpoonup u$ in E, and $u_k \to u$ in $L^2(\Omega)$ and a.e. Taking the limit in (16.7), we obtain

$$(I'(u), v)/2 = \frac{1}{P}(\nabla u, \nabla v) + \frac{\lambda}{P}(u, v) + \int_\Omega \frac{au}{1+au^2} v\,dx = 0, \quad v \in E. \tag{16.12}$$

Thus, u satisfies $I'(u) = 0$. Since $u \in E$, it satisfies

$$(I'(u), u)/2 = \frac{1}{P}(\nabla u, \nabla u) + \frac{\lambda}{P}(u, u) + \int_\Omega \frac{au^2}{1+au^2}\,dx = 0. \tag{16.13}$$

Also, from the limit in (16.8), we have

$$\begin{aligned}\lim \frac{1}{P}\|\nabla u_k\|^2 &= \lim(I'(u_k), u_k)/2 \\ &\quad - \lim[\frac{\lambda}{P}\|u_k\|^2 + \int_\Omega \frac{au_k^2}{1+au_k^2}\,dx] \\ &= -[\frac{\lambda}{P}\|u\|^2 + \int_\Omega \frac{au^2}{1+au^2}\,dx] \\ &= \frac{1}{P}\|\nabla u\|^2.\end{aligned}$$

Consequently, $\nabla u_k \to \nabla u$ in $L^2(\Omega)$. This shows that $I(u_k) \to I(u)$. Hence, $I(u) = c$. □

Lemma 16.3 *If $I'(u) = 0$, then u is a solution of (16.1).*

Proof. From (16.12) we see that

$$|(\nabla u, \nabla v)| \leq C\|v\|, \quad v \in E.$$

From the fact that the functions and Ω are periodic with the same period, it follows that $u \in H^{2,2}(\Omega)$ and satisfies (16.1). (cf. Theorem 11.19). □

Lemma 16.4 *Assume that the function $f(x,t)$ is continuous and satisfies*

$$|f(x,t)| \leq C(|t|+1), \quad x \in \Omega,\ t \in \mathbb{R}. \tag{16.14}$$

Assume in addition that

$$2F(x,t) \geq \lambda_\ell t^2, \quad x \in \Omega,\ |t| \leq \delta \text{ for some } \delta > 0, \tag{16.15}$$

where $F(x,t) = \int_0^t f(x,s)\,ds$. Then the following alternative holds: Either (a) there is an infinite number of functions $y \in E(\lambda_\ell) \setminus \{0\}$ such that

$$-\Delta y = f(x,y),$$

or (b) for each $\rho > 0$ sufficiently small, there is an $\epsilon > 0$ such that

$$G(v) \geq \epsilon, \quad \|v\|_E = \rho,\ v \in N,$$

where

$$G(v) = 2\int_\Omega F(x,v)\,dx - \|\nabla v\|^2.$$

Proof. As before, let N denote the subspace of $L^2(\Omega)$ spanned by the eigenfunctions of $-\Delta$ corresponding to the eigenvalues $\lambda_0, \cdots, \lambda_\ell$, and let $M = N^\perp \cap E$. Thus $E = M \oplus N$.

Then

$$\|\nabla v\|^2 \leq \lambda_\ell \|v\|^2, \quad v \in N \tag{16.16}$$

and

$$\lambda_{\ell+1}\|w\|^2 \leq \|\nabla w\|^2, \quad w \in M. \tag{16.17}$$

Let N' be the set of those functions in N which are orthogonal to $E(\lambda_\ell)$. It is spanned by those eigenfunctions corresponding to $\lambda_0, \ldots, \lambda_{\ell-1}$. For $v \in N$, we write $v = v' + y$, where $v' \in N'$ and $y \in E(\lambda_\ell)$. Since $E(\lambda_\ell)$ is finite dimensional and contained in $L^\infty(\Omega)$, there is a $\rho > 0$ such that

$$\|y\|_E \leq \rho \text{ implies } \|y\|_\infty \leq \delta/2, \quad y \in E(\lambda_\ell), \tag{16.18}$$

where δ is given by (4.5.3). Thus if

$$\|v\|_E \leq \rho \text{ and } |v(x)| \geq \delta, \tag{16.19}$$

then

$$\delta \leq |v(x)| \leq |v'(x)| + |y(x)| \leq |v'(x)| + \delta/2.$$

Hence

$$|v(x)| \leq 2|v'(x)| \tag{16.20}$$

holds for all $x \in \bar{\Omega}$ satisfying (16.19). Thus by (16.14) and (16.15),

$$\begin{aligned} G(v) &\geq \lambda_\ell \int_{|v|<\delta} v^2 dx - C \int_{|v|>\delta} \{|v|^2 + |v|\}\, dx - \|\nabla v\|^2 \\ &\geq \lambda_\ell \|v\|^2 - \|\nabla v\|^2 - C \int_{2|v'|>\delta} \{|v'|^2 + |v'|\}\, dx \\ &\geq \lambda_\ell \|v'\|^2 - \|\nabla v'\|^2 - C' \int_{2|v'|>\delta} |v'|^2 dx \\ &\geq \varepsilon \|v'\|^2 + \left(\frac{\lambda_\ell - \varepsilon}{\lambda_{\ell-1}} - 1\right) \|\nabla v'\|^2 - C' \int_{2|v'|>\delta} |v'|^2 dx \\ &\geq \varepsilon \|v'\|_E^2 - C' \int_{2|v'|>\delta} |v'|^2 dx, \end{aligned}$$

where $\varepsilon < (\lambda_\ell - \lambda_{\ell-1})/(\lambda_{\ell-1} + 1)$. I claim that

$$\int_{2|v_k'|>\delta} |v_k'|^2 dx / \|v_k'\|_E^2 \to 0 \tag{16.21}$$

as $\rho_k = \|v_k'\|_E \to 0$. To show that this is true, assume there is a sequence $\{v_k'\}$ such that

$$\int_{2|v_k'|>\delta} |v_k'|^2 dx / \|v_k'\|_E^2 \geq \epsilon > 0, \tag{16.22}$$

while $\rho_k = \|v_k'\|_E \to 0$. Let $\tilde{v}_k' = v_k'/\rho_k$. Then $\|\tilde{v}_k'\|_E = 1$. Let $\mu_k(x)$ be the characteristic function of the set of those $x \in \Omega$ such that $2|v_k'(x)| \geq \delta$. Then (16.22) becomes

$$\int_\Omega |\tilde{v}_k'|^2 \mu_k(x)\, dx \geq \epsilon. \tag{16.23}$$

But

$$\begin{aligned}\mu_k(x) &= 1 \text{ when } 2|\tilde{v}_k'(x)| \geq \delta/\rho_k \to \infty \\ &= 0 \text{ when } 2|\tilde{v}_k'(x)| < \delta/\rho_k.\end{aligned}$$

Hence $\mu_k(x) \to 0$ a.e. We know that there is a renamed subsequence such that $\tilde{v}_k'$ converges in $L^2(\Omega)$. But

$$\tilde{v}_k'(x)^2 \mu_k(x) \leq \tilde{v}_k'(x)^2,$$

and the right-hand side converges in $L^1(\Omega)$. Since the left-hand side converges a.e. to 0, we see that

$$\int |\tilde{v}_k'(x)|^2 \mu_k(x)\, dx \to 0.$$

Hence (16.22) holds, and we have

$$G(v) \geq \varepsilon \|v'\|_E^2, \quad \|v\|_E \leq \rho, \; v \in N. \tag{16.24}$$

Now suppose alternative *(b)* of the lemma did not hold. Then there would be a sequence such that

$$G(v_k) \to 0, \quad v \in N, \; \|v_k\|_E = \rho. \tag{16.25}$$

If ρ is taken sufficiently small, 16.24 implies that $\|v_k'\|_E \to 0$. Consequently, $\|y_k\|_E \to \rho$. Since $E(\lambda_\ell)$ is finite dimensional, there is a renamed subsequence such that $y_k \to y_0$ in $E(\lambda_\ell)$. Thus we have

$$\|y_0\|_E = \rho, \; G(y_0) = 0, \; |y_0(x)| \leq \delta/2, \quad x \in \Omega.$$

Consequently, (16.15) implies

$$2F(x, y_0(x)) \geq \lambda_\ell \, y_0(x)^2, \quad x \in \Omega. \tag{16.26}$$

Since

$$\int_\Omega \{2F(x, y_0(x)) - \lambda_\ell \, y_0(x)^2\}\, dx = G(y_0) = 0$$

and the integrand is ≥ 0 a.e. by (16.26), we see that

$$2F(x, y_0(x)) \equiv \lambda_\ell \, y_0(x)^2, \quad x \in \Omega.$$

Let $\varphi(x)$ be any function in $C_0^\infty(\Omega)$. Then for $t > 0$ sufficiently small

$$t^{-1}[2F(x, y_0 + t\varphi) - \lambda_\ell \, (y_0 + t\varphi)^2 - 2F(x, y_0) + \lambda_\ell \, y_0^2] \geq 0.$$

Taking the limit as $t \to 0$, we have

$$(f(x, y_0) - \lambda_\ell \, y_0)\varphi(x) \geq 0, \quad x \in \Omega.$$

Since this is true for every $\varphi \in C_0^\infty(\Omega)$, we see that

$$f(x, y_0(x)) \equiv \lambda_\ell \, y_0(x), \quad x \in \Omega.$$

Since $y_0 \in E(\lambda_\ell)$, it follows that

$$-\Delta y_0 = \lambda_\ell \, y_0(x) = f(x, y_0(x)).$$

Since this is true for every $\rho > 0$ sufficiently small, alternative (a) holds. □

Lemma 16.5 *Assume that the function $f(x, t)$ is continuous and satisfies*

$$|f(x, t)| \leq C(|t| + 1), \quad x \in \Omega, \; t \in \mathbb{R}.$$

Assume in addition that

$$2F(x, t) \leq \lambda_{\ell+1} t^2, \quad x \in \Omega, \; |t| \leq \delta \text{ for some } \delta > 0,$$

where $F(x, t) = \int_0^t f(x, s)\, ds$. Then the following alternative holds: Either (a) there is an infinite number of functions $y \in E(\lambda_{\ell+1}) \setminus \{0\}$ such that

$$-\Delta y = f(x, y),$$

or (b) for each $\rho > 0$ sufficiently small, there is an $\epsilon > 0$ such that

$$G(w) \geq \epsilon, \quad \|w\|_E = \rho, \; w \in M,$$

where

$$G(w) = \|\nabla w\|^2 - 2\int_\Omega F(x, w)\, dx.$$

Proof. See the proof of Lemma 16.4. □

16.3 Proof of Theorem 16.1

Assume $P > 0, \lambda < 0$, and $\lambda_\ell + m_1 P \le \sigma < \lambda_{\ell+1} + m_0 P,\ \sigma > \lambda_{\ell+1}$, where $\sigma = -\lambda$. We have

$$PI(u) = \|\nabla u\|^2 - \sigma \|u\|^2 + P \int_\Omega \ln\{1 + a|u|^2\}\, dx. \tag{16.27}$$

Hence,

$$\begin{aligned} PI(v) &\le [\lambda_\ell - \sigma]\|v\|^2 + P \int_\Omega \ln\{1 + a|v|^2\}\, dx \\ &\le [\lambda_\ell - \sigma + m_1 P]\|v\|^2 \le 0, \quad v \in N, \end{aligned}$$

provided $\sigma - m_1 P \ge \lambda_\ell$ since $\ln(1+t) \le t$ for $t \ge 0$.

Take

$$2F(x,t) = \sigma t^2 - P \ln\{1 + at^2\}$$

and

$$f(x,t) = \sigma t - P \frac{at}{1 + at^2}.$$

Then the hypotheses of Lemma 16.5 are satisfied if $\sigma - m_0 P < \lambda_{\ell+1}$ since

$$\begin{aligned} 2F(x,t) &= \sigma t^2 - P \ln\{1 + at^2\} \\ &\le (\sigma - m_0 P)t^2 + m_1^2 P \frac{t^4}{1 + m_0 t^2} \\ &< \lambda_{\ell+1} t^2 \end{aligned}$$

provided $|t| < \delta$ and $m_1^2 P \delta^2 < \lambda_{\ell+1} - \sigma + m_0 P$. Here we used the fact that

$$t^2 - \ln(1 + t^2) \le \frac{t^4}{1 + t^2}, \quad t \in \mathbb{R}. \tag{16.28}$$

Hence, either there is an infinite number of solutions of

$$-\Delta u = f(x,u) = \sigma u - \frac{Pau}{1 + a|u|^2}$$

or there is an $\epsilon > 0$ such that

$$G(w) \ge \epsilon, \quad \|w\|_E = \rho,\ w \in M,$$

where

$$\begin{aligned}G(w) &= \|\nabla w\|^2 - 2\int_\Omega F(x,w)\,dx = PI(w)\\ &= \|\nabla w\|^2 - \sigma\|w\|^2 + P\int_\Omega \ln(1+aw^2).\end{aligned}$$

Since the first alternative gives us what we want, we can assume that the second holds.

Let φ_k be an eigenfunction corresponding to λ_k such that $\|\varphi_k\| = 1$. For $\theta \geq 0$,

$$\begin{aligned}PI(\theta\varphi_{\ell+1}+v) &= \theta^2\|\nabla\varphi_{\ell+1}\|^2 + \|\nabla v\|^2 - \sigma(\theta^2\|\varphi_{\ell+1}\|^2 + \|v\|^2)\\ &\quad + P\int_\Omega \ln(1+a(\theta\varphi_{\ell+1}+v)^2)\\ &\leq (\lambda_{\ell+1}-\sigma)\theta^2 + (\lambda_\ell - \sigma)\|v\|^2\\ &\quad + P\int_\Omega \ln(1+a(\theta\varphi_{\ell+1}+v)^2)\\ &\to -\infty \quad as \quad \theta^2 + \|v\|_E^2 \to \infty,\end{aligned}$$

provided $\lambda_{\ell+1} < \sigma$. Thus we have

$$PI(v) \leq 0, \quad v \in N,$$

$$PI(w) \geq \epsilon, \quad \|w\|_E = \rho,\ w \in M,$$

$$PI(\theta\varphi_{\ell+1}+v) \to -\infty \quad as \quad \theta^2 + \|v\|_E^2 \to \infty.$$

If we take

$$\begin{aligned}A &= \{v \in N : \|v\| \leq R\} \cup \{\theta\varphi_{\ell+1}+v : v \in N, \theta \geq 0, \|\theta\varphi_{\ell+1}+v\| = R\},\\ B &= \partial\mathcal{B}_\rho \cap M,\ 0 < \rho < R,\end{aligned}$$

then A and B link each other by Example 3 of Chap. 1, and we have shown that

$$a_0 = \sup_A PI \leq 0 < \epsilon \leq \inf_B PI. \tag{16.29}$$

We can now apply Theorem 1.4 and Lemmas 16.2 and 16.3 to conclude that (16.1) has a nontrivial solution.

If $0 < \sigma < m_0 P$, we take $N = \{0\},\ M = E$, and reason as follows.

$$2F(x,t) = \sigma t^2 - P\ln\{1+at^2\}$$

$$\leq (\sigma - m_0 P)t^2 + m_1^2 P \frac{t^4}{1 + m_0 t^2}$$
$$< 0 = \lambda_0$$

provided $|t| < \delta$ and $m_1^2 P \delta^2 < m_0 P - \sigma$. Hence, either there is an infinite number of solutions of

$$-\Delta u = f(x, u) = \sigma u - \frac{Pau}{1 + a|u|^2}$$

or there is an $\epsilon > 0$ such that

$$G(w) \geq \epsilon, \quad \|w\|_E = \rho, \; w \in M = E,$$

where

$$G(w) = \|\nabla w\|^2 - 2\int_\Omega F(x, w)\,dx = PI(w)$$
$$= \|\nabla w\|^2 - \sigma\|w\|^2 + P\int_\Omega \ln(1 + aw^2).$$

Since the first alternative gives us what we want, we can assume that the second holds.

On the other hand, for the constants we have,

$$PI(c) \leq [-\sigma c^2 + P\ln(1 + m_1 c^2)]|\Omega| \to -\infty \; as \; |c| \to \infty, \; c \in E(\lambda_1).$$

Since $I(0) = 0$, we can take $A = \{0, \; c\}$, $B = \mathcal{B}_\rho$ with $c > \rho$ large, and apply Example 1 of Chap. 1, Theorem 1.4, and Lemmas 16.2 and 16.3 to conclude that (16.1) has a nontrivial solution.

If $0 < \lambda < -m_0 P$, we have

$$PI(u) = \|\nabla u\|^2 + \lambda\|u\|^2 + P\int_\Omega \ln\{1 + a|u|^2\}\,dx \to \infty \tag{16.30}$$

as $\|u\|_E \to \infty$. Hence, $PI(u)$ has a minimum in E. To see this, we first note that

$$\alpha = \inf_E PI > -\infty.$$

For if u_k is a minimizing sequence, it must be bounded in E. Hence, there is a renamed subsequence such that $u_k \rightharpoonup u$ in E, and $u_k \to u$ in $L^2(\Omega)$ and a.e. Thus

$$\alpha = \inf_E PI \geq \lambda\|u\|^2 + P\int_\Omega \ln\{1 + a|u|^2\}\,dx.$$

Letting $N = \{0\},\ M = E$, we can apply Corollary 3.5 to conclude that there is a sequence satisfying

$$I(u_k) \to c > -\infty, \quad \|I'(u_k)\| \to 0. \tag{16.31}$$

Since the sequence must be bounded, this is equivalent to

$$I(u_k) \to c > -\infty, \quad (1 + \|u_k\|_E)\|I'(u_k)\| \to 0. \tag{16.32}$$

It now follows from Corollary 16.2 that u is a solution of $I'(u) = 0$. Let

$$\psi(t) = \lambda t + P\ln(1 + m_0 t).$$

Then $\psi(0) = 0,\ \psi'(0) \le \lambda + m_0 P < 0$. Hence, there is a constant $c \neq 0$ such that $\psi(c^2) < 0$. Thus,

$$PI(c) \le \psi(c^2)|\Omega| < 0.$$

This shows that the minimum value of PI is < 0. Consequently, the minimum point u is $\neq 0$.

Next, assume $P < 0$ and $\lambda_\ell + m_0 P < \sigma \le \lambda_{\ell+1} + m_1 P,\ \sigma < \lambda_\ell$. In this case,

$$\begin{aligned} PI(w) \ge & [\lambda_{\ell+1} - \sigma + m_1 P]\|w\|^2 \\ & + P\int_\Omega [\ln\{1 + a|w|^2\} - aw^2]\,dx \ge 0, \quad w \in M, \end{aligned}$$

provided $\sigma \le \lambda_{\ell+1} + m_1 P$. Moreover,

$$\sigma t^2 - P\ln\{1 + at^2\} \ge (\sigma - m_0 P)t^2 + P\frac{a^2 t^4}{1 + at^2} > \lambda_\ell t^2$$

provided $|t| < \delta$ and $m_1^2|P|\delta^2 < \sigma - m_0 P - \lambda_\ell$. Hence, either there is an infinite number of solutions of

$$-\Delta u = f(x, u) = \sigma u - \frac{Pau}{1 + a|u|^2}$$

or there is an $\epsilon > 0$ such that

$$G(v) \ge \epsilon, \quad \|v\|_E = \rho, \ v \in N,$$

where

$$G(v) = 2\int_\Omega F(x, v)\,dx - \|\nabla v\|^2 = -PI(v).$$

We also have for $\theta \geq 0$,

$$\begin{aligned} PI(\theta\varphi_\ell + w) &= \theta^2\|\nabla\varphi_\ell\|^2 + \|\nabla w\|^2 - \sigma(\theta^2\|\varphi_\ell\|^2 + \|w\|^2) \\ &\quad + P\int_\Omega \ln(1 + a(\theta\varphi_\ell + w)^2) \\ &\geq (\lambda_\ell - \sigma)\theta^2 + (\lambda_{\ell+1} - \sigma)\|w\|^2 \\ &\quad + P\int_\Omega \ln(1 + a(\theta\varphi_{\ell+1} + w)^2) \\ &\to \infty \quad as \quad \theta^2 + \|w\|_E^2 \to \infty, \end{aligned}$$

provided $\lambda_\ell > \sigma$. Let

$$B = \partial\mathcal{B}_\delta \cap N,$$

and

$$A = \{w \in M : \|w\| \leq R\} \cup \{w + \theta\varphi_\ell : w \in M, \theta \geq 0, \|w + \theta\varphi_\ell\| = R\},$$

where $0 < \delta < R$. By Example 7 of Chap. 1 the sets A, B link each other, and we have shown that

$$a_0 = \sup_A G \leq 0 < \epsilon \leq \inf_B G, \tag{16.33}$$

where $G = -PI$. We can now apply Theorem 1.4 and Lemmas 16.2 and 16.3 to conclude that (16.1) has a nontrivial solution. Theorem 16.1 is completely proved.

16.4 Multiple Lattices

Nonlinear Schrödinger equations arising in optics describing the propagation of a light wave in two induced photonic lattices can be written in the form

$$iU_t + \Delta U = \frac{PU}{1 + |U|^2 + |V|^2}$$

$$iV_t + \Delta V = \frac{PV}{1 + |U|^2 + |V|^2}$$

for the periodic wave functions $U(x,t)$, $V(x,t)$ over a periodic bounded spatial domain $\Omega \subset \mathbb{R}^2$, where P, Q are parameters. To find a steady state solution, we look for solutions of the form

$$U(x, t) = e^{i\lambda t}u(x), \quad V(x, t) = e^{i\lambda t}v(x),$$

where λ is a real constant. This leads to the following system of equations over a periodic domain $\Omega \subset \mathbb{R}^2$:

$$\Delta u = \frac{Pu}{1 + |u|^2 + |v|^2} + \lambda u, \tag{16.34}$$

$$\Delta v = \frac{Qv}{1 + |u|^2 + |v|^2} + \lambda v, \tag{16.35}$$

where P, Q, λ are parameters. The solutions u, v are to be periodic in Ω with the same periods. One wishes to obtain intervals of the parameter λ for which there are nontrivial solutions. This will provide continuous energy spectrum that allows the existence of steady state solutions.

We shall show that there are many intervals of the parameters in which nontrivial solutions exist. Our results are true in any dimension. In stating our results, we shall make use of the following considerations. Let Ω be a bounded periodic domain in $\mathbb{R}^n$, $n \geq 1$. Consider the operator $-\Delta$ on functions in $L^2(\Omega)$ having the same periods as Ω. The spectrum of $-\Delta$ consists of isolated eigenvalues of finite multiplicity:

$$0 = \lambda_0 < \lambda_1 < \cdots < \lambda_\ell < \cdots,$$

with eigenfunctions in $L^\infty(\Omega)$. Let λ_ℓ, $\ell \geq 0$, be one of these eigenvalues, and define

$$N = \bigoplus_{\lambda \leq \lambda_\ell} E(\lambda), \quad M = N^\perp.$$

We shall prove the following.

Theorem 16.6

1. *If* $P > 0$, $\lambda < 0$ *and there is an* $\ell \geq 0$ *such that* $\lambda_\ell \leq |\lambda| - P < \lambda_{\ell+1}$, $|\lambda| > \lambda_{\ell+1}$, *then (16.34), (16.35) has a nontrivial solution.*
2. *If* $P < 0$, $\lambda < 0$ *and there is an* $\ell \geq 0$ *such that* $\lambda_\ell < |\lambda| - P \leq \lambda_{\ell+1}$, $|\lambda| < \lambda_\ell$, *then (16.34), (16.35) has a nontrivial solution.*
3. *If* $0 < -\lambda < P$ *or* $0 < -\lambda < Q$, *then (16.34), (16.35) has a nontrivial solution.*
4. *If* $Q > 0$, $\lambda < 0$ *and there is an* $\ell \geq 0$ *such that* $\lambda_\ell \leq |\lambda| - Q < \lambda_{\ell+1}$, $|\lambda| > \lambda_{\ell+1}$, *then (16.34), (16.35) has a nontrivial solution.*
5. *If* $Q < 0$, $\lambda < 0$ *and there is an* $\ell \geq 0$ *such that* $\lambda_\ell < |\lambda| - Q \leq \lambda_{\ell+1}$, $|\lambda| < \lambda_\ell$, *then (16.34), (16.35) has a nontrivial solution.*

To prove the existence of a nontrivial solution of system (16.34), (16.35), it suffices to obtain a nontrivial solution of either

$$\Delta u = \frac{Pu}{1 + |u|^2} + \lambda u, \tag{16.36}$$

or

$$\Delta v = \frac{Qv}{1 + |v|^2} + \lambda v. \tag{16.37}$$

We make use of this observation.

Remark 16.7 For some values of the parameters, solutions of system (16.34), (16.35) which have both u and v guaranteed to be nontrivial exist. For instance, if (16.36) has a nontrivial solution, then there is a solution $u \neq 0, v \neq 0$ of

$$\Delta u = \frac{Pu}{1 + |u|^2 + |v|^2} + \lambda u, \tag{16.38}$$

$$\Delta v = \frac{Pv}{1 + |u|^2 + |v|^2} + \lambda v. \tag{16.39}$$

The author is unaware if such solutions are desirable from the physical point of view.

If $P \neq Q$, (16.34), (16.35) has a nontrivial solution and either u or v is a constant, then one of them must vanish. If P=Q, then (16.34), (16.35) has a solution with both u and v nontrivial constants iff

$$\frac{P}{\lambda} < -1.$$

Using the fact that a nontrivial solution of (16.34), (16.35) is obtained when u is a nontrivial solution of (16.36) or if v is a nontrivial solution of (16.37), we see that Theorem 16.6 is a consequence of Theorem 16.1 if we take $V(x) = 0$ and $m_0 = m_1 = 1$.

16.5 Notes and Remarks

Wave propagation in nonlinear periodic lattices has been studied by many researchers (cf. e.g., [12, 49, 67, 84, 101, 102, 125, 149, 163, 167, 287–291], and their bibliographies.)

The material in this chapter is from [214, 216]. These equations were studied in [288, 289], where it was shown that there is a continuous energy or wavenumber spectrum that allows the existence of steady state solutions. In particular, they showed that

1. If $P > 0$, there is a constant $\delta > 0$ such that the equation (18.1) has a nontrivial solution provided $0 < -\lambda < \delta$.
2. If $P < 0$ and $0 < \lambda < -P/(1+V_0)$, then the Eq. (18.1) has a nontrivial solution.
3. If $P < 0$ and $\lambda \geq -P/(1+v_0)$, then (18.1) has only the trivial solution.

The system (16.34), (16.35) was studied in [49], where it was shown that:

1. If P, Q, λ are all positive, then the only solution is trivial.
2. If $P < 0$ and $0 < \lambda < -P$, then the system (16.34), (16.35) has a nontrivial solution.
3. If $P > 0$, there is a constant $\delta > 0$ such that the system (16.34), (16.35) has a nontrivial solution provided $0 < -\lambda < \delta$.
4. All of these statements are true if we replace P by Q.

Chapter 17
Radially Symmetric Wave Equations

17.1 Introduction

In this chapter we study periodic solutions of the Dirichlet problem for the semilinear wave equation

$$\Box u := u_{tt} - \Delta u = f(t, x, u), \quad t \in \mathbb{R}, \quad x \in \mathcal{B}_R, \tag{17.1}$$

$$u(t, x) = 0, \quad t \in \mathbb{R}, \quad x \in \partial\mathcal{B}_R, \tag{17.2}$$

$$u(t + T, x) = u(t, x), \quad t \in \mathbb{R}, \quad x \in \mathcal{B}_R, \tag{17.3}$$

where

$$\mathcal{B}_R = \{x \in \mathbb{R}^n : |x| < R\}. \tag{17.4}$$

In this case we have

$$f(t, x, u) = f(t, |x|, u), \quad x \in \mathcal{B}_R.$$

Our basic assumption is that the ratio R/T is rational. Thus, we can write

$$8R/T = a/b, \tag{17.5}$$

where a, b are relatively prime positive integers. We show that

$$n \not\equiv 3 \qquad (\text{mod } (4, a)) \tag{17.6}$$

M. Schechter, *Critical Point Theory*,
https://doi.org/10.1007/978-3-030-45603-0_17

implies that the linear problem corresponding to (17.1)–(17.3) has no essential spectrum. If

$$n \equiv 3 \qquad (\text{mod } (4, a)), \tag{17.7}$$

then the essential spectrum of the linear operator consists of precisely one point

$$\lambda_0 = -(n-3)(n-1)/4R^2. \tag{17.8}$$

This shows that the spectrum has at most one limit point. We consider the nonlinear case for $f(t, r, s)$ satisfying

$$|f(t, r, s)| \leq C(|s| + 1), \quad s \in \mathbb{R}, \ r = |x|. \tag{17.9}$$

Assume that

$$m_- s^2 - W_1(t, r) \leq 2F(t, r, s) \leq m_+ s^2 + W_2(t, r), \tag{17.10}$$

where

$$F(t, r, s) = \int_0^s f(t, r, \sigma)\, d\sigma, \tag{17.11}$$

(m_-, m_+) is contained in the resolvent set and the functions W_1, W_2 are in $L^1(Q, \rho)$ with $Q = [0, T] \times [0, R]$ and $\rho = r^{n-1}$. We also assume that

$$H(t, r, s) := sf(t, r, s) - 2F(t, r, s) \tag{17.12}$$

satisfies

$$\limsup_{|s| \to \infty} H(t, r, s)/|s| \leq h(t, r) < 0. \tag{17.13}$$

Our main theorem is the following.

Theorem 17.1 *If (17.6) holds, then (17.1)–(17.3) has a weak rotationally invariant solution. If (17.7) holds and $m_- \geq \lambda_0$, assume in addition that there is a point $\mu \in (m_-, m_+)$ such that*

$$p(t, r, s) = f(t, r, s) - \mu s \tag{17.14}$$

is nondecreasing in s. If $m_+ \leq \lambda_0$, assume that there is such a point such that $p(t, r, s)$ is nonincreasing in s. Then (17.1)–(17.3) has a weak rotationally invariant solution.

Remark 17.2 The essential spectrum of a self-adjoint operator consists of those points of the spectrum that are not isolated eigenvalues of finite multiplicity. Cf., e.g., [195].

The following theorem will be used in our proof. We believe it is of interest in its own right. It will be proved in the last section together with the main theorem.

Theorem 17.3 *Let N be a closed separable subspace of a Hilbert space E. Let G be a continuously differentiable functional on E such that*

$$v_n = Pu_n \to v \text{ weakly in } E, \quad w_n = (I - P)u_n \to w \text{ strongly in } E$$

implies

$$G'(v_n + w_n) \to G'(v + w) \text{ weakly in } E, \tag{17.15}$$

where P is the projection of E onto N. Assume

$$a_0 := \sup_N G < \infty, \quad b_0 := \inf_M G > -\infty. \tag{17.16}$$

Then there is a sequence $\{u_k\} \subset E$ such that

$$G(u_k) \to c, \ b_0 \leq c \leq a_0, \ G'(u_k) \to 0. \tag{17.17}$$

17.2 The Spectrum of the Linear Operator

In proving Theorem 17.1 we shall need to calculate the spectrum of the linear operator $\Box$ applied to periodic rotationally symmetric functions. Specifically, we shall need the following.

Theorem 17.4 *Let L_0 be the operator*

$$L_0 u = u_{tt} - u_{rr} - r^{-1}(n-1)u_r \tag{17.18}$$

applied to functions $u(t, r)$ in $C^\infty(\bar{Q})$ satisfying

$$u(T, r) = u(0, r), \ u_t(T, r) = u_t(0, r), \quad 0 \leq r \leq R \tag{17.19}$$

$$u(t, R) = u_r(t, 0) = 0, \quad t \in \mathbb{R}, \tag{17.20}$$

where $Q = [0, T] \times [0, R]$. Then L_0 is symmetric on $L^2(Q, \rho)$, where $\rho = r^{n-1}$. Assume that $8R/T = a/b$, where a, b are relatively prime integers (i.e., $(a, b) = 1$). Then L_0 has a self-adjoint extension L having no essential spectrum other than the point $\lambda_0 = -(n-3)(n-1)/4R^2$. If $n \not\equiv 3$ $(mod(4, a))$, then L has no essential *spectrum. If $n \equiv 3$ $(mod(4, a))$, then the essential spectrum of L is precisely the point λ_0.*

Proof. Let $\nu = (n-2)/2$ and let γ be a positive root of $J_\nu(x) = 0$, where J_ν is the Bessel function of the first kind. Set

$$\varphi(r) = J_\nu(\gamma r/R)/r^\nu. \tag{17.21}$$

Then

$$\varphi'' + (n-1)\varphi'/r = (x^2 J_\nu'' + x J_\nu' - \nu^2 J_\nu)/r^{\nu+2} = -\gamma^2 J_\nu / R^2 r^\nu, \tag{17.22}$$

where $x = \gamma r/R$. If

$$\psi(t, r) = \varphi(r) e^{2\pi ikt/T}, \tag{17.23}$$

then

$$L_0\psi = [(\gamma/R)^2 - (2\pi k/T)^2]\psi. \tag{17.24}$$

Let γ_j be the j-th positive root of $J_\nu(x) = 0$, and set

$$\psi_{jk}(t, r) = r^{-\nu} J_\nu(\gamma_j r/R) e^{2\pi ikt/T}. \tag{17.25}$$

Then $\psi_{jk}(t, r)$ is an eigenfunction of L_0 with eigenvalue

$$\lambda_{jk} = (\gamma_j/R)^2 - (2\pi k/T)^2. \tag{17.26}$$

It is easily checked that the functions ψ_{jk}, when normalized, form a complete orthonormal sequence in $L^2(Q, \rho)$. We shall show that the corresponding eigenvalues (17.26) are not dense in $\mathbb{R}$. It will then follow that L_0 has a self-adjoint extension L with spectrum equal to the closure of the set $\{\lambda_{jk}\}$ (cf., e.g., [195]). Now

$$\gamma_j = \beta_j - (\mu - 1)/8\beta_j + O(\beta_j^{-3}) \text{ as } \beta_j \to \infty, \tag{17.27}$$

where

$$\beta_j = \pi(j + \frac{1}{2}\nu - \frac{1}{4}), \quad \mu = 4\nu^2 \tag{17.28}$$

(cf., e.g., [275]). Thus

$$\begin{aligned} \lambda_{jk} R^2 &= [\beta_j - \tau_k - (\mu - 1)/8\beta_j + O(\beta_j^{-3})] \\ &\cdot [\beta_j + \tau_k - (\mu - 1)/8\beta_j + O(\beta_j^{-3})] \\ &= \beta_j^2 - \tau_k^2 - (\mu - 1)/4 + O(\beta_j^{-2}), \end{aligned}$$

where $\tau_k = 2k\pi R/T$. (We may assume $k \geq 0$.) Now

$$\beta_j - \tau_k = \pi(j + \frac{1}{2}\nu - \frac{1}{4} - ak/4b) = \pi[(4j + n - 3)b - ak]/4b. \tag{17.29}$$

Since the expression in the brackets is an integer, we see that either $\beta_j = \tau_k$ or

$$|\beta_j - \tau_k| \geq \pi/4b. \tag{17.30}$$

Thus

$$\lim_{\substack{j,|k|\to\infty \\ \beta_j=\tau_k}} \lambda_{jk} = -(\mu - 1)/4R^2 = \lambda_0 \tag{17.31}$$

and

$$\lim_{\substack{j,|k|\to\infty \\ \beta_j\neq\tau_k}} |\lambda_{jk}| = \infty. \tag{17.32}$$

If $n - 3$ is not a multiple of $(4, a)$, then

$$\beta_j - \tau_k = \pi[(4j + n - 3) - ak/b]/4 \tag{17.33}$$

can never vanish. To see this, note that if $(b, k) \neq b$, then ak/b is not an integer. Hence $\beta_j \neq \tau_k$. If $b = (b, k)$, then

$$(n - 3) \neq ak' - 4j \quad \forall j,\ k' = k/b. \tag{17.34}$$

Thus in this case we always have $\beta_j \neq \tau_k$ and $|\lambda_{jk}| \to \infty$ as $j, k \to \infty$. On the other hand, if $n \equiv 3\ (mod(4, a))$, then there is an infinite number of positive integers j, k' such that

$$n - 3 = ak' - 4j. \tag{17.35}$$

Hence, the point λ_0 is a limit point of eigenvalues. Consequently, it is in $\sigma_e(L)$. This completes the proof. □

17.3 The Nonlinear Case

We now turn to the problem solving (17.1)–(17.3). If one is searching for rotationally invariant solutions, the problem reduces to

$$Lu = f(t, r, u), \quad u \in D(L), \tag{17.36}$$

where L is the self-adjoint extension of the operator L_0 given in Theorem 17.4. Under the hypotheses of that theorem the spectrum of L is discrete. We assume that $f(t, r, s)$ is a Carathéodory function on $Q \times \mathbb{R}$ such that

$$|f(t, r, s)| \leq C(|s| + 1), \quad s \in \mathbb{R}. \tag{17.37}$$

We have the following.

Theorem 17.5 *Let $f(t, r, s)$ satisfy (17.10), (17.13), (17.37), and assume the hypotheses of Theorem 17.4. If*

$$n \not\equiv 3 \ (mod(4, a)) \tag{17.38}$$

make no further assumptions. If

$$n \equiv 3 \ (mod(4, a)) \tag{17.39}$$

and $m_- \geq \lambda_0$, assume in addition that there is a point $\mu \in (m_-, m_+)$ such that

$$p(t, r, s) = f(t, r, s) - \mu s \tag{17.40}$$

is nondecreasing in s. If $m_+ \leq \lambda_0$, assume that there is such a point such that $p(t, r, s)$ is nonincreasing in s. Then (17.36) has at least one weak solution.

Proof. Since μ is in the resolvent set of L, there is a $\delta > 0$ such that

$$|\lambda_{jk} - \mu| \geq \delta \quad \forall\ j, k, \tag{17.41}$$

where the λ_{jk} are given by (17.26). Each $u \in L^2(Q, \rho)$ can be expanded in the form

$$u = \sum \alpha_{jk} \psi_{jk}(t, r), \tag{17.42}$$

where the ψ_{jk} are given by (17.25). Let N_0 be the subspace of those $u \in L^2(Q, \rho)$ for which $\alpha_{jk} = 0$ if $\beta_j \neq \tau_k$ (cf. the Proof of Theorem 17.4). For $u \in N_0$

$$u = \sum_{\beta_j = \tau_k} \alpha_{jk} \psi_{jk}(t, r), \tag{17.43}$$

where summation is taken over those j, k for which $\beta_j = \tau_k$. Let E be the subspace of $L^2(Q, \rho)$ consisting of those u for which

$$\|u\|_E^2 = \sum |\lambda_{jk} - \mu| |\alpha_{jk}|^2 \tag{17.44}$$

is finite. With this norm, E becomes a separable Hilbert space. Note that $E \subset D(|L|^{1/2})$ and the embedding of $E \ominus N_0$ into $L^2(Q, \rho)$ is compact (we use (17.32) for this purpose). Let

$$G(u) = ([L - \mu]u, u) - 2\int\int_Q P(t, r, u)\rho dtdr, \quad u \in E \tag{17.45}$$

where

$$P(t, r, s) = \int_0^s p(t, r, \sigma)d\sigma, \tag{17.46}$$

and the scalar product is that of $L^2(Q, \rho)$. One checks readily that G is a C^1 functional on E with

$$(G'(u), v)/2 = ([L - \mu]u, v) - (p(u), v), \quad u, v \in E, \tag{17.47}$$

where we write $p(u)$ in place of $p(t, r, u)$. This shows that u is a weak solution of (17.36) iff $G'(u) = 0$. Let N be the subspace of E spanned by the ψ_{jk} corresponding to those $\lambda_{jk} < \mu$ and let M denote the subspace of E spanned by the rest. Thus $M = N^\perp$ in E and

$$\|u\|_E^2 = \sum(\mu - \lambda_{jk})|\alpha_{jk}|^2, \quad u \in N. \tag{17.48}$$

and

$$\|u\|_E^2 = \sum(\lambda_{jk} - \mu)|\alpha_{jk}|^2, \quad u \in M. \tag{17.49}$$

Thus

$$\begin{aligned} G(v) &= ([L - \mu]v, v) - 2\int\int_Q P(t, r, v)\rho\, dtdr \\ &\leq (m_- - \mu)\|v\|^2 + (\mu - m_-)\|v\|^2 + B_1 \leq B_1, \quad v \in N, \end{aligned}$$

where

$$B_j = \int\int_Q W_j(t, r)\rho\, dtdr.$$

Also,

$$G(w) \geq (m_+ - \mu)\|w\|^2 - (m_+ - \mu)\|w\|^2 - B_2 \geq -B_2, \quad w \in M. \tag{17.50}$$

Assume first that (17.38) holds. Then $N_0 = \{0\}$. In this case, if $\{u_k\} \subset E$ is a sequence converging weakly to u in E, then $\{u_k\}$ has a renamed subsequence which converges strongly in $L^2(Q, \rho)$ and a.e. in Q. This follows from the fact that the embedding of E in $L^2(Q, \rho)$ is compact because of (17.32). Now by (17.47),

$$(G'(u_k), v)/2 = (w_k, v)_E - (v_k, v)_E - \mu(u_k, v) - (p(u_k), v), \quad v \in E, \tag{17.51}$$

where $u_k = v_k + w_k$, $v_k \in N$, $w_k \in M$. Clearly, the first three terms on the right converge. Concerning the last, note that

$$\begin{aligned} |[p(u_k) - p(u)]v| &\leq C(|u_k| + |u| + 1)|v| \\ &\leq C|u_k v| + C(|u| + 1)|v| \\ &\leq C(u_k^2 + v^2 + (|u| + 1)|v|), \end{aligned}$$

which converges in $L^1(Q, \rho)$. Thus $G'(u_k) \to G'(u)$ weakly in E. Hence all of the hypotheses of Theorem 17.3 are satisfied, and we can conclude that there is a sequence $\{u_k\}$ satisfying (17.17). Write $u_k = v_k + w_k$, where $v_k \in N$, $w_k \in M$. Since

$$G(u) = ([L - \mu]u, u) - 2\int\int_Q P(t, r, u)\rho\, dtdr, \quad u \in E \tag{17.52}$$

and

$$(G'(u), v)/2 = ([L - \mu]u, v) - (p(u), v), \quad u, v \in E, \tag{17.53}$$

we have,

$$\int\int_Q H(t, r, u)\rho\, dtdr = (G'(u), u) - 2G(u). \tag{17.54}$$

Hence,

$$|\int\int_Q H(t, r, u_k)\rho\, dtdr| = o(\|u_k\|_E). \tag{17.55}$$

Suppose $\rho_k = \|u_k\|_E \to \infty$. Let $\tilde{u}_k = u_k/\rho_k$. Then $\|\tilde{u}_k\|_E = 1$. In this case, it has a renamed subsequence $\{u_k\} \subset E$ converging weakly to u in E, strongly in $L^2(Q, \rho)$, and a.e. in Q. This implies

$$\begin{aligned} \limsup \int\int_Q H(t, r, u_k)\rho\, dtdr/\rho_k &= \limsup \int\int_Q [H(t, r, u_k)/|u_k|]|\tilde{u}_k|\rho\, dtdr \\ &\leq \int\int_Q h(t, r)|\tilde{u}|\rho\, dtdr < 0, \end{aligned}$$

contrary to (17.55), unless $\tilde{u} = 0$ a.e.

On the other hand,

$$(G'(u_k), w_k - v_k)/2 = \|u_k\|_E^2 - (p(u_k), w_k - v_k). \tag{17.56}$$

Consequently,

$$1 \leq \liminf C \int\int_Q |\tilde{u}_k(\tilde{w}_k - \tilde{v}_k)|\rho\, dtdr = C \int\int_Q |\tilde{u}(\tilde{w} - \tilde{v})|\rho\, dtdr. \tag{17.57}$$

Thus, $\tilde{u} \neq 0$ a.e. This contradiction shows that

$$\|u_k\|_E \leq C. \tag{17.58}$$

Consequently, there is a renamed subsequence which converges weakly in E, a.e. in Q, and strongly in $L^2(Q, \rho)$. Taking the limit in

$$(G'(u_k), v)/2 = ([L - \mu]u_k, v) - (p(u_k), v), \tag{17.59}$$

we obtain a weak solution of (17.36). This completes the proof of this case.

Next assume that (17.39) holds and $\mu < \lambda_0$ with $p(s)$ nonincreasing. Then $\dim(N \cap N_0) < \infty$. If $\{u_k\} \subset E$ is a sequence such that $v_k = Pu_k \to v = Pu$ weakly in N and $w_k = (I - P)u_k \to w = (I - P)u$ strongly in M, where P is the projection of E onto N, then $\{u_k\}$ has a renamed subsequence which converges strongly in $L^2(Q, \rho)$. The reason is that $\{v_k\}$ has such a subsequences because the embedding of $E \ominus N_0$ in $L^2(Q, \rho)$ is compact. Thus $G'(u_n) \to G'(u)$ weakly in E. Hence all of the hypotheses of Theorem 17.3 are satisfied, and we can conclude that there is a sequence $\{u_k\}$ satisfying (17.17). Write $u_k = v_k + w_k + y_k$, where $v_k \in N$, $w_k \in M \ominus N_0$, $y_k \in N_0$. Suppose $\rho_k = \|u_k\|_E \to \infty$. Let $\tilde{u}_k = u_k/\rho_k$. Then $\|\tilde{u}_k\|_E = 1$. The argument given above implies that (17.55) is violated unless $\tilde{u} = 0$ a.e.

To see that this is not the case, note that

$$\begin{aligned}(G'(u_k), w_k + y_k - v_k) &= \|u_k\|_E^2 - (p(u_k), w_k + y_k - v_k)\\ &= \|u_k\|_E^2 - (p(u_k), u_k) - 2(p(u_k), v_k)\\ &= \|u_k\|_E^2 - (p(u_k) - p(0), u_k) - 2(p(u_k), v_k) - (p(0), u_k)\\ &\geq \|u_k\|_E^2 - 2(p(u_k), v_k) - (p(0), u_k).\end{aligned}$$

Dividing by ρ_k^2 and taking the limit, we find that

$$1 \leq C\|\tilde{v}\|,$$

showing that $\tilde{u} \neq 0$. Since this violates (17.55), we can conclude that $\|u_k\|_E$ is bounded and has a renamed subsequence which converges weakly in E and such that $u'_k = v_k + w_k$ converges strongly in $L^2(Q, \rho)$. Now

$$(G'(u_k), y_k - y)/2 = ([L - \mu](y_k - y), y_k - y) \quad (17.60)$$
$$- (p(u_k) - p(u'_k + y), y_k - y)$$
$$- (p(u'_k + y) - p(u), y_k - y)$$
$$+ ([L - \mu]y - p(u), y_k - y),$$

where $y_k \to y$ weakly in E and $L^2(Q, \rho)$ and $u'_k \to u'$ weakly in E and strongly in $L^2(Q, \rho)$. By hypothesis

$$(p(u_k) - p(u'_k + y), y_k - y) \leq 0 \quad (17.61)$$

if $\mu < \lambda_0$. Moreover,

$$(G'(u_k), y_k - y) \to 0$$
$$(p(u'_k + y) - p(u), y_k - y) \to 0$$

and

$$([L - \mu]y - p(u), y_k - y) \to 0. \quad (17.62)$$

Hence

$$\|y_k - y\|_E^2 = ([L - \mu](y_k - y), y_k - y) \leq o(1), \quad k \to \infty. \quad (17.63)$$

This shows that $y_k \to y$ in E, and the proof proceeds as before. If $\lambda_0 < \mu$, we apply Theorem 17.3 to $-G(u)$ and come to the same conclusion. In this case, the inequality in (17.61) is reversed. This completes the proof. □

Proof of Theorem 17.3 Assume that there is no sequence satisfying (17.17). Then there is a positive number δ such that

$$\|G'(u)\| \geq 2\delta \quad (17.64)$$

whenever u belongs to the set

$$E_1 = \{u \in E : b_0 - 2\delta \leq G(u) \leq a_0 + 2\delta\}. \quad (17.65)$$

Since N is separable, we can norm it with a norm $|v|_w$ satisfying

$$|v|_w \leq \|v\|, \quad v \in N \quad (17.66)$$

and such that the topology induced by this norm is equivalent to the weak topology of N on bounded subsets of N (see Chap. 5). For $u \in E$, we write $u = v + w$, where $v \in N$, $w \in M = N^{\perp}$, and take

$$|u|_w^2 = |v|_w^2 + \|w\|^2. \tag{17.67}$$

Then clearly

$$|u|_w \leq \|u\|, \quad u \in E, \tag{17.68}$$

and convergence of a bounded sequence $u_n = v_n + w_n$ with respect to this norm means that v_n converges weakly in N and w_n converges strongly in M. We denote E equipped with this norm by E_w. For $u \in E_1$, let $q(u) = G'(u)/\|G'(u)\|$. Then by (17.64)

$$(G'(u), q(u)) \geq 2\delta, \quad u \in E_1. \tag{17.69}$$

Let

$$\begin{aligned} T &= (a_0 - b_0 + 4\delta)/\delta, \\ \mathcal{B}_R &= \{u \in E : \|u\| < R\}, \\ R &= 2T + \delta, \\ B &= \bar{\mathcal{B}}_R \cap E_1. \end{aligned} \tag{17.70}$$

For each $u \in B$ there is a E_w neighborhood $W(u)$ of u such that

$$(G'(h), q(u)) > \delta, \quad h \in W(u) \cap B. \tag{17.71}$$

For otherwise there would be a sequence $\{h_k\} \subset B$ such that

$$|h_k - u|_w \to 0 \text{ and } (G'(h_k), q(u)) \leq \delta. \tag{17.72}$$

Since B is bounded in E, $Ph_k \to Pu$ weakly in N and $(I - P)h_k \to (I - P)u$ strongly in M. Hence, by hypothesis,

$$(G'(h_k), q(u)) \to (G'(u), q(u)) \geq 2\delta$$

in view of (17.69). This contradicts (17.72). Let B_w be the set B with the inherited topology of E_w. It is a metric space and $W(u) \cap B$ is an open set in this space. Thus $\{W(u) \cap B\}$, $u \in B$, is an open covering of the paracompact space B_w. Consequently, there is a locally finite refinement $\{W_\tau\}$ of this cover. For each τ there is an element u_τ such that $W_\tau \subset W(u_\tau)$. Let $\{\psi_\tau\}$ be a partition of unity subordinate to this covering. Each ψ_τ is locally Lipschitz continuous with respect to the norm $|u|_w$ and consequently with respect to the norm of E. Let

$$Y(u) = \sum \psi_\tau(u) q(u_\tau), \quad u \in B. \tag{17.73}$$

Then $Y(u)$ is locally Lipschitz continuous with respect to both norms. Moreover,

$$\|Y(u)\| \leq \sum \psi_\tau(u) \|q(u_\tau)\| \leq 1 \tag{17.74}$$

and

$$(G'(u), Y(u)) = \sum \psi_\tau(u)(G'(u), h(u_\tau)) \geq \delta, \quad u \in B. \tag{17.75}$$

Let $Q = \mathcal{B}_{T+\delta} \cap N$. For $u \in \bar{Q} \cap E_1$, let $\sigma(t)u$ be the solution of

$$\sigma'(t) = -Y(\sigma(t)), \quad t \geq 0, \ \sigma(0) = u. \tag{17.76}$$

Note that $\sigma(t)u$ will exist as long as $\sigma(t)u$ is in B. Moreover, it is continuous in (u, t) with respect to both topologies.

Next we note that if $u \in \bar{Q} \cap E_1$ and $\sigma(t)u \in B$, then

$$\begin{aligned} dG(\sigma(t)u)/dt &= (G'(\sigma), \sigma') \\ &= -(G'(\sigma), Y(\sigma)) \leq -\delta. \end{aligned} \tag{17.77}$$

Hence if $\sigma(t)u \in B$ for $0 \leq t \leq T$, then

$$G(\sigma(T)u) \leq G(u) - \delta T \leq b_0 - 4\delta. \tag{17.78}$$

But if $\sigma(s)u$ exists for $0 \leq s \leq t < T$, then $\sigma(t)u \in B$. To see this note that

$$u - \sigma(t)u = z_t(u) := \int_0^t Y(\sigma(s)u)ds. \tag{17.79}$$

By (17.74)

$$\|z_t(u)\| \leq t.$$

Consequently

$$\|\sigma(t)u\| \leq \|u\| + t < R. \tag{17.80}$$

Thus $\sigma(t)u \in B$. We can now conclude that for each $u \in \bar{Q} \cap E_1$ there is a $t < T$ such that $\sigma(s)u$ exists for $0 \leq s \leq t$ and $G(\sigma(t)u) \leq b_0 - \delta$. Let

$$T_u := \inf\{t \geq 0 : G(\sigma(t)u) \leq b_0 - \delta\}, u \in \bar{Q} \cap E_1. \tag{17.81}$$

Then $\sigma(t)u$ exists for $0 \leq t \leq T_u < T$. Moreover, T_u is continuous in u. Define

$$\begin{aligned} \sigma_1(t)u &= \sigma(t)u, \quad 0 \leq t \leq T_u \\ &= \sigma(T_u)u, \quad T_u \leq t \leq T \end{aligned}$$

for $u \in \bar{Q} \cap E_1$. For $u \in \bar{Q} \setminus E_1$, define $\sigma_1(t)u = u, \quad 0 \leq t \leq T$. Then $\sigma_1(t)u$ is continuous in (u, t), and

$$G(\sigma_1(T)u) \leq b_0 - \delta, \quad u \in \bar{Q}. \tag{17.82}$$

Let

$$K = \{(u, t) : u = \hat{\sigma}(t)v, \ v \in \bar{Q}, \ t \in [0, T]\}.$$

Then K is a compact subset of $E_w \times \mathbb{R}$. To see this, let (u_k, t_k) be any sequence in K. Then $u_k = \sigma(t_k)v_k$, where $v_k \in \bar{Q}$. Since Q is bounded, there is a subsequence such that $v_k \to v_0$ weakly in E and $t_k \to t_0$ in $[0, T]$. Since $\bar{Q}$ is convex and bounded, v_0 is in $\bar{Q}$ and $|v_k - v_0|_w \to 0$. Since $\hat{\sigma}(t)$ is continuous in $\tilde{E} \times \mathbb{R}$, we have

$$u_k = \hat{\sigma}(t_k)v_k \rightharpoonup \hat{\sigma}(t_0)v_0 = u_0 \in K.$$

Each $u_0 \in B$ has a neighborhood $W(u_0)$ in E_w and a finite dimensional subspace $S(u_0)$ such that $Y(u) \subset S(u_0)$ for $u \in W(u_0) \cap B$. Since $\sigma_1(t)v$ is continuous in (v, t), for each (v_0, t_0) in K there are a neighborhood $W(v_0, t_0) \subset E_w$ and a finite dimensional subspace $S(v_0, t_0) \subset E_w$ such that $\hat{z}_t(v) \subset S(v_0, t_0)$ for $(v, t) \in W(v_0, t_0)$, where

$$\hat{z}_t(u) := u - \hat{\sigma}(t)u = \begin{cases} \int_0^t Y(\hat{\sigma}(s)u)ds, & u \in \hat{E}, \\ 0, & u \notin \hat{E}. \end{cases} \tag{17.83}$$

Since K is compact, there is a finite number of points $(v_j, t_j) \subset K$ such that $K \subset W = \cup W(v_j, t_j)$. Let S be a finite dimensional subspace of E_w containing all the $S(v_j, t_j)$ and such that $Q \cap S \neq \{0\}$. Let

$$\varphi(v, t) = P\sigma_1(t)v, \quad v \in \bar{Q}, \ 0 \leq t \leq T. \tag{17.84}$$

We note that $\varphi(v, t)$ maps $\bar{Q} \cap S \times [0, T]$ into S since $P\sigma_1(t)v = v - P\hat{z}_t(v)$, and $P\hat{z}_t(v)$ is in S when v is in $\bar{Q}$. Let $\varphi_t(v) = \varphi(v, t), \ (v, t) \in K$. Then

$$\varphi_t(v) \neq 0, \quad v \in \partial(Q \cap S) = \partial Q \cap S, \ 0 \leq t \leq T. \tag{17.85}$$

For if $\varphi(v,t) = 0$, then $\sigma_1(t)v \in M$. This implies $G(\sigma_1(t)v) \geq b_0$ by (17.16). But (17.77) and (17.82) imply that $G(\sigma_1(t)v) < b_0$ for $t > 0$. Since $0 \notin \partial Q$ by hypothesis, $\varphi_0(v) = v \neq 0$. Thus (17.85) holds. Consequently the Brouwer degree $d(\varphi_t, Q \cap S, 0)$ can be defined. Since $\varphi_t(v)$ is continuous, we have

$$d(\varphi_T, Q \cap S, 0) = d(\varphi_0, Q \cap S, 0) = d(I, Q \cap S, 0) = 1.$$

Hence there is a $v \in Q$ such that $P\sigma_1(T)v = 0$. Consequently, $\sigma_1(T)v \in M$. In view of (17.16), this implies

$$G(\sigma_1(T)u) \geq b_0,$$

contradicting (17.82). This completes the proof. □

17.4 Notes and Remarks

Many authors have studied the one-dimensional periodic-Dirichlet problem for the semilinear wave equation

$$u_{tt} - u_{xx} = f(t,x,u), \quad t \in \mathbb{R},\ x \in (0,\pi),$$

$$u(t,x) = 0, \quad t \in \mathbb{R},\ x = 0,\ x = \pi,$$

$$u(t + 2\pi, x) = u(t,x), \quad t \in \mathbb{R},\ x \in (0,\pi)$$

A basic problem in this one-dimensional case is that the null space N of the linear part $\Box u = u_{tt} - u_{xx}$ is infinite dimensional. On the other hand, $\Box$ has a compact inverse on the orthogonal complement of N. In contrast to this, the higher dimensional periodic-Dirichlet problem for the semilinear wave equation has the additional difficulty that $\Box$ does not have a compact inverses on $N^\perp$. In fact, it has a sequence of eigenvalues of infinite multiplicities stretching to ∞. This complication causes all of the methods used to solve the one-dimensional case to fail.

Beginning with Smiley [236], some authors have examined the radially symmetric problem (17.1)–(17.3) (cf. [23–25, 35, 80, 151, 154, 192, 236, 237]). This allows one to reduce the problem to

$$u_{tt} - u_{rr} - r^{-1}(n-1)u_r = f(t,r,u),$$

$$u(T,r) = u(0,r),\ u_t(T,r) = u_t(0,r), \quad 0 \leq r \leq R,$$

$$u(t,R) = u_r(t,0) = 0, \quad t \in \mathbb{R}.$$

The complications for this problem depend on the values of R and T. Only in [192] were all possible rational values of R/T considered.

In [23, 25, 154, 236] the hypotheses included inequalities of the form

$$p \leq \liminf \frac{f(u)}{u} \leq \limsup \frac{f(u)}{u} \leq q.$$

In [25] the authors examine radially symmetric solutions to the problem

$$u_{tt} - \Delta u + g(u) = f(t, x),$$

$$u(t + T, \cdot) = u(t, \cdot),$$

where x belongs to a bounded ball B in $\mathbb{R}^n$ with radius R, u satisfies the homogeneous Dirichlet boundary conditions on ∂B, and R/T is rational. The existence of at least one weak solution is proved provided that g is asymptotically linear and the behavior of $g(u)/u$ for u tending to $\pm\infty$ is suitably related to the eigenvalues of the operator $Lv = v_{tt} - \Delta v$, $v(t + T, \cdot) = v(t, \cdot)$.

In [35] irrational values of R/T are considered.

In [192] we proved Theorem 17.1 under the assumption

$$|p(t, r, s)| \leq C(|s|^{\theta} + 1), \quad s \in \mathbb{R}, \tag{17.86}$$

holding for some $\theta < 1$. This assumption is a far greater restriction than (17.9) and (17.13).

What distinguishes our presentation from the results of others is that we cover all rational values of R/T and our hypotheses are given in terms of the primitive

$$P(t, r, s) = \int_0^s p(t, r, \sigma)\, d\sigma, \tag{17.87}$$

of $p(t, r, s)$ rather than the function $p(t, r, s)$ itself.

Chapter 18
Multiple Solutions

18.1 Introduction

A typical characteristic of nonlinear problems is the fact that many times there are multiple solutions. Usually it is difficult to obtain any solutions, let alone more than one. In general, it is much harder to determine when there are even two.

In this chapter we consider the following problem:

$$\begin{cases} -\ddot{u}(t) + B(t)u = \mu \nabla V(t, u), \\ \\ u(T) - u(0) = \dot{u}(T) - \dot{u}(0) = 0, \end{cases} \tag{18.1}$$

where B is a nonnegative symmetric matrix valued function and μ is a positive real parameter. In particular, first we require a suitable behavior of the potential $V(t, \cdot)$ near zero in order to establish the existence of a positive interval of parameters for which problem (18.1) admits at least one nontrivial solution (see Theorem 18.3). Then, assuming in addition that $V(t, \cdot)$ satisfies certain conditions at infinity, a second nontrivial solution is assured (see Theorems 18.4–18.6). The multiplicity results are obtained by combining a linking theorem for functionals depending on a parameter with a monotonicity trick.

18.2 Variational Setting and Preliminary Results

We will assume the following conditions on the matrix valued function B in $I = [0, T]$.

$(B1)$ *$B(t) = \big(b_{ij}(t)\big)$ is a symmetric matrix with each $b_{ij}(t) \in L^1(I)$.*

M. Schechter, *Critical Point Theory*,
https://doi.org/10.1007/978-3-030-45603-0_18

$(B2)$ *There exists a function $\gamma \in L^1(I)$ such that $\gamma \geq 0$, $\gamma \not\equiv 0$ and*

$$B(t)x \cdot x \geq \gamma(t)|x|^2$$

for every $x \in \mathbb{R}^n$ and a.e. t in I.

Thus

$$\gamma(t)|x|^2 \leq B(t)x \cdot x \leq \Lambda(t)|x|^2,$$

for almost every $t \in I$ and every $x \in \mathbb{R}^n$, where $\Lambda(t) \in L^1(I)$. Let H be the Sobolev space of functions $u \in L^2(I, \mathbb{R}^n)$ having a weak derivative $\dot{u} \in L^2(I, \mathbb{R}^n)$. It is well known that H, endowed with the norm

$$\|u\|_H := \left(\int_0^T |u(t)|^2 \, dt + \int_0^T |\dot{u}(t)|^2 \, dt \right)^{1/2},$$

is a Hilbert space, compactly embedded in $C^0(I, \mathbb{R}^n)$ and $L^\infty(I, \mathbb{R}^n)$.

Because of conditions $(B1)$ and $(B2)$, it is possible to introduce on H the following inner product

$$d(u, v) = \int_0^T B(t)u(t) \cdot v(t) \, dt + \int_0^T \dot{u}(t) \cdot \dot{v}(t) \, dt,$$

for every $u, v \in H$. The norm induced by $d(u, v)$ is

$$d(u)^{1/2} := \left(\int_0^T B(t)u(t) \cdot u(t) \, dt + \int_0^T |\dot{u}(t)|^2 \, dt \right)^{1/2}.$$

In fact, we have the following:

Lemma 18.1 *$d(\cdot)^{1/2}$ is equivalent to the norm of H. In particular, there is a constant $c_0 > 0$ such that*

$$\|x\|_\infty^2 \leq c_0 d(x), \quad x \in H. \tag{18.2}$$

Proof. We need to only consider the case $n = 1$. Clearly,

$$d(u) \leq \int_I \Lambda(t)\|u\|_\infty^2 + \int_I |\dot{u}(t)|^2 \leq C\|u\|_H^2.$$

Moreover, there is an $\varepsilon > 0$ such that $\varepsilon\|u\|_H^2 \leq d(u)$. Otherwise, there would be a sequence u_k such that $\|u_k\|_H = 1$ and $d(u_k) \to 0$. Thus

$$\|u_k\|_H := \left(\int_0^T |u_k(t)|^2\,dt + \int_0^T |\dot{u}_k(t)|^2\,dt\right)^{1/2} = 1,$$

and

$$d(u_k)^{1/2} := \left(\int_0^T B(t)u_k(t)\cdot u_k(t)\,dt + \int_0^T |\dot{u}_k(t)|^2\,dt\right)^{1/2} \to 0.$$

Consequently, there would be a renamed subsequence converging weakly in H, strongly in $L^\infty(I)$ and $L^2(I)$, and a.e. in I to a limit $u(t) \in H$. This would imply that $\int_I |\dot{u}_k(t)|^2 \to 0$, $\int_I \gamma(t)|u_k(t)|^2 \to 0$ and $\int_I |u_k(t)|^2 \to 1$. Since $u_k \to u$ in $L^\infty(I)$, these imply $\int_I \gamma(t)|u(t)|^2 = 0$ and $\int_I |u(t)|^2 = 1$. Since

$$(u_k, \dot{v}) + (\dot{u}_k, v) = 0, \quad v \in H,$$

we have

$$(u, \dot{v}) = 0, \quad v \in H.$$

Thus $\dot{u}(t) = 0$, and $u(t) \equiv \mathit{constant}$. Since $\gamma \not\equiv 0$, this provides a contradiction. □

The assumptions imply that there is an extension $\mathcal{D}$ of the operator

$$\mathcal{D}_0 x = -\ddot{x}(t) - B(t)x(t)$$

having a discrete, countable spectrum consisting of isolated eigenvalues of finite multiplicity with a lower bound $\lambda_0 > 0$

$$0 < \lambda_0 < \lambda_1 < \lambda_2 < \ldots < \lambda_l < \ldots \tag{18.3}$$

(cf. Lemma 10.18). The fact that $\lambda_0 > 0$ follows from the observation that $\mathcal{D}u = \lambda u$ implies $d(u) = \lambda\|u\|^2$ showing that $\lambda = 0$ implies $u = 0$.

A solution of problem (18.1) is any function $u_0 \in C^1(I, \mathbb{R}^n)$ such that $\dot{u}_0$ is absolutely continuous, and satisfies

$$-\ddot{u}_0 + B(t)u_0 = \mu\nabla_u V(t, u_0) \quad \text{a.e. in } I,$$

and

$$u_0(T) - u_0(0) = \dot{u}_0(T) - \dot{u}_0(0) = 0.$$

It follows that if we put $\lambda = 1/\mu$, a critical point of the functional

$$G_\lambda(u) = \lambda d(u) - 2\int_I V(t,u)\,dt, \quad 0 < \lambda < \infty,$$

is a solution of (18.1) where the system takes the form

$$\lambda \mathcal{D}u(t) = \nabla_u V(t, u(t)). \tag{18.4}$$

We introduced the parameter λ to make use of the monotonicity trick.

Define

$$G_\lambda(u) = \lambda I(u) - J(u), \quad \lambda > 0,$$

where

$$I(u) = d(u), \quad J(u) = 2\int_I V(t,u)\,dt.$$

Assume one of the two following alternatives holds.

(H$_1$) $I(u) \geq 0$ for all $u \in H$ and $I(u) + |J(u)| \to \infty$ as $\|u\| \to \infty$.

(H$_2$) $I(u) \leq 0$ for all $u \in H$ and $|I(u)| + |J(u)| \to \infty$ as $\|u\| \to \infty$.

Moreover assume that

(H$_3$) There are sets A, B such that A links B and

$$a_0 := \sup_A G_\lambda \leq b_0 := \inf_B G_\lambda$$

for each $\lambda > 0$. Moreover

$$a(\lambda) := \inf_{\Gamma\in\Phi} \sup_{\substack{0\leq s\leq 1\\ u\in A}} G_\lambda(\Gamma(s)u)$$

is finite for each $\lambda > 0$.

The following was proved in Chap. 4 (cf. Corollary 4.13).

Theorem 18.2 *Assume that* **(H$_1$)***(or* **(H$_2$)***) and* **(H$_3$)** *hold. Then for almost all* $\lambda \in \Lambda = (0,\infty)$ *there exists a bounded sequence* $u_k(\lambda) \in H$ *such that*

$$\|G'_\lambda(u_k)\| \to 0, \quad G_\lambda(u_k) \to a(\lambda) \quad \text{as } k \to \infty.$$

18.3 Statement of the Theorems

Theorem 18.3 *Assume:*

1. *There are a function $b(t) \in L^1(I)$ and positive constants m and $\theta < 2$ such that*

$$2V(t, x) \leq b(t)|x|^\theta, \quad |x| \leq m, \ x \in \mathbb{R}^n.$$

2. *There is a constant $M > K_0 = c_0 m^{\theta-2}\|b\|_1$ such that*

$$\liminf_{c \to 0} 2 \int_I V(t, c\varphi)/c^2 \|\varphi\|_2^2 > M\lambda_0, \tag{18.5}$$

where φ is an eigenfunction of $\mathcal{D}$ corresponding to the first eigenvalue λ_0 and c_0 satisfies (18.2).

Then the system (18.4) has a nontrivial solution u_λ satisfying

$$d(u_\lambda) < m^2/c_0, \quad G_\lambda(u_\lambda) < 0$$

for each $\lambda \in (K_0, M)$.

Theorem 18.4 *Assume that hypotheses (1) and (2) of Theorem 18.3 are satisfied in addition to*

$$\liminf_{c \to \infty} 2 \int_I V(t, c\varphi)/c^2 \|\varphi\|_2^2 > M\lambda_0. \tag{18.6}$$

Then the system (18.4) has two nontrivial solutions u_λ, v_λ satisfying

$$d(u_\lambda) < m^2/c_0, \ G_\lambda(u_\lambda) < 0, \quad G_\lambda(v_\lambda) > 0$$

for almost all $\lambda \in (K_0, M)$.

Theorem 18.5 *Assume that hypotheses (1) and (2) of Theorem 18.3 are satisfied. Moreover,*

3. *The function V is such that*

$$V(t, x)/|x|^2 \to \infty, \quad \text{as } |x| \to \infty, \tag{18.7}$$

uniformly with respect to t.

4. *There is a function $W(t) \in L^1(I)$ such that*

$$2V(t, x) - 2V(t, rx) + (r^2 - 1)x \cdot \nabla_x V(t, x) \leq W(t), \quad t \in I, \ x \in \mathbb{R}^n, \ r \in [0, 1]. \tag{18.8}$$

Then the system (18.4) has two nontrivial solutions u_λ, v_λ satisfying

$$d(u_\lambda) < m^2/c_0, \; G_\lambda(u_\lambda) < 0, \quad G_\lambda(v_\lambda) > 0$$

for each $\lambda \in (K_0, M)$.

Theorem 18.6 *The conclusions of Theorem 18.5 hold if we replace Hypothesis (18.8) with the following:*

There are a constant C and a function $W(t) \in L^1(I)$ such that

$$H(t, \theta x) \leq C(H(t, x) + W(t)), \quad 0 \leq \theta \leq 1, \; t \in I, \; x \in \mathbb{R}^n,$$

where

$$H(t, x) = \nabla_x V(t, x) \cdot x - 2V(t, x).$$

18.4 Proofs of the Theorems

Before giving the proofs, we shall prove a few lemmas.

Lemma 18.7 *If (18.8) holds, then*

$$\int_I [2V(t, u) - 2V(t, ru) + (r^2 - 1)u \cdot \nabla_u V(t, u)] \leq C, \quad u \in H, r \in [0, 1], \tag{18.9}$$

where the constant C does not depend on u, r.

Proof. This follows from (18.8) if we take $u = x$. □

Lemma 18.8 *If $u \in H$ satisfies $G'_\lambda(u) = 0$ for some $\lambda > 0$, then there is a constant C independent of u, λ, r such that*

$$G_\lambda(ru) - G_\lambda(u) \leq C \tag{18.10}$$

for all $r \in [0, 1]$.

Proof. From $G'_\lambda(u) = 0$ one has that

$$(G'_\lambda(u), g)/2 = \lambda(\mathcal{D}u, g) - \int_I g \cdot \nabla_u V(t, u) = 0$$

for every $g \in H$. Take

$$g = (1 - r^2)u.$$

Then we have

$$\begin{aligned}
G_\lambda(ru) - G_\lambda(u) =& \lambda(r^2-1)(\mathcal{D}u, u) \\
&+ \int_I [2V(t,u) - 2V(t,ru)]\,dt \\
&= \int_I [2V(t,u) - 2V(t,ru) + ((r^2-1)u \cdot \nabla_u V(t,u)]\,dt \\
&\le C
\end{aligned}$$

by Lemma 18.7. □

Proof of Theorem 18.3 Fix $\lambda \in (K_0, M)$, put $r^2 = m^2/c_0$, and define

$$\mathcal{B}_r = \{u \in H : d(u) \le r^2\}, \quad \partial\mathcal{B}_r = \{u \in H : d(u) = r^2\}.$$

We claim that

$$\inf_{u \in \partial\mathcal{B}_r} G_\lambda(u) > 0. \tag{18.11}$$

Indeed, let $\delta > 0$ be such that $K_0 < K_0 + \delta < \lambda < M$. Then for every $u \in \partial\mathcal{B}_r$ one has

$$G_\lambda(u) \ge \lambda d(u) - \int_I b(t)|u(t)|^\theta \ge \lambda m^2/c_0 - m^\theta \|b\|_1 \ge \delta m^2/c_0,$$

and (18.11) holds. On the other hand, from (18.5), for fixed $\varepsilon \in \big(0, M\lambda_0 - 2\int_I V(t,c\varphi)/c^2\|\varphi\|_2^2\big)$, there exists a $\bar\sigma > 0$ such that

$$2\int_I V(t,c\varphi)/c^2\|\varphi\|_2^2 > M\lambda_0 + \varepsilon$$

for every $|c| < \bar\sigma$. Hence, for c sufficiently small one has $c\varphi \in \mathcal{B}_r$, as well as

$$\begin{aligned}
G_\lambda(c\varphi) &= c^2\|\varphi\|_2^2(\lambda\lambda_0 - 2\int_I V(t,c\varphi)/c^2\|\varphi\|_2^2) \\
&\le c^2\|\varphi\|_2^2(M\lambda_0 - 2\int_I V(t,c\varphi)/c^2\|\varphi\|_2^2) \\
&\le -c^2\|\varphi\|_2^2 \cdot \varepsilon < 0.
\end{aligned}$$

For each λ let $\mu(\lambda) = \inf_{\mathcal{B}_r} G_\lambda$. Then $-\infty < \mu(\lambda) < 0$. There is a minimizing sequence $(u_k) \subset \mathcal{B}_r$ such that $G_\lambda(u_k) \to \mu(\lambda)$. Consequently, there is a renamed subsequence such that $u_k \rightharpoonup u \in H$ and $u_k \to u \in L^\infty(I)$. Thus

$$\lambda d(u_k) \to \mu(\lambda) + 2\int_I V(t,u)\,dt.$$

Also $\lambda d(u) \leq \liminf \lambda d(u_k) = \mu(\lambda) + 2\int_I V(t,u)\,dt$, namely $G_\lambda(u) \leq \mu(\lambda) < 0$ and $u \notin \partial\mathcal{B}_r$. Hence, u is in the interior of $\mathcal{B}_r$ and we have $G'_\lambda(u) = 0$. □

Proof of Theorem 18.4 First observe that we defined

$$I(u) = d(u), \quad J(u) = 2\int_I V(t,u)\,dt$$

for every $u \in H$. Since $I(u) \geq 0$ for all $u \in H$, it is clear that $(\mathbf{H_1})$ holds. Moreover, as in the proof of Theorem 18.3, we take $r^2 = m^2/c_0$. Then

$$\nu(\lambda) = \inf_{\partial\mathcal{B}_r} G_\lambda > 0, \quad \lambda \in (K_0, M).$$

By hypothesis, there are c_1, c_2 such that $c_1\varphi \in \mathcal{B}_r$ and $c_2\varphi \notin \mathcal{B}_r$ with $G_\lambda(c_i\varphi) < 0$, $i = 1, 2$. The set $A = (c_1\varphi, c_2\varphi)$ links $B = \partial\mathcal{B}_r$ (cf. Example 1 of Section 3.2). Applying Theorem 18.2, for almost every λ we obtain a bounded sequence $(y_k) \subset H$ such that

$$G_\lambda(y_k) \to a(\lambda) := \inf_{\Gamma\in\Phi} \sup_{0\leq s\leq 1,\, u\in A} G_\lambda(\Gamma(s)u) \geq \nu(\lambda), \quad G'_\lambda(y_k) \to 0.$$

Since the sequence is bounded, there is a renamed subsequence such that $y_k \rightharpoonup y \in H$ and $y_k \to y \in L^\infty(I)$. Since $G'_\lambda(y_k) \to 0$, we have

$$\lambda d(y_k, v) - \int_I \nabla V(t, y_k)v(t) \to 0.$$

In the limit this gives $G'_\lambda(y) = 0$. We also have $\lambda d(y_k) \to \int_I \nabla V(t,y)y = \lambda d(y)$. Consequently, we have $G_\lambda(y_k) = \lambda d(y_k) - 2\int_I V(t,y_k) \to \lambda d(y) - 2\int_I V(t,y) = G_\lambda(y)$ showing that $G_\lambda(y) = a(\lambda) \geq \nu(\lambda) > 0$. The proof is completed by taking u_λ as the solution obtained by Theorem 18.3 and setting $v_\lambda = y$. □

18.5 The Remaining Proofs

Proof of Theorem 18.5 Note that (18.7) implies (18.6). By Theorem 18.4, for a.e. $\lambda \in (K_0, M)$, there exists a solution u_λ such that $G'_\lambda(u_\lambda) = 0, \quad G_\lambda(u_\lambda) = a(\lambda) \geq \nu(\lambda) > 0$. Let λ satisfy $K_0 < \lambda < M$. Choose $\lambda_n \to \lambda,\ \lambda_n > \lambda$ such that there exists a sequence u_n satisfying

$$G'_{\lambda_n}(u_n) = 0, \quad G_{\lambda_n}(u_n) = a(\lambda_n) \geq \nu(\lambda_n) > 0.$$

Therefore,

$$\int_I \frac{2V(t,u_n)}{d(u_n)}dt < M.$$

Now we prove that $\{u_n\}$ is bounded in H. If $\|u_n\|_H \to \infty$, let $\tilde{u}_n = u_n/d^{1/2}(u_n)$. Then $d(\tilde{u}_n) = 1$ and there is a renamed subsequence such that $\tilde{u}_n \to \tilde{u}$ weakly in H, strongly in $L^\infty(I)$ and a.e. in I. Let $\Omega_0 \subset I$ be the set where $\tilde{u} \neq 0$. Then $|u_n(t)| \to \infty$ for $t \in \Omega_0$. Note that (18.7) and the continuity of V imply the existence of $\beta \in \mathbb{R}$ such that

$$V(t,x) \geq \beta$$

for every $(t,x) \in I \times \mathbb{R}^n$. If Ω_0 had positive measure, we would have

$$\begin{aligned} M > \int_I \frac{2V(t,u_n)}{d(u_n)}dt &= \int_{\Omega_0} \frac{2V(t,u_n)}{d(u_n)} + \int_{I\backslash\Omega_0} \frac{2V(t,u_n)}{d(u_n)}dt \\ &\geq \int_{\Omega_0} \frac{2V(t,u_n)}{|u_n|^2}|\tilde{u}_n|^2 dt + \int_{I\backslash\Omega_0} \frac{2\beta}{d(u_n)}dt \to \infty \end{aligned}$$

as $n \to \infty$. This shows that $\tilde{u} = 0$ a.e. in I. Hence, $\tilde{u}_n \to 0$ in $L^\infty(I)$. For any $s > 0$ and $h_n = s\tilde{u}_n$, we have

$$\int_I V(t,h_n)\,dt \to \int_I V(t,0)\,dt. \tag{18.12}$$

Take $r_n = s/d^{1/2}(u_n) \to 0$. By Lemma 18.8

$$G_{\lambda_n}(r_n u_n) - G_{\lambda_n}(u_n) \leq C. \tag{18.13}$$

Hence,

$$G_{\lambda_n}(s\tilde{u}_n) \leq C + G_{\lambda_n}(u_n) = C + a(\lambda_n) \leq C + a(M). \tag{18.14}$$

But

$$\begin{aligned} G_{\lambda_n}(s\tilde{u}_n) &= \lambda_n s^2(\mathcal{D}\tilde{u}_n, \tilde{u}_n) - 2\int_I V(t, s\tilde{u}_n) \\ &\geq s^2\lambda d(\tilde{u}_n) - 2\int_I V(t, s\tilde{u}_n) \\ &\to \lambda s^2 - \int_I V(t,0)\,dt \end{aligned}$$

by (18.12). This implies

$$\limsup G_{\lambda_n}(s\tilde{u}_n) = \infty \quad as\ s \to \infty,$$

contrary to (18.14).

This contradiction shows that $\|u_n\|_H \le C$. Then there is a renamed subsequence such that $u_n \to u$ weakly in H, strongly in $L^\infty(I)$, and a.e. in I. It now follows that for the bounded renamed subsequence,

$$G'_\lambda(u_n) \to 0, \quad G_\lambda(u_n) \to \lim_{n\to\infty} a(\lambda_n) \ge \nu(\lambda).$$

We can now follow the proof of Theorem 18.4 to obtain the desired solution. □

Proof of Theorem 18.6 We follow the proof of Theorem 18.5 until we conclude that $\tilde{u}_n \to 0$ in $L^\infty(I)$ as a consequence of the fact that we assume that $\|u_n\|_H \to \infty$. We define $\theta_n \in [0, 1]$ by

$$G_{\lambda_n}(\theta_n u_n) = \max_{\theta\in[0,1]} G_{\lambda_n}(\theta u_n).$$

For any $c > 0$ and $h_n = c\tilde{u}_n$, we have

$$\int_I V(t, h_n)\,dt \to \int_I V(t, 0)\,dt \le 0.$$

Thus, for every fixed $c > 0$, if n is large enough one has that $0 < c/d^{1/2}(u_n) < 1$ and

$$G_{\lambda_n}(\theta_n u_n) \ge G_{\lambda_n}((c/d^{1/2}(u_n))u_n) = G_{\lambda_n}(c\tilde{u}_n) = c^2\lambda_n d(\tilde{u}_n) - 2\int_I V(t, h_n)\,dt,$$

so that

$$\liminf_{n\to\infty} G_{\lambda_n}(\theta_n u_n) \ge c^2\lambda,$$

namely, $\lim_{n\to\infty} G_{\lambda_n}(\theta_n u_n) = \infty$. If there is a renamed subsequence such that $\theta_n = 1$ for every n, then

$$G_{\lambda_n}(u_n) \to \infty. \tag{18.15}$$

If $0 \le \theta_n < 1$ for all n, then we have $(G'_{\lambda_n}(\theta_n u_n), \theta_n u_n) = 0$. Indeed, if one defines $h(\theta) = G_{\lambda_n}(\theta u_n)$ for every $\theta \in [0, 1]$, one has

$$\frac{d}{d\theta}h(\theta) = (G'_{\lambda_n}(\theta u_n), u_n).$$

Hence, if $\theta_n = 0$, then $(G'_{\lambda_n}(\theta_n u_n), \theta_n u_n) = 0 \cdot \frac{d}{d\theta}h(0) = 0$. Otherwise, if $0 < \theta_n < 1$, since $h(\theta_n) = \max_{\theta \in [0,1]} h(\theta)$, one has

$$(G'_{\lambda_n}(\theta_n u_n), \theta_n u_n) = \theta_n \cdot \frac{d}{d\theta}h(\theta_n) = 0.$$

Therefore,

$$\begin{aligned}\int_I H(t, \theta_n u_n)\, dt &= \int_I \Big(\nabla V(t, \theta_n u_n)\theta_n x_n - 2V(t, \theta_n u_n)\Big)\, dt \\ &= G_{\lambda_n}(\theta_n u_n) - \frac{1}{2}(G'_{\lambda_n}(\theta_n u_n), \theta_n u_n) \\ &= G_{\lambda_n}(\theta_n u_n) \to \infty.\end{aligned}$$

By hypothesis,

$$\begin{aligned}G_{\lambda_n}(u_n) &= \int_I H(t, u_n) \\ &\geq \int_I H(t, \theta_n u_n)\, dt/C - \int_I W(t)\, dt \to \infty.\end{aligned}$$

Thus, (18.15) holds in any case. But

$$G_{\lambda_n}(u_n) = a(\lambda_n) \leq a(M) < \infty.$$

This contradiction shows that $\|u_n\|_H \leq C$. It now follows that for a renamed subsequence,

$$G'_\lambda(u_n) \to 0, \quad G_\lambda(u_n) \to \lim_{n\to\infty} a(\lambda_n) \geq \nu(\lambda).$$

We can now follow the proof of Theorem 18.4 to obtain the desired solution. □

18.6 Some Examples

Here we show that the assumptions required in the main theorems are naturally satisfied in many simple and meaningful cases.

For simplicity, in what follows, we suppose that $n = 1$, $I = [0, \pi]$, and $B(t) \equiv 1$ for all $t \in I$ while $\alpha, \beta \in L^1(I)$ are two positive functions. A direct computation shows that the eigenvalues of $\mathcal{D}$, with periodic boundary conditions, are

$$\lambda_l = 4l^2 + 1.$$

Hence, $\lambda_0 = 1$ and the corresponding eigenfunctions are constants.

Example 18.9 Put

$$V(t,x) = \alpha(t)|x|^\theta$$

for every $t \in I$, $x \in \mathbb{R}$, with $1 \le \theta < 2$. Then all the assumptions of Theorem 18.3 are satisfied. Indeed, condition (1) holds with $b(t) = 2\alpha(t)$ and every $m > 0$. Moreover, if $\varphi(t) = k$ for every $t \in I$, with $k \in \mathbb{R} \setminus \{0\}$, one has $\|\varphi\|_2^2 = k^2\pi$ and

$$2\int_I V(t, c\varphi)/c^2\|\varphi\|_2^2 = \|b\|_1|ck|^{\theta-2}/\pi.$$

Thus (2) is verified, since $\liminf_{c\to 0} 2\int_I V(t, c\varphi)/c^2\|\varphi\|_2^2 = +\infty$.

Finally, observe that in this case the interval of parameters for which the conclusions of Theorem 18.3 hold is $(0, +\infty)$.

Example 18.10 Let $g : \mathbb{R} \to \mathbb{R}$ be a positive and continuously differentiable function such that

$$L = \lim_{x\to\infty} g(x) > c_0\pi$$

and

$$2g(1) + g'(1) = 2g(-1) - g'(-1) = \theta,$$

where $1 \le \theta < 2$. Put

$$F(x) = \begin{Bmatrix} |x|^\theta, & |x| \le 1, \\ x^2 g(x), & |x| > 1. \end{Bmatrix}.$$

Then, the function

$$V(t,x) = \alpha(t)F(x)$$

for every $t \in I$, $x \in \mathbb{R}$, satisfies all the assumptions of Theorem 18.4. Indeed, arguing as in the previous example, we see that conditions (1) and (2) hold with $m = 1$. Moreover, for $|c|$ large enough one has

$$2\int_I V(t, c\varphi)/c^2\|\varphi\|_2^2 = 2\|\alpha\|_1 g(ck)/\pi.$$

Hence,

$$\liminf_{c\to\infty} 2\int_I V(t, c\varphi)/c^2\|\varphi\|_2^2 = 2\|\alpha\|_1 L/\pi > 2\|\alpha\|_1 c_0 = K_0,$$

and condition (18.6) holds.

Example 18.11 Assume that $\alpha,\ \beta \in L^\infty(I)$ and put

$$V(t,x) = \alpha(t)|x|^\theta + \beta(t)|x|^\tau$$

for every $t \in I$, $x \in \mathbb{R}$ with $1 \le \theta < 2 < \tau$. Then all the assumptions of Theorem 18.5 are satisfied. Indeed, condition (1) holds with $b(t) = 2(\alpha(t) + \beta(t)$ and $m = 1$. Moreover, if $\varphi(t) = k$ for every $t \in I$, with $k \in \mathbb{R} \setminus \{0\}$, one has

$$2\int_I V(t, c\varphi)/c^2\|\varphi\|_2^2 = 2(\|\alpha\|_1|ck|^{\theta-2} + \|\beta\|_1|ck|^{\tau-2})/\pi.$$

Hence

$$\liminf_{c\to 0} 2\int_I V(t, c\varphi)/c^2\|\varphi\|_2^2 = +\infty$$

and (2) is verified. It is an easy matter to verify that condition (3) holds. Finally, if

$$\mathcal{V}_r(t,x) = 2V(t,x) - 2V(t,rx) + (r^2-1)x\cdot\nabla_x V(t,x)$$

for every $t \in I$, $x \in \mathbb{R}$, and $r \in [0,1]$, then, exploiting the choice of θ and τ and observing that $2 - \tau + \tau r^2 - 2r^\tau \le 0$, there exists $C > 0$ independent from t, x, and r, such that

$$\begin{aligned}\mathcal{V}_r(t,x) &= 2\alpha(t)|x|^\theta(1-r^\theta) + 2\beta(t)|x|^\tau(1-r^\tau)\\ &\quad + (r^2-1)[\alpha(t)\theta|x|^\theta + \beta(t)\tau|x|^\tau]\\ &= \alpha(t)|x|^\theta(2-\theta+\theta r^2 - 2r^\theta)\\ &\quad + \beta(t)|x|^\tau(2-\tau+\tau r^2-2r^\tau) < C,\end{aligned}$$

namely (18.8) holds.

We conclude with a further example that points out how Theorem 18.5 applies to functions that do not satisfy the well-known Ambrosetti–Rabinowitz condition.

Example 18.12 Let $\alpha \in L^\infty(I)$ and put

$$V(t,x) = \alpha(t)|x|^2\ln^2|x|$$

for all $t \in I$ and $x \in \mathbb{R}$ (with the meaning $V(t,0) = 0$). Since

$$\lim_{x\to 0} |x|^{2-\theta}\ln^2|x| = 0$$

for every $0 < \theta < 2$, it is clear that condition (1) holds with $b(t) = \alpha(t)$ and m small enough. Moreover, if as usual $\varphi(t) = k$ for $t \in I$ and $k \in \mathbb{R} \setminus \{0\}$, one has

$$\liminf_{c\to 0} 2\int_I V(t, c\varphi)/c^2\|\varphi\|_2^2 = \liminf_{c\to 0} 2\|\alpha\|_1 \ln^2 |ck|/\pi = +\infty,$$

that is, (2) is verified. It is simple to check that (3) holds. Finally, if $\mathcal{V}_r$ is defined as in the previous example, for $r \in (0, 1]$ one has

$$\begin{aligned}
\mathcal{V}_r(t, x) &= 2\alpha(t)|x|^2 \left[\ln^2 |x| - r^2 \ln^2 |rx| + (r^2 - 1)(\ln^2 |x| + \ln |x|)\right] \\
&= 2\alpha(t)|x|^2 \left[-r^2 \ln^2 r - 2r^2 \ln r \ln |x| + r^2 \ln |x| - \ln |x|\right] \\
&\leq 2\alpha(t)|x|^2 \ln |x| \left[r^2 - 1 - 2r^2 \ln r\right].
\end{aligned}$$

Since $r^2 - 1 - 2r^2 \ln r \leq 0$ for every $r \in [0, 1]$, there exists a constant $C > 0$ independent of t, x, and r such that

$$\mathcal{V}_r(t, x) < C.$$

For $r = 0$ one has

$$\mathcal{V}_0(t, x) = -2\alpha(t)|x|^2 \ln |x|.$$

Thus, in every case, (18.8) holds.

18.7 Notes and Remarks

We follow the approach of [39]. There we assumed the following:

$(B1)$ *$B(t) = \big(b_{ij}(t)\big)$ is a symmetric matrix with each $b_{ij}(t) \in L^\infty(I)$.*
$(B2)$ *There exists a function $\gamma \in L^\infty(I)$ such that $\gamma \geq 0$, $\gamma \not\equiv 0$ and*

$$B(t)x \cdot x \geq \gamma(t)|x|^2$$

for every $x \in \mathbb{R}^n$ and a.e. t in I.

Here we show that this was not necessary. One need to only assume that each $b_{ij}(t) \in L^1(I)$ and $\gamma \in L^1(I)$.

Previous work can be found in [38] and [40].

Bibliography

1. Adams RA. Sobolev spaces. Cambridge: Academic; 1975.
2. Alama S, Li YY. On "multibump" bound states for certain semilinear elliptic equations. Indiana Univ Math J. 1992;41(4):983–1026.
3. Ambrosetti A, Coti Zelati V. Periodic solutions of singular Lagrangian systems. Boston: Birkhäuser; 1993.
4. Ambrosetti A, Rabinowitz PH. Dual variational methods in critical point theory and applications. J Func Anal. 1973;14:349–381.
5. Angenent S. The shadowing lemma for elliptic PDE. In: Dynamics of infinite-dimensional systems (Lisbon, 1986). NATO Advanced science institutes series F: computer and systems sciences, vol. 37. Berlin: Springer; 1987. pp. 7–22.
6. Antonacci F. Existence of periodic solutions of Hamiltonian systems with potential indefinite in sign. Nonlinear Anal. 1997;29:1353–1364.
7. Antonacci F, Magrone P. Second order nonautonomous systems with symmetric potential changing sign. Rend Mat Appl. 1998;18:367–379.
8. Arcoya D, Orsina L. Landesman-Lazer conditions and quasilinear elliptic equations. Nonlinear Anal. 1997;28(10):1623–1632.
9. Bahri A, Berestycki H. Existence of forced oscillations for some nonlinear differential equations. Comm Pure Appl Math. 1984;37:403–442.
10. Barletta G, Livrea R. Existence of three periodic solutions for a non-autonomous second order system. Matematiche (Catania) 2002;57:205–215.
11. Barletta G, Papageorgiou N. Nonautonomous second order periodic systems: existence and multiplicity of solutions. J Nonlin Convex Anal. 2007;8:373–390.
12. Bartal G, Manela O, Cohen O, Fleischer JW, Segev M. Observation of second-band vortex solitons in 2D photonic lattices. Phys Rev Lett. 2005;95:053904.
13. Bartolo P, Benci V, Fortunato D. Abstract critical applications to some nonlinear problems with "strong" resonance at infinity. Nonlinear Anal. TMA 1983;7:981–1012.
14. Bartsch T, Ding Y. On a nonlinear Schrödinger equation with periodic potential. Math Ann. 1999;313:15–37.
15. Bartsch T, Ding Y. Homoclinic solutions of an infinite-dimensional Hamiltonian systems. Math Z. 2001;240:289–310.
16. Bartsch T, Ding Y. Critical point theory with applications to asymptotically linear wave and beam equations. Nonl Anal. TMA 2001;44:727–748.
17. Bartsch T, Wang Z. Existence and multiplicity results for some superlinear elliptic problems on $\mathbb{R}^N$. Comm. PDE 1995;20:1725–1741.

M. Schechter, *Critical Point Theory*,
https://doi.org/10.1007/978-3-030-45603-0

18. Bartsch T, Willem M. Periodic solutions of nonautonomous Hamiltonian systems with symmetries. J Reine Angew Math. 1994;451:149–159.
19. Bartsch T, Pankov A, Wang Z-Q. Nonlinear Schrödinger equations with steep potential well. Commun Contemp Math. 2001;3(4):549–569.
20. Benci V. Some critical point theorems and applications. Comm Pure Appl Math. 1980;33:147–172.
21. Benci V, Rabinowitz PH. Critical point theorems for indefinite functionals. Invent Math. 1979;52:241–273.
22. Ben-Naoum AK. The periodic problem for a spherical wave equation. (English summary) Ital J Pure Appl Math. 2004;15:9–18.
23. Ben-Naoum AK, Berkovits J. On the existence of periodic solutions for semilinear wave equation on a ball in R^n with the space dimension n odd. Nonlinear Anal. 1995;24(2):241–250.
24. Ben-Naoum AK, Mawhin J. The periodic-Dirichlet problem for some semilinear wave equations. J Diff Eq. 1992;96:340–354.
25. Ben-Naoum AK, Mawhin J. Periodic solutions of some semilinear wave equations on balls and on spheres, Top Meth Nonlinear Anal. 1993;1:113–137.
26. Ben Naoum AK, Troestler C, Willem M. Existence and multiplicity results for homogenous second order differential equations, J Diff Eq. 1994;112:269–249.
27. Berestycki H, Lions P-L. Nonlinear scalar field equations. I. Existence of a ground state. Arch Ration Mech Anal. 1983;82(4):313–345.
28. Berestycki H, Capuzzo-Dolcetta I, Nirenberg L. Variational methods for indefinite superlinear homogeneous elliptic problems. Nonlinear Diff Eq Appl. 1995;2:553–572.
29. Berger MS. On periodic solutions of second order Hamiltonian systems. I. J Math Anal Appl. 1970;29:512–522.
30. Berger MS. Periodic solutions of second order dynamical systems and isoperimetric variational problems. Amer J Math. 1971;93:1–100.
31. Berger MS. On a family of periodic solutions of Hamiltonian systems. J. Diff Eq. 1971;10:17–26.
32. Berger MS. Autonomous perturbations of some Hamiltonian systems, all of whose solutions are periodic. In: Proceedings of the Conference on Mathematics Research Center, Naval Research Laboratory, Washington, 1971. New York: Academic; 1971. pp. 351–357.
33. Berger MS, Schechter M. On the solvability of semilinear gradient operator equations. Adv Math. 1977;25:97–132.
34. Bergh J, Löfström J. Interpolation spaces. Berlin: Springer; 1976.
35. Berkovits J, Mawhin J. Diophantine approximation, Bessel functions and radially symmetric periodic solutions of semilinear wave equations in a ball. Trans Amer Math Soc. 2001;353:5041–5055.
36. Bertotti ML. Multiplicity of subharmonic solutions of forced Hamiltonian systems near an equilibrium. Manuscripta Math. 1989;64:389–402.
37. Bonanno G, Livrea R. Periodic solutions for a class of second-order Hamiltonian systems. Electron J Diff Eq. 2005;115:13.
38. Bonanno G, Livrea R. Multiple periodic solutions for Hamiltonian systems with not coercive potential. J Math Anal Appl. 2010;363:627–638.
39. Bonanno G, Livrea R, Schechter M. Multiple solutions of Hamiltonian systems. In: Infinite dimensional morse theory and multiple solution problems. Progress in Nonlinear Differential Equations and Their Applications, vol. 6. Boston: Birkhäuser; 1993.
40. Bonanno G, Livrea R, Schechter M. Some notes on a superlinear second order Hamiltonian system. Manuscripta Math. 2017;154:59–77.
41. Bouchala J, Drábek P. Strong resonance for some quasilinear elliptic equations. J Math Anal Appl. 2000;245(1):7–19.
42. Brezis H. Periodic solutions of nonlinear vibrating strings and duality principles. Bull Amer Math Soc. (N.S.) 1983;8(3);409–426.

43. Brézis H, Coron J. Periodic solutions of nonlinear wave equations and Hamiltonian systems. Amer J Math. 1981;103:559–570.
44. Brezis H, Nirenberg L. Remarks on finding critical points. Comm Pure Appl Math. 1991;44:939–964.
45. Castro A. Hammerstein integral equations with indefinite kernel. Int J Math Math Sci. 1978;1(2):187–201.
46. Cerami G. An existence criterion for the critical points on unbounded manifolds. (Italian) Istit Lombardo Acad Sci Lett Rend A 1978/1979;112(2):332–336.
47. Chabrowski J, Szulkin A. On a semilinear Schrödinger equation with critical Sobolev exponent. Proc Amer Math Soc. 2002;130(1):85–93.
48. Chang KC. Infinite dimensional Morse theory and multiple solution problems. Boston: Birkhauser; 1993.
49. Chen S, Lei Y. Existence of steady-state solutions in a nonlinear photonic lattice model. J Math Phys. 2011;52(6):063508.
50. Chen G, Ma S. Periodic solutions for Hamiltonian systems without Ambrosetti-Rabinowitz and spectrum 0. J Math Anal Appl. 2011;379:842–851.
51. Chen W, Mills DL. Gap solitons and the nonlinear optical response of superlattices. Phys Rev Lett. 1989;62:1746–1749.
52. Chen S, Tang C. Periodic and subharmonic solutions of a class of superquadratic Hamiltonian systems. J Math Anal Appl. 2004;297:267–284.
53. Chen S, Wu X, Zhao F. New existence and multiplicity theorems of periodic solutions for non-autonomous second order Hamiltonian systems, Math Comput Model. 2007;46:550–556.
54. Chen J, Yang M, Zhao F. A periodic solution for a second-order asymptotically linear Hamiltonian system. Nonlinear Anal. 2009;11:4021–4026.
55. Clark DC. Periodic solutions of variational systems of ordinary differential equations. J Diff Eq. 1978;28:354–368.
56. Clarke FH. A classical variational principle for periodic Hamiltonian trajectories. Proc Amer Math Soc. 1979;76:186–188.
57. Clarke FH. Periodic solutions to Hamiltonian inclusions. J Diff Eq. 1981;40:1–6.
58. Clarke FH. Periodic solutions of Hamiltonian equations and local minima of the dual action. Trans Amer Math Soc. 1985;287:239–251.
59. Clarke FH, Ekeland I. Solutions périodiques, de période donnée, des équations hamiltoniennes. C R Acad Sci Paris Sér. A-B 1978;287:1013–1015.
60. Clarke FH, Ekeland I. Oscillations entretenues de systèmes Hamiltoniens non linéaires. C R Acad Sci Paris 1979;289:A393–A395.
61. Clarke FH, Ekeland I. Nonlinear oscillations and boundary value problems for Hamiltonian systems. Arch Rational Mech Anal. 1982;78:315–333.
62. Conley C, Zehnder E. Morse-type index theory for flows and periodic solutions for Hamiltonian equations. Comm Pure Appl Math. 1984;37:207–253.
63. Cordaro G. Three periodic solutions to an eigenvalue problem for a class of second-order Hamiltonian systems. Abstr Appl Anal. 2003;18:1037–1045.
64. Cordaro G, Rao G. Three periodic solutions for perturbed second-order Hamiltonian systems. J Math Anal Appl. 2009;359:780–785.
65. Costa DG. On a class of elliptic systems in $\mathbb{R}^N$. Electron J Diff Eq. 1994;7:1–14.
66. Costa DG, Magalhaes CA. Un probleme non-quadratique a l'Infini. C R Acad Sci Paris T Ser I 1992;315:1059–1062.
67. Costa DG, Magalhaes CA. A variational approach to subquadratic perturbations of elliptic systems. J Diff Eq. 1994;111(1):103–122.
68. Costa DG, Magalhaes CA. Variational elliptic problems which are nonquadratic at infinity. Nonlinear Anal. 1994;23(11):1401–1412.
69. Coti Zelati V. Periodic solutions of dynamical systems with bounded potential. J Diff Eq. 1987;67:400–413.
70. Coti Zelati V, Li S, Wu S. Periodic solutions for a class of singular nonautonomous second order system in a potential well, Variational methods in nonlinear analysis. Basel: Gordon and Breach; 1995. pp. 67–76.

71. de Figueiredo DG, Felmer PL. On superquadratic elliptic systems. Trans Amer Math Soc. 1994;343(1):99–116.
72. de Figueiredo DG, Mitidieri E. A maximum principle for an elliptic system and applications to semilinear problems. SIAM J Math Anal. 1986;17:836–849.
73. Deng X, Yang R. Periodic solutions for some Hamiltonian systems. J Appl Math Comput. 2009;31:309–315.
74. Ding Y, Girardi M. Periodic and homoclinic solutions to a class of Hamiltonian systems with the potentials changing sign. Dynam Syst Appl. 1993;2:131–145.
75. Ding Y, Girardi M. Periodic solutions for a class of symmetric and subquadratic Hamiltonian systems. Math Comput Modell. 1996;23:59–71.
76. Ding Y, Lee C. Periodic solutions of Hamiltonian systems. SIAM J Math Anal. 2000;32:555–571.
77. Ding W-Y, Ni W-M. On the existence of positive entire solutions of a semilinear elliptic equation. Arch Rat Mech Anal. 1986;91:183–308.
78. Ding W, Qian D. Periodic solutions for sublinear systems via variational approach. Nonlinear Anal Real World Appl. 2010;11:2603–2609.
79. Ding Y, Szulkin A. Bound states for semilinear Schrödinger equations with sign-changing potential. Calc Var Partial Diff Eq. 2007;29(3):397–419.
80. Ding Y, Li S, Willem M. Periodic solutions of symmetric wave equations. J Diff Eq. 1998;145(2):217–241.
81. Dong Y. Index theory, nontrivial solutions, and asymptotically linear second-order Hamiltonian systems. J Diff Eq. 2005;214:233–255.
82. Drábek P, Robinson SB. Resonance problems for the p-Laplacian. J Funct Anal. 1999;169(1):189–200.
83. Dunford N, Schwartz JT. Linear operators, part I, Interscience. Hoboken: Wiley; 1967.
84. Efremidis NK, Sears S, Christodoulides DN. Discrete solitons in photorefractive optically-induced photonic lattices. Phys Rev Lett. 2000;85:1863–1866.
85. Ekeland I, Hofer H. Periodic solutions with prescribed minimal period for convex autonomous Hamiltonian systems. Invent Math. 1985;81:155–188.
86. Ekeland I, Hofer H. Subharmonics for convex nonautonomous Hamiltonian systems. Comm Pure Appl Math. 1987;40:1–36.
87. Ekeland I, Ghoussoub N. Certain new aspects of the calculus of variations in the large. Bull Amer Math Soc. 2002;39:207–265.
88. Ekeland I, Temam R. Convex analysis and variational analysis. Amsterdam: North-Holland; 1976.
89. e Silva EADB. Linking theorems and applications to semilinear elliptic problems at resonance. Nonlinear Anal. TMA 1991;16:455–477.
90. e Silva EADB. Existence and multiplicity of solutions for semilinear elliptic systems. NoDEA 1994:1:339–363.
91. e Silva EADB. Nontrivial solutions for noncooperative elliptic systems at resonance. In: Proceedings of the Conference on USA-Chile Workshop on Nonlinear Analysis (Vi–a del Mar-Valparaiso, 2000) (Electronic). Electronic Journal of Differential Equations, vol. 6, Southwest Texas State University, San Marcos, 2001. pp. 267–283.
92. Faraci F. Three periodic solutions for a second order nonautonomous system. J Nonlinear Convex Anal. 2002;3:393–399.
93. Faraci F. Multiple periodic solutions for second order systems with changing sign potential. J Math Anal Appl. 2006;319:567–578.
94. Faraci F, Iannizzotto A. A multiplicity theorem for a perturbed second-order non-autonomous system. Proc Edinb Math Soc. 2006; 49(2):267–275.
95. Faraci F, Livrea G. Infinitely many periodic solutions for a second-order nonautonomous system. Nonlinear Anal. 2003;54:417–429.
96. Fei G. On periodic solutions of superquadratic Hamiltonian systems. Electron J Diff Eq. 2002;8:12.

97. Fei G, Kim S, Wang T. Periodic solutions of classical Hamiltonian systems without Palais-Smale condition. J Math Anal Appl. 2002;267:665–678.
98. Felmer P. Periodic solutions of 'superquadratic' Hamiltonian systems. J Diff Eq. 1993;102:188–207.
99. Felmer P, Silva EAD. Homoclinic and periodic orbits for Hamiltonian systems. Ann Scuola Norm Sup Pisa Cl Sci. 1998;26:285–301.
100. Feng J, Han Z. Periodic solutions to differential systems with unbounded or periodic nonlinearities. J Math Anal Appl. 2006;323:1264–1278.
101. Fleischer JW, Segev M, Efremidis NK, Christodolides DN. Observation of two-dimensional discrete solitons in optically induced nonlinear photonic lattices. Nature 2003;422:147–149.
102. Fleischer JW, Bartal G, Cohen O, et al. Observation of vortex-ring 'discrete' solitons in 2D photonic lattices. Phys Rev Lett. 2004;92:123904.
103. Fonda A, Mawhin J. Iterative and variational methods for the solvability of some semilinear equations in Hilbert spaces. J Diff Eq. 1992;98:355–375.
104. Fonda A, Mawhin J. An iterative method for the solvability of semilinear equations in Hilbert spaces and applications. In: Wiener-Hale, editor. Proceedings of the Conferences on Differential Equations, Edinburgh, Texas; 1991, Longman; 1992. pp. 126–132.
105. Furtado MF, Silva EAB. Double resonant problems which are locally non-quadratic at infinity. In: Proceedings of the Conference on USA-Chile Workshop on Nonlinear Analysis (Vi-a del Mar-Valparaiso, 2000) (Electronic). Electronic Journal of Differential Equations, vol. 6, Southwest Texas State University, San Marcos; 2001. pp. 155–171.
106. Furtado MF, Maia LA, Silva EAB. Solutions for a resonant elliptic system with coupling in $\mathbb{R}^N$. Comm Partial Diff Eq. 2002;27(7–8):1515–1536.
107. Furtado MF, Maia LA, Silva EAB. On a double resonant problem in $\mathbb{R}^N$. Diff Int Eq. 2002;15(11):1335–1344.
108. Ghoussoub N. Duality and perturbation methods in critical point theory, Cambridge tracts in mathematics, vol. 107. Cambridge: Cambridge University Press; 1993.
109. Giannoni F. Periodic solutions of dynamical systems by a saddle point theorem of Rabinowitz. Nonlinear Anal. 1989;13:707–719.
110. Girardi M, Matzeu M. Existence and multiplicity results for periodic solutions of superquadratic Hamiltonian systems where the potential changes sign. NoDEA Nonlinear Diff Eq Appl. 1995;2:35–61.
111. Girardi M, Matzeu M. On periodic solutions of the system $x(t)+b(t)(V1(x(t))+V2(x(t)))0$ where $b(\cdot)$ changes sign and $V1$, $V2$ have different superquadratic growths. In: Variational and local methods in the study of Hamiltonian systems, Trieste, 1994. River Edge: World Scientific Publishing; 1995. pp. 65–76.
112. Gordon WB. A theorem on the existence of periodic solutions to Hamiltonian systems with convex potential. J Diff Eq. 1970;10:324–335.
113. Guo Z, Xu Y. Existence of periodic solutions to second-order Hamiltonian systems with potential indefinite in sign. Nonlinear Anal. 2002;51:1273–1283.
114. Han Z. 2π-periodic solutions to ordinary differential systems at resonance. Acta Math Sinica (Chinese) 2000;43:639–644.
115. Hu S, Papageorgiou N. Nontrivial solutions for superquadratic nonautonomous periodic systems. Topol Methods Nonlinear Anal. 2009;34:327–338.
116. Hulshof J, Vander Vorst RCAM. Differential systems with strongly indefinite variational structure. J Funct Anal. 1993;114:32–58.
117. He X, Wu X. Periodic solutions for a class of nonautonomous second order Hamiltonian systems. J Math Anal Appl. 2008;341:1354–1364.
118. Jeanjean L. On the existence of bounded Palais-Smale sequences and application to a Landesman-Lazer type problem set on $\mathbf{R}^N$. Proc Roy Soc Edinburgh 1999;A129:787–809.
119. Jeanjean L. Local conditions insuring bifurcation from the continuous spectrum. Math Z. 1999;232:651–664.
120. Jiang M. Subharmonic solutions of second order subquadratic Hamiltonian systems with potential changing sign. J Math Anal Appl. 2000;244:291–303.

121. Jiang M. Periodic solutions of second order superquadratic Hamiltonian systems with potential changing sign I. J Diff Eq. 2005;219:323–341.
122. Jiang M. Periodic solutions of second order superquadratic Hamiltonian systems with potential changing sign II. J Diff Eq. 2005;219:342–362.
123. Jiang Q, Tang C. Periodic and subharmonic solutions of a class of subquadratic second-order Hamiltonian systems. J Math Anal Appl. 2007;328:380–389.
124. Jiang J, Tang X. Existence and multiplicity of periodic solutions for a class of second-order Hamiltonian systems. Comput Math Appl. 2010;59:3646–3655.
125. Kelley JL. General topology. New York: Van Nostrand Reinhold; 1955.
126. Kryszewski W, Szulkin A. Generalized linking theorems with an application to semilinear Schrodinger equation. Adv Diff Eq. 1998;3:441–472.
127. Kryszewski W, Szulkin A. Infinite-dimensional homology and multibump solutions. J Fixed Point Theory Appl. 2009;5(1):1–35.
128. Kuchment P. The mathematics of photonic crystals. In: Mathematical modeling in optical science. Frontiers in applied mathematics and statistics, vol. 22. Philadelphia: SIAM; 2001. pp. 207–272.
129. Kyritsi S, Papageorgiou N. On superquadratic periodic systems with indefinite linear part. Nonlinear Anal. 2010;72:946–954.
130. Lassoued L. Periodic solutions of a nonlinear second order system. In: Variational methods (Progress in nonlinear differential equations and their applications), Paris, 1988. Boston: Birkhauser; 1990. pp. 455–468.
131. Lassoued L. Solutions périodiques d'un système différentiel non linéaire du second ordre avec changement de signe. Ann Mat Pura Appl. 1990;156:73–111.
132. Lassoued L. Periodic solutions of a second order superquadratic system with a change of sign in the potential. J Diff Eq. 1991;93:1–18.
133. Lazer A, McKenna P. On steady-state solutions of a system of reaction-diffusion equations from biology. Nonlinear Anal TMA 1982;6:523–530.
134. Li Y. Remarks on a semilinear elliptic equation on RN. J Diif Eq. 1988;74:34–49.
135. Li S. Periodic solutions of nonautonomous second order systems with superlinear terms. Diff Int Eq. 1992;5:1419–1424.
136. Li S, Szulkin A. Periodic solutions of an asymptotically linear wave equation. Topol Method Nonlinear Anal. 1993;1(2):211–230.
137. Li S, Szulkin A. Periodic solutions for a class of nonautonomous Hamiltonian systems. J Diff Eq. 1994;112:226–238.
138. Li S, Szulkin A. Periodic solutions for a class of nonautonomous wave equations. Diff Int Eq. 1996;9(6):1197–1212.
139. Li S, Willem M. Applications of local linking to critical point theory. J Math Anal Appl. 1995;189:6–32.
140. Li G, Yang J. Asymptotically linear elliptic systems. (English summary) Comm Partial Diff Eq. 2004;29(5–6):925–954.
141. Li S, Zou W. Infinitely many solutions for Hamiltonian systems. J Diff Eq. 2002;186:141–164.
142. Li YQ, Wang ZQ, Zeng J. Ground states of nonlinear Schrödinger equations with potentials. (English, French summary) Ann Inst H Poincaré Anal Non Linéaire 2006;23(6):829–837.
143. Long YM. Multiple solutions of perturbed superquadratic second order Hamiltonian systems. Trans Amer Math Soc. 1989;311:749–780.
144. Long YM. Periodic solutions of superquadratic Hamiltonian systems with bounded forcing terms. Math Z. 1990;203:453–467.
145. Long YM. Nonlinear oscillations for classical Hamiltonian systems with bi-even subquadritic potentials. Nonlinear Anal. 1995;24:1665–1671.
146. Luan SX, Mao AM. Periodic solutions of nonautonomous second order Hamiltonian systems. Acta Math Sin. (English) 2005;21:685–690.
147. Ma S. Computations of critical groups and periodic solutions for asymptotically linear Hamiltonian systems. J Diff Eq. 2010;248:2435–2457.

148. Ma J, Tang C. Periodic solutions for some nonautonomous second-order systems. J Math Anal Appl. 2002;275:482–494.
149. Martin H, Eugenieva ED, Chen Z. Discrete solitons and soliton-induced dislocations in partially coherent photonic lattices. Phys Rev Lett. 2004;92:123902.
150. Mawhin J. Periodic solutions of nonlinear dispersive wave equations. In: Albrecht, Kirchgässner, Collatz, editors. Constructive methods for nonlinear BVPs and nonlinear oscillations. Basel: Birkhäuser; 1979. pp. 102–109.
151. Mawhin J. Nonlinear functional analysis and periodic solution of semilinear wave equation. In: Lakshmikantham, editor. Nonlinear phenomena in mathematical sciences. Cambridge: Academic; 1982. pp. 671–681.
152. Mawhin J. Periodic oscillations of nonlinear wave systems. In: Mitropolsky, editor. Ninth International Conference on Nonlinear Oscillations, Kiev 1981, vol. 1. Kiev: Naukova Dumka; 1984. pp. 47–53.
153. Mawhin J. Forced second order conservative systems with periodic nonlinearity. Ann Inst Poincaré Anal Non Linéare 1989;6:415–434.
154. Mawhin J. Periodic solutions of some semilinear wave equations and systems: a survey. Chaos Soliton Fract. 1995;5(9):1651–1669.
155. Mawhin J, Ward JR. Asymptotic nonuniform non-resonance conditions in the periodic-Dirichlet problem for semilinear wave equations. Ann Mat Pura Appl. 1983;135(4):85–97.
156. Mawhin J, Ward JR. Nonuniform non-resonance conditions in the Periodic-Dirichlet problem for semi-linear wave equations with jumping nonlinearities. J Nigerian Math Soc. 1984;3:41–55.
157. Mawhin J, Willem M. Critical points of convex perturbations of some indefinite quadratic forms and semilinear boundary value problems at resonance. Ann Inst H Poincaré Anal Non Linearé 1986;3:431–453.
158. Mawhin J, Willem M. Critical point theory and Hamiltonian systems. Berlin: Springer; 1989.
159. Meng Q, Tang XH. Solutions of a second-order Hamiltonian system with periodic boundary conditions. Commun Pure Appl Anal. 2010;9:1053–1067.
160. Miyagaki OH, Souto MAS. Superlinear problems without Ambrosetti and Rabinowitz growth condition. J Diff Eq. 2008;245(12):3628–3638.
161. Motreanu D, Motreanu V, Papageorgiou N. Periodic solutions for nonautonomous systems with nonsmooth quadratic or superquadratic potential. Topol Methods Nonlinear Anal. 2004;24:269–296.
162. Motreanu D, Motreanu V, Papageorgiou N. Nonautonomous resonant periodic systems with indefinite linear part and a nonsmooth potential. Commun Pure Appl Anal. 2011;10:1401–1414.
163. Neshev DN, Alexander TJ, Ostrovskaya EA, et al. Observation of discrete vortex solitons in optically induced photonic lattices. Phys Rev Lett. 2004; 92:123903.
164. Nirenberg L. Variational and topological methods in nonlinear problems. Bull Amer Math Soc. 1981;4:267–302.
165. Omana W, Willem M. Homoclinic orbits for a class of Hamiltonian systems. Diff Int Eq. 1992;5:1115–1120.
166. Ou Z, Tang C. Periodic and subharmonic solutions for a class of superquadratic Hamiltonian systems. Nonlinear Anal. 2004;58:245–258.
167. Pankov A. Periodic nonlinear Schrödinger equation with application to photonic crystals. Milan J Math. 2005;73:259–287.
168. Perera K. One-sided resonance for quasilinear problems with asymmetric nonlinearities. Abstr Appl Anal. 2002;7(1):53–60.
169. Perera K, Schechter M. Morse index estimates in saddle point theorems without a finite-dimensional closed loop. Indiana Univ Math J. 1988;47(3):1083–1095.
170. Perera K, Schechter M. Type II regions between curves of the Fučik spectrum and critical groups. Topol Methods Nonlinear Anal. 1998;12(2):227–243.
171. Perera K, Schechter M. A generalization of the Amann-Zehnder theorem to nonresonance problems with jumping nonlinearities. NoDEA Nonlinear Diff Eq Appl. 2000;7(4):361–367.

172. Perera K, Schechter M. The Fučík spectrum and critical groups. Proc Amer Math Soc. 2001;129(8):2301–2308 (Electronic).
173. Perera K, Schechter M. Critical groups in saddle point theorems without a finite dimensional closed loop. Math Nachr. 2002;243:156–164.
174. Perera K, Schechter M. Double resonance problems with respect to the Fučík spectrum. Indiana Univ Math J. 2003;52(1):1–17.
175. Perera K, Schechter M. Sandwich pairs in p-Laplacian problems. Topol Method Nonlinear Anal. 2007;29(1):29–34.
176. Perera K, Schechter M. Flows and critical points. NoDEA Nonlinear Diff Eq Appl. 2008;15(4–5):495–509.
177. Perera K, Schechter M. Sandwich pairs for p-Laplacian systems. J Math Anal Appl. 2009;358(2):485–490.
178. Pipan J, Schechter M. Non-autonomous second order Hamiltonian systems. J Diff Eq. 2014;257(2):351–373.
179. Rabinowitz PH. On subharmonic solutions of Hamiltonian systems. Comm Pure Appl Math. 1980;33:609–633.
180. Rabinowitz PH. Minimax methods in critical point theory with applications to differential equations. In: CBMS Regional Conference Series in Mathematics. vol. 65. Providence: American Mathematical Society; 1986.
181. Reed M, Simon B. Methods of modern mathematical physics, vol. 4. Cambridge: Academic; 1978.
182. Schechter M. Operator methods in quantum mechanics. New York: Elsevier; 1981.
183. Schechter M. Spectra of partial differential operators. North-Holland series in applied mathematics and mechanics, vol. 14. 2nd ed. Amsterdam: North-Holland; 1986. xiv+310. ISBN: 0-444-87822-X.
184. Schechter M. A variation of the mountain pass lemma and applications. J London Math Soc. 1991;44:491–502.
185. Schechter M. A bounded mountain pass lemma without the (PS) condition and applications. Trans Amer Math Soc. 1992;331:681–703.
186. Schechter M. A generalization of the saddle point method with applications. Ann Polon Math. 1992;57(3):269–281.
187. Schechter M. New saddle point theorems. In: Generalized functions and their applications (Varanasi, 1991). New York: Plenum; 1993. pp. 213–219.
188. Schechter M. The intrinsic mountain pass. Pacific J Math. 1995;171:529–544.
189. Schechter M. Infinite-dimensional linking. Duke Math J. 1998;94(3):573–595.
190. Schechter M. New linking theorems. Rend Sem Mat Univ Padova. 1998;99:255–269.
191. Schechter M. Critical point theory with weak-to-weak linking. Comm Pure Appl Math. 1998;51(11–12):1247–1254.
192. Schechter M. Rotationally invariant periodic solutions of semilinear wave equations. Abstr Appl Anal. 1998;3:171–180.
193. Schechter M. Linking methods in critical point theory. Boston: Birkhauser; 1999.
194. Schechter M. Periodic solutions of semilinear higher dimensional wave equations. Chaos Solitons Fractals 2001;12(6):1029–1034.
195. Schechter M. Principles of functional analysis. In: Graduate Studies in Mathematics, vol. 36. 2nd ed. Providence: American Mathematical Society; 2002. xxii+425 pp.
196. Schechter M. An introduction to nonlinear analysis. Cambridge Studies in Advanced Mathematics, vol. 95. Cambridge: Cambridge University Press; 2004.
197. Schechter M. Saddle point techniques. Nonlinear Anal. 2005;63:699–711.
198. Schechter M. Periodic non-autonomous second-order dynamical systems. J Diff Eq. 2006;223(2):290–302.
199. Schechter M. Periodic solutions of second-order nonautonomous dynamical systems. Bound Value Probl. 2006:Art. ID 25104, 9.
200. Schechter M. The use of Cerami sequences in critical point theory. Abstr Appl Anal. 2007;2007: Art. ID 58948, 28.

201. Schechter M. Semilinear problems in unbounded domains. J Math Anal Appl. 2007; 327(2):1143–1154.
202. Schechter M. Sandwich pairs in critical point theory. Trans Amer Math Soc. 2008;360(6):2811–2823.
203. Schechter M. Strong sandwich pairs. Indiana Univ Math J. 2008;57(3):1105–1131.
204. Schechter M. Infinite-dimensional sandwich pairs. Pacific J Math. 2008;235:73–88.
205. Schechter M. Systems of sandwich pairs. J Fixed Point Theory Appl. 2009;5(1):145–155.
206. Schechter M. Minimax systems and critical point theory. Boston: Birkhauser; 2009.
207. Schechter M. Monotonicity methods for infinite dimensional sandwich systems. Discrete Contin Dyn Syst. 2010;28(2):455–468.
208. Schechter M. Nonautonomous second order Hamiltonian systems. Pacific J Math. 2011;251:431–452.
209. Schechter M. The role of core functions in semilinear elliptic problems. Q J Math. 2011;62(3):717–735.
210. Schechter M. Custom monotonicity methods. Math Scand. 2011;109(1):133–146.
211. Schechter M. Noncooperative elliptic systems. Z Angew Math Phys. 2011;62(4):649–666.
212. Schechter M. Superlinear Schrödinger operators. J Funct Anal. 2012;262:2677–2694.
213. Schechter M. Ground state solutions of superlinear Schrödinger equations. Comm Appl Nonlinear Anal. 2012;19(1):91–97.
214. Schechter M. Steady state solutions for Schrödinger equations governing nonlinear optics. J Math Phys. 2012;53:043504, 8.
215. Schechter M. Monotonicity methods for noncooperative elliptic systems. Complex Var Elliptic Eq. 2012;57(11):1163–1178.
216. Schechter M. Photonic lattices. J Math Phys. 2013;54:061502, 7.
217. Schechter M. Sandwich systems. J Math Anal Appl. 2013;399(1):88–99.
218. Schechter M. Ground state solutions for non-autonomous dynamical systems. J Math Phys. 2014;55(10):101504, 13.
219. Schechter M. Periodic second order superlinear Hamiltonian systems J Math Anal Appl. 2015;426(1):546–562.
220. Schechter M. Nonlinear Schrödinger operators with zero in the spectrum. Z Angew Math Phys. 2015;66(5):2125–2141.
221. Schechter M. Carathéodory functions in partial differential equations. In: Contributions in mathematics and engineering. Cham: Springer; 2016. pp. 535–549.
222. Schechter M. Homoclinic solutions of nonlinear second-order Hamiltonian systems. Ann Mat Pura Appl. 2016;195(5):1665–1683.
223. Schechter M. Global solutions of nonlinear Schrödinger equations. Calc Var Partial Diff Eq. 2017;56(2):Art. 40, 17.
224. Schechter M. Infinite dimensional linking, No. 961116. Mathematics department preprint series, UCI.
225. Schechter M. Weak linking No. 970202. Mathematics department preprint series, UCI.
226. Schechter M, Tintarev K. Pairs of critical points produced by linking subsets with applications to semilinear elliptic problems. Bull Soc Math Belg. 1992;44:249–261.
227. Schechter M, Zou W. Double linking theorem and multiple solutions. J Funct Anal. 2003;205(1):37–61.
228. Schechter M, Zou W. Weak linking theorems and Schrödinger equations with critical Sobolev exponent. ESAIM Control Optim Calc Var. 2003;9:601–619 (Electronic).
229. Schechter M, Zou W. Weak linking. Nonlinear Anal. 2003;55(6):695–706.
230. Schechter M, Zou W. Super-linear problems. Pacific J Math. 2004;214(1):145–160.
231. Schechter M, Zou W. Semi-classical bound states of Schrödinger equations. Math Proc Cambridge Philos Soc. 2014;156(1):167–181.
232. Shi S. Ekeland's variational principle and the mountain pass lemma. Acta Math Sinica (N.S.) 1985;1(4):348–355.
233. Shilgba LK. Periodic solutions of non-autonomous second order systems with quasisubadditive potential. J Math Anal Appl. 1995;189:671–675.

234. Shilgba LK. Existence result for periodic solutions of a class of Hamiltonian systems with super quadratic potential. Nonlinear Anal. 2005;63:565–574.
235. Sirakov B. Existence and multiplicity of solutions of semi-linear elliptic equations in RN. Calc Var. 2000;11:119–142.
236. Smiley MW. Time periodic solutions of nonlinear wave equations in balls. In: Oscillations, bifurcation an chaos (Toronto, 1986). Proceedings of the Conference on Canadian Mathematical Society; 1987. pp. 287–297.
237. Smiley MW. Eigenfunction methods and nonlinear hyperbolic boundary value problems at resonance. J Math Anal Appl. 1987;122:129–151.
238. Smiley MW. On the existence of smooth breathers for nonlinear wave equations. J Diff Eq. 1992;96(2):295–317.
239. Struwe M. The existence of surfaces of constant mean curvature with free boundaries. Acta Math. 1988;160:19–64.
240. Struwe M. Variational methods. 2nd ed. Berlin: Springer; 1996.
241. Szulkin A, Weth T. Ground state solutions for some indefinite variational problems. J Funct Anal. 2009;257(12):3802–3822.
242. Szulkin A, Zou W. Infinite dimensional cohomology groups and periodic solutions of asymptotically linear Hamiltonian systems. J Diff Eq. 2001;174:369–371.
243. Tang CL. Periodic solutions of nonautonomous second order systems with γ -quasisubadditive potential. J Math Anal Appl. 1995;189:671–675.
244. Tang C. Periodic solutions of non-autonomous second order systems. J Math Anal Appl. 1996;202:465–469.
245. Tang C. An existence theorem of nontrivial solutions of semilinear equations in reflexive Banach spaces and its applications. Acad Roy Belg Bull Cl. Sci. 1996;7(6):87–100.
246. Tang CL. Periodic solutions of nonautonomous second order systems with sublinear nonlinearity. Proc Amer Math Soc. 1998;126:3263–3270.
247. Tang C. Existence and multiplicity of periodic solutions for nonautonomous second order systems. Nonlinear Anal. 1998;32:299–304.
248. Tang C, Tao Z. Periodic and subharmonic solutions of second-order Hamiltonian systems. J Math Anal Appl. 2004;293:435–445.
249. Tang C, Wu X. Periodic solutions of a class of non-autonomous second-order systems. J Math Anal Appl. 1999;236:227–235.
250. Tang CL, Wu XP. Periodic solutions for second order systems with not uniformly coercive potential. J Math Anal Appl. 2001;259(2):386–397.
251. Tang CL, Wu XP. Periodic solutions for a class of nonautonomous subquadratic second order Hamiltonian systems. J Math Anal Appl. 2002;275(2):870–882.
252. Tang C, Wu X. Periodic solutions of nonautonomous second-order systems Hamiltonian systems with even-typed potentials. Nonlinear Anal. 2003;55:759–769.
253. Tang CL, Wu XP. Notes on periodic solutions of subquadratic second order systems. J Math Anal Appl. 2003;285(1):8–16.
254. Tang C, Wu X. A note on periodic solutions of nonautonomous second-order systems. Proc Amer Math Soc. 2004;132:1295–1303.
255. Tang C, Wu X. Some critical point theorems and their applications to periodic solution for second order Hamiltonian systems. J Diff Eq. 2010;248:660–692.
256. Tang XH, Xiao L. Existence of periodic solutions to second-order Hamiltonian systems with potential indefinite in sign. Nonlinear Anal. 2008;69:3999–4011.
257. Tang C, Ye Y. Periodic solutions for some nonautonomous second order Hamiltonian systems. J Math Anal Appl. 2008;344:462–471.
258. Tang C, Ye Y. Periodic and subharmonic solutions for a class of superquadratic second order Hamiltonian systems. Nonlinear Anal. 2009;71:2298–2307.
259. Tang XH, Zhang Q. New existence of periodic solutions for second order non-autonomous Hamiltonian systems. J Math Anal Appl. 2010;369:357–367.
260. Tao Z, Wu S, Yan S. Periodic solutions for a class of superquadratic Hamiltonian systems. J Math Anal Appl. 2007;331:152–158.

261. Thews K. T-periodic solutions of time dependent Hamiltonian systems with a potential vanishing at infinity. Manuscripta Math. 1981;33:327–338.
262. Tintarev K. Isotopic linking and critical points of functionals. Proceedings of the second world congress of nonlinear analysts, Part 7 (Athens, 1996). Nonlinear Anal. 1997;30(7):4145–4149.
263. Tintarev K. Solutions to elliptic systems of Hamiltonian type in R^N. Electron J Diff Eq. 1999;1999(29):11 pp.
264. Troestler C, Willem M. Nontrivial solution of a semilinear Schrödinger equation. Comm Partial Diff Eq. 1996;21(9–10):1431–1449.
265. Wang Y, Zhang L. A note on existence of a unique periodic solution of non-autonomous second-order Hamiltonian systems. J Franklin Inst. 2010;347:781–794.
266. Wang Z, Zhang J. Periodic solutions for nonautonomous second order Hamiltonian systems with sublinear nonlinearities. Bound Value Probl. 2011;2011 1:23, 14.
267. Wang Z, Zhang J, Zhang Z. Periodic solutions of second order non-autonomous Hamiltonian systems with local superquadratic potential. Nonlinear Anal. 2009;70:3672–3681.
268. Ward JR. Periodic solutions of perturbed conservative systems. Proc Amer Math Soc. 1978;72:281–285.
269. Ward JR. The existence of periodic solutions for nonlinearly perturbed conservative systems. Nonlinear Anal. 1979;5:697–705.
270. Watson G. A treatise on the theory of Bessel functions. Cambridge: University Press; 1922.
271. Willem M. Oscillations forcées systèmes hamiltoniens. Beseancon: Public Sémin Analy Non Linéarie, University; 1981.
272. Willem M. Periodic oscillations of odd second order Hamiltonian systems. Boll Un Mat Ital. 1984;3:293–304.
273. Willem M. Periodic solutions of convex Hamiltonian systems. Bull Soc Math Belg. 1984;36:11–22.
274. Willem M. Subharmonic oscillations of convex Hamiltonian systems. Nonlinear Anal. 1985;9:1303–1311.
275. Willem M. Minimax theorems. Boston: Birkharser; 1996.
276. Willem M, Zou W. On a Schrödinger equation with periodic potential and spectrum point zero. Indiana Univ Math J. 2003;52(1):109–132.
277. Willem M, Zou W. On a semilinear Dirichlet problem and a nonlinear Schrödinger equation with periodic potential. Preprint.
278. Wu X. Periodic solutions for nonautonomous second-order systems with bounded nonlinearity. J Math Anal Appl. 1999;230:135–141.
279. Wu XP, Tang CL. Periodic solutions of a class of nonautonomous second order systems. J Math Anal Appl. 1999;236:227–235.
280. Wu X, Zhao F. Periodic solutions for a class of non-autonomous second order systems. J Math Anal Appl. 2004;296:422–434.
281. Wu X, Zhao F. Saddle point reduction method for some non-autonomous second order systems. J Math Anal Appl. 2004;291:653–665.
282. Wu X, Zhao F. Existence and multiplicity of nonzero periodic solution with saddle point character for some nonautonomous second order systems. J Math Anal Appl. 2005;305:588–595.
283. Xiao Y. Periodic solutions for a class of nonlinear Hamiltonian systems. Acta Math Sinica 2008;51:1205–1212.
284. Xu X. Periodic solutions for non-autonomous Hamiltonian systems possessing superquadratic potentials. Nonlinear Anal. 2002;51:941–955.
285. Yang Y. Solitons in field theory and nonlinear analysis. New York: Springer; 2001.
286. Yang R. Periodic solutions of some non-autonomous second order Hamiltonian systems. Nonlinear Anal. 2008;69:2333–2338.
287. Yang M. Ground state solutions for a periodic Schrödinger equation with superlinear nonlinearities. Nonlinear Anal. 2010;72(5):2620–2627.

288. Yang Y, Zhang R. Steady state solutions for nonlinear Schrödinger equation arising in optics. J Math Phys. 2009;50:053501-9.
289. Yang Y, Zhang R. Erratum. Steady state solutions for nonlinear Schrödinger equation arising in optics, [J. Math. Phys. 50 (2009) 053501-9] J Math Phys. 2010;51:049902.
290. Yang J, Bezryadina A, Chen Z, Makasyuk I. Observation of two-dimensional lattice vector solitons. Opt Lett. 2004;29:1656.
291. Yang J, Makasyuk I, Bezryadina A, Chen Z. Dipole and quadrupole solitons in optically induced two-dimensional photonic lattices: theory and experiment. Stud Appl Math. 2004;113:389–412.
292. Yang M, Chen W, Ding Y. Solutions for periodic Schrödinger equation with spectrum zero and general superlinear nonlinearities. J Math Anal Appl. 2010;364(2):404–413.
293. Zevin AA. Some conditions for the existence and stability of periodic oscillations in nonlinear nonautonomous Hamiltonian systems. J Appl Math Mech. 1985;48:451–458.
294. Zevin AA. Nonlocal criteria for the existence and stability of periodic oscillations in autonomous Hamiltonian systems. J Appl Math Mech. 1987;50:45–51.
295. Zhang S. Periodic solutions for some second order Hamiltonian systems. Nonlinearity 2009;22:2141–2150.
296. Zhang X, Zhou Y. Periodic solutions of non-autonomous second order Hamiltonian systems. J Math Anal Appl. 2008;345:929–933.
297. Zhang X, Zhou Y. On periodic solutions of non-autonomous second order Hamiltonian systems. Appl Math. 2010;55:373–384.
298. Zhao P, Zhou W, Zhong C. The existence of three nontrivial solutions of a class of elliptic systems. Nonlinear Anal. Ser. A Theory Method. 2002;49(3):431–443.
299. Zou W. Multiple solutions for asymptotically linear elliptic systems. J Math Anal Appl. 2001; 255(1):213–229.

Index

M. Schechter, *Critical Point Theory*,
https://doi.org/10.1007/978-3-030-45603-0

The manufacturer's authorised representative in the EU is Springer Nature Customer Service Centre GmbH, Europaplatz 3, 69115 Heidelberg, Germany. If you have any concerns regarding our products, please contact ProductSafety@springernature.com

Printed and bound by CPI Group (UK) Ltd, Croydon, CR0 4YY

30/07/2026

02181693-0012